Mental Evolution in Animals

With a Posthumous Essay on Instinct by Charles Darwin

George John Romanes

Cambridge
University Press

CAMBRIDGE UNIVERSITY PRESS

Cambridge, New York, Melbourne, Madrid, Cape Town,
Singapore, São Paolo, Delhi, Tokyo, Mexico City

Published in the United States of America by Cambridge University Press, New York

www.cambridge.org
Information on this title: www.cambridge.org/9781108037877

This edition first published 1883
This digitally printed version 2011

ISBN 978-1-108-03787-7 Paperback

	Products of Emotional Development.	
50		50
49		49
48		48
47		47
46		46
45		45
44		44
43		43
42		42
41		41
40		40
39		39
38		38
37		37
36		36
35		35
34		34
33		33
32		32
31		31
30		30
29		29
28	Shame, Remorse, Deceitfulness, Ludicrous.	28
27	Revenge, Rage.	27
26	Grief, Hate, Cruelty, Benevolence.	26
25	Emulation, Pride, Resentment, Æsthetic love of ornament, Terror.	25
24	Sympathy.	24
23		23
22	Affection.	22
21	Jealousy, Anger, Play.	21
20	Parental affection, Social feelings, Sexual selection, Pugnacity, Industry, Curiosity.	20
19	Sexual emotions without sexual selection.	19
18	Surprise, Fear.	18
17		17
16		16
15		15
14		14
13		13
12		12
11		11
10		10
9		9
8		8
7		7
6		6
5		5
4		4
3		3
2		2
1		1

CAMBRIDGE LIBRARY COLLECTION

Books of enduring scholarly value

Life Sciences

Until the nineteenth century, the various subjects now known as the life sciences were regarded either as arcane studies which had little impact on ordinary daily life, or as a genteel hobby for the leisured classes. The increasing academic rigour and systematisation brought to the study of botany, zoology and other disciplines, and their adoption in university curricula, are reflected in the books reissued in this series.

Mental Evolution in Animals

George John Romanes (1848–94), considered by *The Times* to be 'the biological investigator upon whom in England the mantle of Mr. Darwin has most conspicuously descended', wrote this influential work on the evolution of the mental faculties of animals in 1883. The two scientists were close friends, and Darwin gave Romanes his notes on psychology to use in his studies. Much of the book is devoted to instinct, and contained in the appendix is a posthumous essay by Darwin on the subject, originally intended for a later edition of *On the Origin of Species*. Romanes' method of using anecdotal evidence over empirical research has been criticised, but this book stands as an influential work in the history of evolutionary biology; it was followed in 1888 by his *Mental Evolution in Man* (also reissued in this series), which discussed some of the most important issues of nineteenth-century evolutionary psychology.

Cambridge University Press has long been a pioneer in the reissuing of out-of-print titles from its own backlist, producing digital reprints of books that are still sought after by scholars and students but could not be reprinted economically using traditional technology. The Cambridge Library Collection extends this activity to a wider range of books which are still of importance to researchers and professionals, either for the source material they contain, or as landmarks in the history of their academic discipline.

Drawing from the world-renowned collections in the Cambridge University Library, and guided by the advice of experts in each subject area, Cambridge University Press is using state-of-the-art scanning machines in its own Printing House to capture the content of each book selected for inclusion. The files are processed to give a consistently clear, crisp image, and the books finished to the high quality standard for which the Press is recognised around the world. The latest print-on-demand technology ensures that the books will remain available indefinitely, and that orders for single or multiple copies can quickly be supplied.

The Cambridge Library Collection will bring back to life books of enduring scholarly value (including out-of-copyright works originally issued by other publishers) across a wide range of disciplines in the humanities and social sciences and in science and technology.

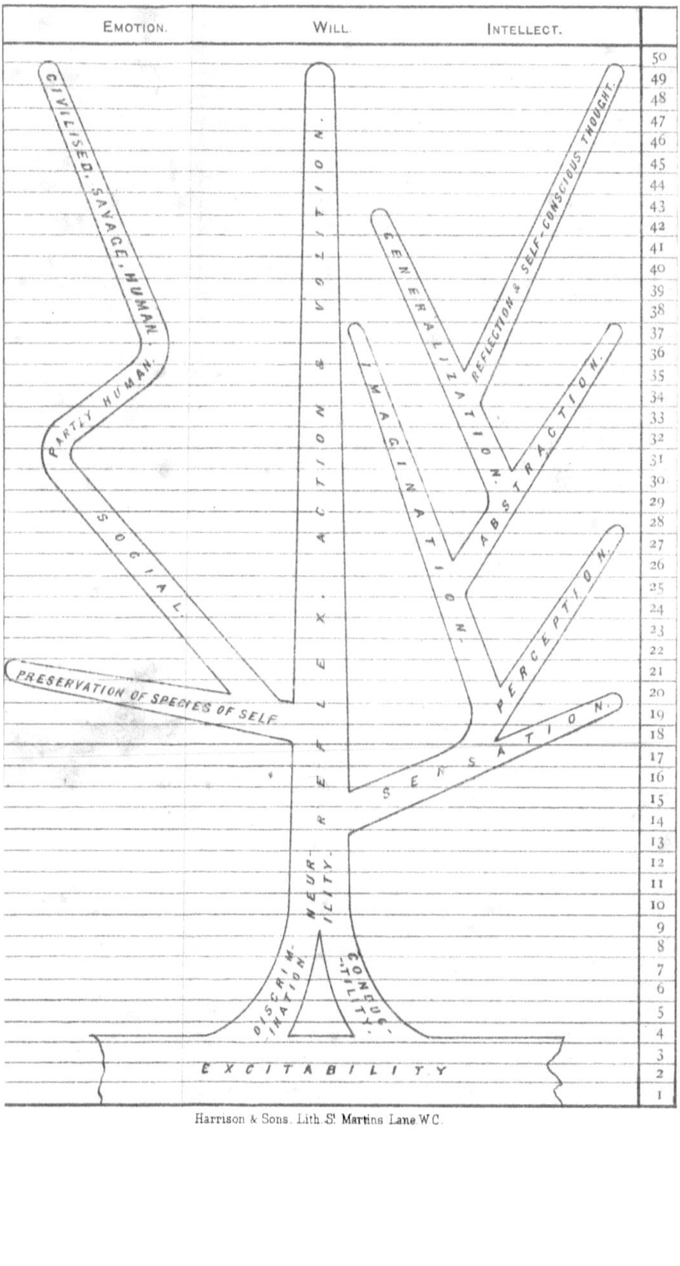

Harrison & Sons. Lith. St Martins Lane W.C.

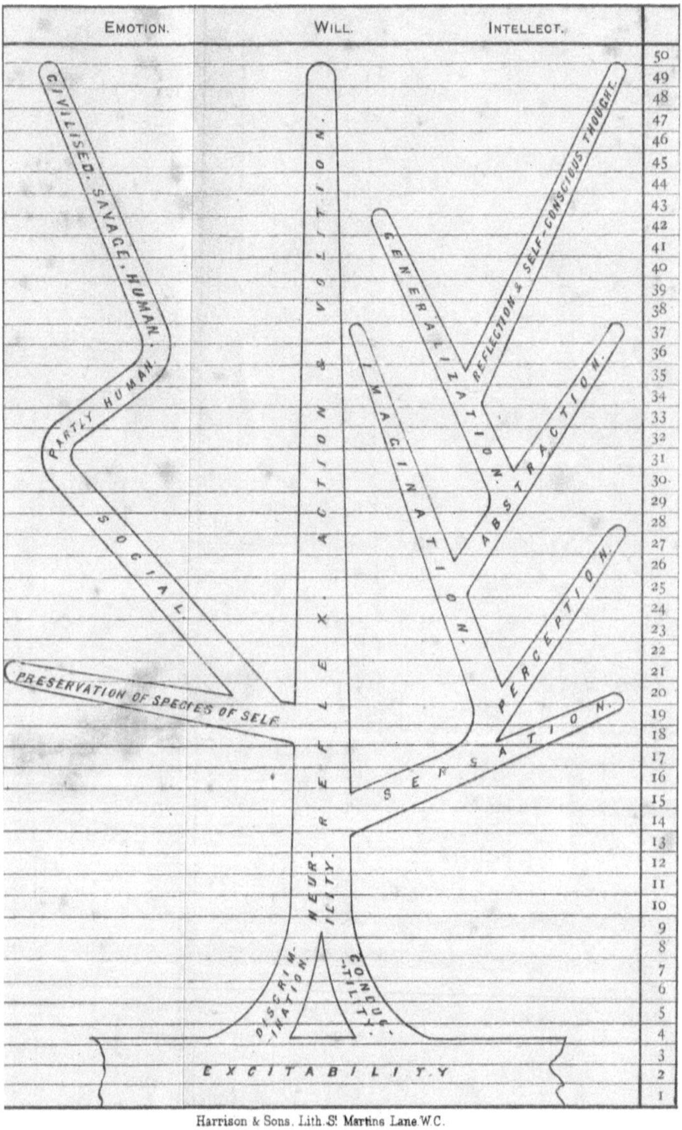

Harrison & Sons. Lith. St. Martins Lane W.C.

Products of Intellectual Development.	The Phychological Scale.	Psycogenesis of Man.	
			50
			49
			48
			47
			46
			45
			44
			43
			42
			41
			40
			39
			38
			37
			36
			35
			34
			33
			32
			31
			30
			29
Indefinite morality.	Anthropoid Apes and Dog.	15 months.	28
Use of tools.	Monkeys, and Elephant	12 months.	27
Understanding of mechanisms.	Carnivora, Rodents, and Ruminants	10 months.	26
Recognition of Pictures, Understanding of words, Dreaming.	Birds.	8 months.	25
Communication of ideas.	Hymenoptera.	5 months.	24
Recognition of persons.	Reptiles and Cephalopods.	4 months.	23
Reason.	Higher Crustacia.	14 weeks.	22
Association by similarity.	Fish and Batrachia.	12 weeks.	21
Recognition of offspring, Secondary instincts.	Insects and Spiders.	10 weeks.	20
Association by contiguity.	Mollusca.	7 weeks.	19
Primary instincts.	Larvæ of Insects, Annelida	3 weeks.	18
Memory.	Echinodermata.	1 week.	17
Pleasures and pains. <- Consciousness ->		Birth.	16
	Cœlenterata.		15
Nervous adjustments.			14
			13
	Unknown animals,		12
Partly nervous adjustments.	probably Cœlenterata.		11
	perhaps extinct.	Embryo.	10
			9
			8
Non-nervous adjustments.	Unicellular organisms.		7
			6
			5
			4
Protoplasmic movements.	Protoplasmic organisms.	Ovum and	3
			2
		Spermatozoa	1

MENTAL EVOLUTION IN ANIMALS.

BY

GEORGE JOHN ROMANES, M.A., LL.D., F.R.S.,

ZOOLOGICAL SECRETARY OF THE LINNEAN SOCIETY.

WITH A

POSTHUMOUS ESSAY ON INSTINCT.

BY

CHARLES DARWIN, M.A., LL.D., F.R.S.

LONDON:

KEGAN PAUL, TRENCH, & CO., 1, PATERNOSTER SQUARE.

1883.

HARRISON AND SONS,
PRINTERS IN ORDINARY TO HER MAJESTY,
ST. MARTIN'S LANE.

PREFACE.

It will be observed that the title of this volume is "Mental Evolution in Animals." The reasons which have led me to depart from my intention (as expressed in the Preface of "Animal Intelligence") to devote the present essay to mental evolution in man as well as in animals, are given in the introductory chapter.

It may appear that in the following pages a somewhat disproportionate amount of space has been allotted to the treatment of Instinct; but, looking to the confusion which prevails with reference to this important branch of psychology in the writings of our leading authorities, I have deemed it desirable to consider the subject exhaustively.

It is, I think, desirable briefly to explain the circumstances under which I have been enabled to produce so much hitherto unpublished material from the MSS of the late Mr. Darwin, and also to state the extent to which I have availed myself of such of this unpublished material as came into my hands. As I have already explained, in the Preface of "Animal Intelligence," Mr. Darwin himself gave me all his MSS relating to psychological subjects, with the request that I should publish any parts of them that I chose in my works on Mental Evolution. But after his death I felt that the circumstances with reference to this kind offer were changed, and that I should scarcely be justified in appropriating so much material, the value of which had become enhanced. I therefore published at the Linnean Society, and with the consent of Mr. Darwin's family, as much of this material as could be published in a consecutive form; this is the chapter which was intended for the "Origin of Species," and which, for the sake of reference, I have added as an Appendix to my present work. For the rest, the numerous disjointed paragraphs and notes which I found among the MSS I have woven into the text of this book, feeling on the one hand that they were not so well suited to appear as a string of disconnected passages, and on the other hand that it was desirable to publish them somewhere. I have gone through all the MSS carefully, and have arranged so as to introduce every passage in them of any importance which I find to have been hitherto unpublished. In no case have I found any reason to suppress a passage, so that the quotations which I have given may be collectively regarded as a full supplementary publication of all that Mr. Darwin has written in the domain of psychology. In order to facilitate reference, I have given in the Index, under Mr. Darwin's name, the numbers of all the pages in this work where the quotations in question occur.

18, Cornwall Terrace,
Regent's Park, London, N.W.,
November, 1883.

CONTENTS.

INTRODUCTION.

In the family of the sciences Comparative Psychology may claim nearest kinship with Comparative Anatomy; for just as the latter aims at a scientific comparison of the bodily structures of organisms, so the former aims at a similar comparison of their mental structures.* Moreover, in the one science as in the other, the first object is to analyze all the complex structures with which each has respectively to deal. When this analysis, or dissection, has been completed for as great a number of cases as circumstances permit, the next object is to compare with one another all the structures which have been thus analyzed; and, lastly, the results of such comparison supply, in each case alike, the basis for the final object of these sciences, which is that of classifying, with reference to these results, all the structures which have been thus examined.

In actual research these three objects are prosecuted, not successively, but simultaneously. Thus it is not necessary in either case that the final object—that of classification—should wait for its commencement upon the completion of the dissection or analysis of every organism or every mental structure that is to be found upon the earth. On the contrary, the *comparison* in each case begins with the facts that are first found to be *comparable*, and is afterwards progressively extended as knowledge of additional facts becomes more extensive.

Now each of the three objects which I have named affords

* The word "structure" is used in a metaphorical sense when applied to mind, but the usage is convenient.

in its pursuit many and varied points of interest, which are quite distinct from any interest that may be felt in the attainment of the ultimate end—Classification. Thus, for example, the study of the human hand as a *mechanism* has an interest apart from all considerations touching the comparison of its structure with that of the corresponding member in other animals; and, similarly, the study of the psychological faculties in any given animal has an interest apart from all considerations touching their comparison with the corresponding faculties in other animals. Again, just as the comparison of separate bodily members throughout the animal series has an interest apart from any question concerning the classification of animal bodies to which such comparison may ultimately lead, so the study of separate psychical faculties throughout the animal series (including, of course, mankind) has an interest quite distinct from any question concerning the classification of animal minds to which such comparison may ultimately lead. Lastly, around and outside all the objects of these sciences as such, there lies the broad expanse of General Thought, into which these sciences, in all their stages, throw out branches of inference. It is needless to say that of late years the interest with which the unprecedented growth of these branches is watched has become so universal and intense, that it may be said largely to have absorbed the more exclusive sources of interest which I have enumerated.

With the view of furthering these various lines of interest, I have undertaken a somewhat laborious enquiry, part of which has already been published in the International Scientific Series, and a further instalment of which is contained in the present volume. The two works, therefore, "Animal Intelligence" and "Mental Evolution in Animals," although published separately, are really one; and they have been divided only for the following reasons. In the first place, to have produced the whole as one volume would have been to present a book, if not of inconvenient bulk, at least quite out of keeping with the size of all the other books in the same series. Moreover, the subject-matter of each work, although intimately related to that of the other, is nevertheless quite distinct. The first is a compendium of facts relating to Animal Intelligence, which, while necessary as a basis for the present essay, is in itself a separate and distinct treatise, intended to meet the interest already alluded to as attaching to this subject for its own sake; while the second treatise, although based upon the former, has to deal with a wider range of subject-matter.

It is evident that, in entering upon this wider field, I shall frequently have to quit the narrower limits of direct observation within which my former work was confined; and it is chiefly because I think it desirable clearly to distinguish between the objects of Comparative Psychology as a science, and any inferences or doctrines which may be connected with its study, that I have made so complete a partition of the facts of animal intelligence from the theories which I believe these facts to justify.

So much, then, for the reasons which have led to the form of these essays, and the relations which I intend the one to bear to the other. I may now say a few words to indicate the structure and scope of the present essay.

Every discussion must rest on some basis of assumption; every thesis must have some hypothesis. The hypothesis which I shall take is that of the truth of the general theory of Evolution: I shall assume the truth of this theory so far as I feel that all competent persons of the present day will be prepared to allow me. I must therefore first define what degree of latitude I suppose to be thus conceded.

I take it for granted, then, that all my readers accept the doctrine of Organic Evolution, or the belief that all species of plants and animals have had a derivative mode of origin by way of natural descent; and, moreover, that one great law or method of the process has been natural selection, or survival of the fittest. If anyone grants this much, I further assume that he must concede to me the *fact*, as distinguished from the *manner* and *history* of Mental Evolution, throughout the whole range of the animal kingdom, with the exception of man. I assume this because I hold that if the doctrine of Organic Evolution is accepted, it carries with it, as a necessary corollary, the doctrine of Mental Evolution, at all events as far as the brute creation is concerned. For throughout the brute creation, from wholly unintelligent animals to the most highly intelligent, we can trace one continuous gradation; so that if we already believe that all specific forms of animal life have had a derivative origin, we cannot refuse to believe that all the mental faculties which these various forms present must likewise have had a derivative origin. And, as a matter of fact, we do not find anyone so unreasonable as to maintain, or even to suggest, that if the evidence of Organic Evolution is accepted, the evidence of Mental Evolution, within the limits which I have named, can consistently be rejected. The one body of evidence therefore serves as a pedestal to the other, such that in the absence of the former the latter would have no *locus standi* (for no one could well dream of Mental Evolution were it not for the evidence of Organic Evolution, or of the transmutation of species); while the presence of the former irresistibly suggests the necessity of the latter, as the logical structure for the support of which the pedestal is what it is.

It will be observed that in this statement of the case I have expressly excluded the psychology of man, as being a department of comparative psychology with reference to which I am not entitled to assume the principles of Evolution. It seems needless to give my reasons for this exclusion. For it is notorious that from the hour when Mr. Darwin and Mr. Wallace simultaneously propounded the theory which has exerted so enormous an influence on the thought of the present century, the difference between the views of these two joint originators of the theory has since been shared by the ever-increasing host of their disciples. We all know what that difference is. We all know that while Mr. Darwin believed the facts of human psychology to admit of being explained by the general laws of Evolution, Mr. Wallace does not believe these facts to admit of being thus explained. Therefore, while the followers of Mr. Darwin maintain that all organisms whatsoever are alike products of a natural genesis, the followers of Mr. Wallace maintain that a distinct exception must be made to this general statement in the case of the human organism; or at all events in the case of the human mind. Thus it is that the great school of evolutionists is divided into two sects; according to one the mind of man has been slowly evolved from lower types of psychical existence, and according to the other the mind of man, not having been thus evolved, stands apart, *sui generis,* from all other types of such existence.

Now assuredly we have here a most important issue, and as it is one the discussion of which will constitute a large element of my work, it is perhaps desirable that I should state at the outset the manner in which I propose to deal with it.

The question, then, as to whether or not human intelligence has been evolved from animal intelligence can only be dealt with scientifically by comparing the one with the other, in order to ascertain the points wherein they agree and the points wherein they differ. Now there can be no doubt that when this is done, the difference between the mental faculties of the most intelligent animal and the mental faculties of the lowest savage is seen to be so vast, that the hypothesis of their being so nearly allied as Mr. Darwin's teaching implies, appears at first sight absurd. And, indeed, it is not until we have become convinced that the theory of Evolution can alone afford an explanation of the facts of human anatomy, that we are prepared to seek for a similar explanation of the facts of human psychology. But wide as is the difference between the mind of a man and the mind of a brute, we must remember that the question is one, not as to degree, but as to kind; and therefore that our task, as serious enquirers after truth, is calmly and honestly to examine the character of the difference which is presented, in order to determine whether it is really beyond the bounds of rational credibility that the enormous interval which now separates these two divisions of mind can ever have been bridged over, by numberless intermediate gradations, during the untold ages of the past.

While writing the first chapters of the present volume, I intended that the latter half of it should be devoted to a consideration of this question, and therefore in "Animal Intelligence" I said that such would be the case. But as the work proceeded it soon became evident that a full treatment of this question would require more space than could be allowed in a single volume, without seriously curtailing both the consideration of this question itself and also that of Mental Evolution, as this is exhibited in the animal kingdom. I therefore determined on restricting the present essay to a consideration of Mental Evolution in Animals, and on reserving for subsequent publication all the material which I have collected bearing on Mental Evolution in Man. I cannot yet say how long it will be before I can feel that I am justified in publishing my researches concerning this branch of my subject; for the more that I have investigated it, the more have I found that it grows, as it were, in three dimensions—in depth, width, and complexity. But at whatever time I shall be able to publish the third and final instalment of my work, it will of course rest upon the basis supplied by the present essay, as this rests upon the basis supplied by the previous one.

It being understood, then, that the present essay is restricted to a consideration of Mental Evolution in Animals,

I should like to have it also understood that it is further restricted to the psychology as distinguished from the philosophy of the subject. In a short and independent essay, published elsewhere,* I have already stated my views concerning the more important questions of philosophy into which the subject-matter of psychology is so apt to dip; but here it is only needful to emphasize the fact that these two strata of thought, although assuredly in juxtaposition, are no less assuredly distinct. My present enquiry belongs only to the upper stratum, or to the science of psychology as distinguished from any theory of knowledge. I am in no wise concerned with "the transition from the object known to the knowing subject," and therefore I am in no wise concerned with any of the philosophical theories which have been propounded upon this matter. In other words, I have everywhere to regard mind as an object and mental modifications as phenomena; therefore I have throughout to investigate the process of Mental Evolution by what is now generally and aptly termed the historical method. I cannot too strongly impress upon the memory of those who from previous reading are able to appreciate the importance of the distinction, that I thus intend everywhere to remain within the borders of psychology, and nowhere to trespass upon the grounds of philosophy.

On entering so wide a field of enquiry as that whose limits I have now indicated, it is indispensable to the continuity of advance that we should be prepared, where needful, to supplement observation with hypothesis. It therefore seems desirable to conclude this Introduction with a few words both to explain and to justify the method which in this matter I intend to follow.

It has already been stated that the sole object of this work is that of tracing, in as scientific a manner as possible, the probable history of Mental Evolution, and therefore, of

* *Nineteenth Century*, December, 1882.

course, of enquiring into the causes which have determined it. So far as observation is available to guide us in this enquiry, I shall resort to no other assistance. Where, however, from the nature of the case, observation fails us, I shall proceed to inference. But though I shall use this method as sparingly as possible, I am aware that criticism will often find valid ground to object—'It is all very well to map out the supposed genesis of the various mental faculties in this way, but we require some definite experimental or historical proof that the genesis in question actually did take place in the order and manner that you infer.'

Now, in answer to this objection, I have only to say that no one can have a more lively appreciation than myself of the supreme importance of experimental or historical verification, in all cases where the possibility of such verification is attainable. But in cases where such verification is not attainable, what are we to do? We may clearly do either of two things. We may either neglect to investigate the subject at all, or we may do our best to investigate it by employing the only means of investigation which are at our disposal. Of these two courses there can be no doubt which is the one that the scientific spirit prompts. The true scientific spirit desires to examine everything, and if in any case it is refused the best class of instruments wherewith to conduct the examination, it will adopt the next best that are available. In such cases science clearly cannot be forwarded by neglecting to use these instruments, while her cause may be greatly advanced by using them with care. This is proved by the fact that, in the science of psychology, nearly all the considerable advances which have been made, have been made, not by experiment, but by observing mental phenomena and reasoning from these phenomena deductively. In such cases, therefore, the true scientific spirit prompts us, not to throw away deductive reasoning where it is so frequently the only instrument available, but rather to carry it with us, and to use it as not abusing it.

And this, as I have said, is what I shall endeavour to do. No one can regret more than myself that the most interesting of all regions of scientific enquiry should happen to be the one in which experiment, or inductive verification, is least of all applicable; but such being the case, we must take the case as we find it, use deductive reasoning where we clearly see that it is the only instrument available, but use it to as limited an extent as the nature of our subject permits.

ERRATUM.

Page 145, *for* Conceptualism *read* Realism.

MENTAL EVOLUTION IN ANIMALS.

CHAPTER I.

THE CRITERION OF MIND.

THE subject of our enquiry being Mental Evolution, it is desirable to begin by understanding clearly what we mean by Mind,* and then defining the conditions under which known Mind is invariably found to occur. In this chapter, therefore, I shall deal with what I take to be the Criterion of Mind, and shall then pass on in the next chapter to a consideration of the objective conditions under which alone Mind is observed to exist.

It is obvious, then, to start with, that by Mind we may mean two very different things, according as we contemplate it in our own individual selves, or as manifested by other beings. For if I contemplate my own mind, I have an immediate cognizance of a certain flow of thoughts and feelings, which are the most ultimate things—and, indeed, the only things—of which I am cognizant. But if I contemplate Mind in other persons or organisms, I can have no such immediate cognizance of their thoughts and feelings; I can only *infer* the existence of such thoughts and feelings from the activities of the persons or organisms which appear to manifest them. Thus it is that by Mind we may mean either that which is subjective or that which is objective. Now throughout the present work we shall have to consider Mind as an object; and therefore it is well to remember that our only instrument of analysis is the observation of activities

* It was necessary in my work on *Animal Intelligence* briefly to touch on this question; therefore the parts of the analysis which are common to the two works I shall render as much as possible in the same words.

which we infer to be prompted by, or associated with, mental antecedents or accompaniments analogous to those of which we are directly conscious in our own subjective experience. That is to say, starting from what I know subjectively of the operations of my own individual mind, and of the activities which in my own organism these operations seem to prompt, I proceed by analogy to infer from the observable activities displayed by other organisms, the fact that certain mental operations underlie or accompany these activities.

From this statement of the case it will be apparent that our knowledge of mental activities in any organism other than our own is neither subjective nor objective. That it is not subjective I need not wait to show. That it is not objective may be rendered obvious by a few moments' reflection. For it is evident that mental activities in other organisms can never be to us objects of direct knowledge; as I have just said, we can only infer their existence from the objective sources supplied by observable activities of such organisms. Therefore all our knowledge of mental activities other than our own really consists of an inferential interpretation of bodily activities—this interpretation being founded on our subjective knowledge of our own mental activities. By inference we project, as it were, the known patterns of our own mental chromograph on what is to us the otherwise blank screen of another mind; and our only knowledge of the processes there taking place is really due to such a projection of our own subjectively. This matter has been well and clearly presented by the late Professor Clifford, who has coined the exceedingly appropriate term *eject* (in contradistinction to *subject* and *object*), whereby to designate the distinctive character of a mind (or mental process) other than our own in its relation to our own. I shall therefore adopt this convenient term, and speak of all our possible knowledge of other minds as ejective.

Now in this necessarily ejective method of enquiry, what is the kind of activities that we are entitled to regard as indicative of Mind? I certainly do not so regard the flowing of a river or the blowing of a wind. Why? First, because the subjects are too remote in kind from my own organism to admit of my drawing any reasonable analogy between them and it; and, secondly, because the activities which they present are invariably of the same kind under the same circumstances: they therefore offer no evidence of that which I deem the distinctive character of my own mind as such—Consciousness. In other words, two conditions require to be satisfied before we even begin to imagine that observable activities are indicative of mind; the activities must be displayed by a living organism, and they must be of a kind to suggest the presence of consciousness. What then is to be taken as the criterion of consciousness? Subjectively, no criterion is either needful or possible; for to me, individually, nothing can be more ultimate than my own consciousness, and, therefore, my consciousness cannot admit of any criterion having a claim to a higher certainty. But, ejectively, some such criterion is required, and as my consciousness cannot come within the territory of a foreign consciousness, I can only appreciate the latter through the agency of ambassadors—these ambassadors being, as I have now so frequently said, the observable activities of an organism. The next question, therefore, is, What activities of an organism are to be taken as indicative of consciousness? The answer that comes most readily is,—All activities that are indicative of Choice; wherever we see a living organism apparently exerting intentional choice, we may infer that it is conscious choice, and, therefore, that the organism has a mind. But physiology shows that this answer will not do; for, while not disputing whether there is any mind without the power of conscious choice, physiology, as we shall see in the next chapter, is very firm in denying that all apparent choice is due to mind. The host of reflex actions is arrayed against the proposition, and, in view of such non-mental, though apparently intentional adjustments, we find the necessity for some test of the choice-element as real or fictitious. The only test we have is to ask whether the adjustments displayed are invariably the same under the same circumstances of stimulation. The only distinction between adjustive movements due to reflex action, and adjustive movements accompanied by mental perception, consists in the former depending on inherited mechanisms within the nervous system being so constructed as to effect *particular* adjustive movements in response to *particular* stimulations, while the latter are independent of any such inherited adjustment of special mechanisms to the exigencies of special circumstances. Reflex actions, under the influence of their appropriate stimuli, may be compared to the actions of a machine under the manipulations of an operator: when certain springs of action are touched by certain stimuli, the whole machine is thrown into appropriate action; there is no room for choice, there is no room for uncertainty; but, as surely as any of these inherited mechanisms is affected by the stimulus with reference to which it has been constructed to act, so surely will it act in precisely the same way as it always has acted. But the case with conscious mental adjustment is quite different. For, without going into the question concerning the relation of Body and Mind, or waiting to ask whether cases of mental adjustment are not really quite as *mechanical* in the sense of being the necessary result or correlative of a chain of psychical sequences due to a physical stimulation, it is enough to point to the variable and incalculable character of mental adjustments as distinguished from the constant and foreseeable character of reflex adjustments. All, in fact, that in an objective sense we can mean by a mental adjustment, is an adjustment of a kind that has not been definitely fixed by heredity as the only adjustment possible in the given circumstances of stimulation. For, were there no alternative of adjustment, the case, in an animal at least, would be indistinguishable from one of reflex action.

It is, then, adaptive action by a living organism in cases where the inherited machinery of the nervous system does not furnish data for our prevision of what the adaptive action must necessarily be—it is only in such cases that we recognize the element of mind. In other words, ejectively considered, the distinctive element of mind is consciousness, the test of consciousness is the presence of choice, and the evidence of choice is the antecedent uncertainty of adjustive action between two or more alternatives. To this analysis it is, however, needful to add that, although our only criterion of mind is antecedent uncertainty of adjustive action, it does not follow that all adjustive action in which mind is concerned should be of an antecedently uncertain character; or, which is the same thing, that because some such action may be of an antecedently certain character, we should on this account regard it as non-mental. Many adjustive actions which we recognize as mental are, nevertheless, seen beforehand to be, under the given circumstances, inevitable; but analysis would show that such is only the case when we have in view agents whom we already, and from independent evidence, regard as mental.

In positing the evidence of Choice as my objective (or ejective) criterion of Mind, I do not think it necessary to enter into any elaborate analysis of what constitutes this evidence. In a subsequent chapter I shall treat fully of what I call the physiology or objective aspect of choice; and then it will be seen that from the gradual manner in which choice, or the mind-element, arises, it is not practically possible to draw a definite line of demarcation between choosing and non-choosing agents. Therefore, at this stage of the enquiry I prefer to rest in the ordinary acceptation of the term, as implying a distinction which common sense has always drawn, and probably always will draw, between mental and non-mental agents. It cannot be correctly said that a river chooses the course of its flow, or that the earth chooses an ellipse wherein to revolve round the sun. And similarly, however *complex* the operations may be of an agent recognized as non-mental—such, for instance, as those of a calculating machine—or however impossible it may be to *predict* the result of its actions, we never say that such operations or actions are due to choice; we reserve this term for operations or actions, however simple and however easily the result may be foreseen, which are performed, either by agents who in virtue of the non-mechanical nature of these actions prove themselves to be mental, or by agents already recognized as mental—*i.e.*, by agents who have already proved themselves to be mental by performing other actions of such a non-mechanical or unforeseeable nature as we feel assured can only be attributed to choice. And there can be no reasonable doubt that this common-sense distinction between choosing and non-choosing agents is a valid one. Although it may be difficult or impossible, in particular cases, to decide to which of the two categories this or that being should be assigned, this difficulty does not affect the validity of the classification—any more, for instance, than the difficulty of deciding whether Limulus should be classified with the crabs or with the scorpions affects the validity of the classification which marks off the group Crustacea from the group Arachnida. The point is that, notwithstanding special difficulties in assigning this or that being to one or the other class, the psychological classification which I advocate resembles the zoological classification which I have cited; it is a valid classification, inasmuch as it recognizes a distinction where there is certainly something to distinguish. For even if we take the most mechanical view of mental processes that is possible, and suppose that conscious intelligence plays no part whatever in determining action, there still remains the fact that such conscious intelligence *exists*, and that prior to certain actions it is always affected in certain ways. Therefore, even if we suppose that the state of things is, so to speak, accidental, and that the actions in question would always be performed in precisely the same way whether or not they were thus connected with consciousness, it would still remain desirable, for scientific purposes, that a marked distinction should be drawn between cases of activity that proceed without, and those that proceed with this remarkable association with consciousness. As the phenomena of subjectivity are facts at any rate no less real than those of objectivity, if it is found that some of the latter are invariably and faithfully mirrored in those of the former, such phenomena, for this reason alone, deserve to be placed in a distinct scientific category, even though it were proved that the mirror of subjectivity might be removed without affecting any of the phenomena of objectivity.

Without, therefore, entertaining the question as to the connexion between Body and Mind, it is enough to say that under any view concerning the nature of this connexion, we are justified in drawing a distinction between activities which are accompanied by feelings, and activities which, so far as we can see, are not so accompanied. If this is allowed, there seems to be no term better fitted to convey the distinction than the term Choice; agents that are able to *choose* their actions are agents that are able to *feel* the stimuli which determine the choice.

Such being our Criterion of Mind, it admits of being otherwise stated, and in a more practically applicable manner, in the following words which I quote from "Animal Intelligence:"—"It is, then, adaptive action by a living organism in cases where the inherited machinery of the nervous system does not furnish data for our prevision of what the adaptive action must necessarily be—it is only here that we recognize the objective evidence of mind. The criterion of mind, therefore, which I propose, and to which I shall adhere throughout the present volume, is as follows:—Does the organism learn to make new adjustments, or to modify old ones, in accordance with the results of its own individual experience? If it does so, the fact cannot be merely due to reflex action in the sense above described; for it is impossible that heredity can have provided in advance for innovations upon or alterations of its machinery during the lifetime of a particular individual."

Two points have to be observed with regard to this criterion, in whichever verbal form we may choose to express it. The first is that it is not rigidly exclusive either, on the one hand, of a possibly mental character in apparently non-mental adjustments, or, conversely, of a possibly non-mental character in apparently mental adjustments. For it is certain that failure to learn by individual experience is not always conclusive evidence against the existence of mind; such failure may arise merely from an imperfection of memory, or from there not being *enough* of the mind-element present to make the adjustments needful to meet the novel circumstances. Conversely, it is no less certain that some parts of our own nervous system, which are not concerned in the phenomena of consciousness, are nevertheless able in some measure to learn by individual experience. The nervous apparatus of the stomach, for instance, is able in so considerable a degree to adapt the movements of that organ to the requirements of its individual experience, that were the organ an organism we might be in danger of regarding it as dimly intelligent. Still there is no evidence to show that non-mental agents are ever able in any considerable measure thus to simulate the adjustments performed by mental ones; and therefore our criterion, in its practical application, has rather to be guarded against the opposite danger of denying the presence of mind to agents that are really mental. For, as I observed in "Animal Intelligence," "it is clear that long before mind has advanced sufficiently far in the scale of development to become amenable to the test in question, it has probably begun to dawn as nascent subjectivity. In other words, because a lowly organized animal does *not* learn by its own individual experience, we may not therefore conclude that in performing its natural or ancestral adaptations to appropiate stimuli, consciousness, or the mind-element, is wholly absent; we can only say that this element, if present, reveals no evidence of the fact. But, on the other hand, if a lowly organized animal *does* learn by its own individual experience, we are in possession of the best available evidence of conscious memory leading to intentional adaptation. Therefore, our criterion applies to the upper limit of non-mental action, not to the lower limit of mental."

Or, again adopting the convenient terminology of Clifford, we must always remember that we can never know the mental states of any mental beings other than ourselves as *objects;* we can only know them as *ejects,* or as ideal projections of our own mental states. And it is from this broad fact of psychology that the difficulty arises in applying our criterion of mind to particular cases—especially among the lower animals. For if the evidence of mind, or of being capable of choice, must thus always be ejective as distinguished from objective, it is clear that the cogency of the evidence must diminish as we recede from minds inferred to be like our own, towards minds inferred to be not so like our own, passing in a gradual series into not-minds. Or, otherwise stated, although the evidence derived from ejects is practically regarded as good in the case of mental organizations inferred to be closely analogous to our own, this evidence clearly ceases to be trustworthy in the ratio in which the analogy fails; so that when we come to the case of very low animals—where the analogy is least—we feel uncertain whether or not to ascribe to them any ejective existence. But I must again insist that this fact—which springs immediately out of the fundamental isolation of the individual mind—is no argument against my criterion of mind as the best criterion available; it tends, indeed, to show that no better criterion can be found, for it shows the hopelessness of seeking such.

The other point which has to be noted with regard to this criterion is as follows. I again quote from "Animal Intelligence:"—

"Of course to the sceptic this criterion may appear unsatisfactory, since it depends, not on direct knowledge, but on inference. Here, however, it seems enough to point out, as already observed, that it is the best criterion available; and, further, that scepticism of this kind is logically bound to deny evidence of mind, not only in the case of the lower animals, but also in that of the higher, and even in that of men other than the sceptic himself. For all objections which could apply to the use of this criterion of mind in the animal kingdom, would apply with equal force to the evidence of any mind other than that of the individual objector. This is obvious, because, as I have already observed, the only evidence we can have of objective mind is that which is furnished by objective activities; and, as the subjective mind can never become assimilated with the objective so as to learn by direct feeling the mental processes which there accompany the objective activities, it is clearly impossible to satisfy any one who may choose to doubt the validity of inference, that in any case, other than his own, mental processes ever do accompany objective activities.

"Thus it is that philosophy can supply no demonstrative refutation of idealism, even of the most extravagant form. Common-sense, however, universally feels that analogy is here a safer guide to truth than the sceptical demand for impossible evidence; so that if the objective existence of other organisms and their activities is granted—without which postulate comparative psychology, like all the other sciences, would be an unsubstantial dream—common sense will always and without question conclude that the activities of organisms other than our own, when analogous to those activities of our own which we know to be accompanied by certain mental states, are in them accompanied by analogous mental states."

CHAPTER II.

THE STRUCTURE AND FUNCTIONS OF NERVE-TISSUE.

HAVING thus arrived at the best available Criterion of Mind considered as an eject, we have now to pass on to the topic which has already been propounded, viz., to a consideration of the objective conditions under which known mind is invariably found to occur.

Mind, then, so far as human experience extends, is only certainly known to occur in association with living organisms, and, still more particularly, in association with a peculiar kind of tissue which does not occur in all organisms, and even in those in which it does occur never constitutes more than an exceedingly small percentage of their bulk. This peculiar tissue, so sparingly distributed through the animal kingdom, and presenting the unique characteristic of being associated with mind, is, of course, the nervous tissue. It therefore devolves upon us, first of all, to contemplate the structure and the functions of this tissue, as far as it is needful for the purposes of our subsequent discussion that these should be clearly understood.

Throughout the animal kingdom nerve-tissue is invariably present in all species whose zoological position is not below that of the Hydrozoa. The lowest animals in which it has hitherto been detected are the Medusæ, or jelly-fishes, and from them upwards its occurrence is, as I have said, invariable. Wherever it does occur its fundamental structure is very much the same, so that whether we meet with nerve-tissue in a jelly-fish, an oyster, an insect, a bird, or a man, we have no difficulty in recognizing its structural units as everywhere more or less similar. These structural units are microscopical cells and microscopical fibres. (Figs. 1, 2.)

The fibres proceed to and from the cells, so serving to

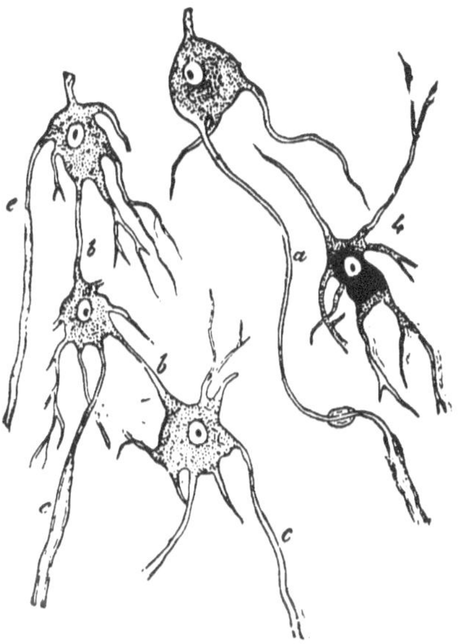

FIG. 1.—Motor Nerve Cells connected by intercellular processes (*b*, *b*), and giving origin to outgoing fibres (*c*, *c*, *c*, and *a*). 4. Multipolar cell containing much pigment around nucleus. Diagrammatic. (Vogt.)

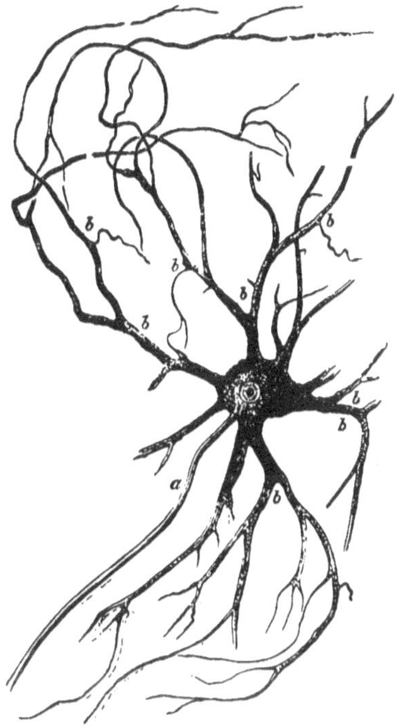

FIG. 2.—Multipolar Ganglion Cell from anterior grey matter of Spinal Cord of Ox. *a*, Axis cylinder process; *b*, branched processes, magnified 150 diameters. (Deiters.)

connect the cells with one another, and also with distant parts of the animal body. The function of the fibres is that of conducting stimuli or impressions (represented by molecular or invisible movements) to and from the nerve-cells, while the function of the cells is that of originating those of the impressions which are conducted by the fibres outwards. Those of the impressions which are conducted by the fibres inwards, or towards the cells, are originated by stimuli affecting the nerve-fibre in any part of its length; such stimuli may be contact with other bodies or pressure arising therefrom (mechanical stimuli), sudden elevations of temperature (thermal stimuli), molecular changes in the nerve-substance produced by irritants (chemical stimuli), effects of electrical disturbance (electrical stimuli), or lastly, the passage of a molecular disturbance from any other nerve-fibre with which the one in question may be connected.

Nerve-cells are usually found collected together in aggregates, which are called ganglia, to and from which large bundles of nerve-fibres come and go. These rope-like clusters of nerve-fibres constitute the white threads and strings which we recognize as nerves when we dissect an animal. (See Fig. 3.) The relation of the clusters of fibres to the cluster of cells is now such as to supply the anatomical condition to the performance of a physiological

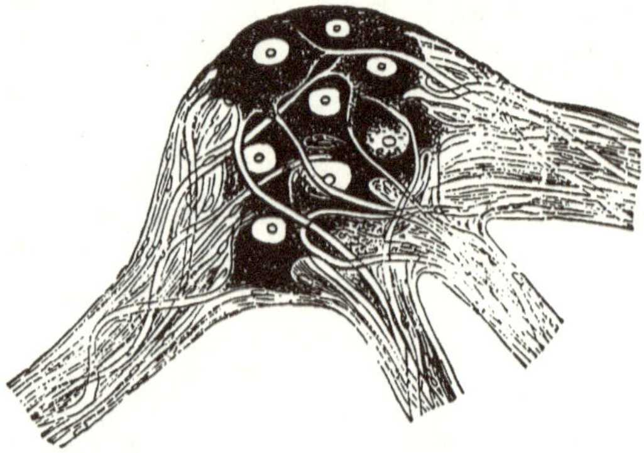

FIG. 3.—Small Sympathetic Ganglion (human) with Multipolar Cells. Magnified about 400 diameters. (Leydig.)

process, which is termed Reflex Action. If we suppose the left-hand bundle of fibres represented in the woodcut to be prolonged and to terminate in a sensory surface, while the other three bundles, when likewise prolonged, terminate in a group of muscles, then a stimulus falling upon the sensory surface would cause a molecular disturbance to travel along the left-hand or in-going nerve to the ganglion; on reaching the ganglion this disturbance would cause the ganglion to discharge its influence into the right-hand or outgoing nerves, which would then conduct this disturbance into the group of muscles and cause them to contract. This process is called reflex action, because the original stimulus falling upon the sensory surface does not pass in a direct line to its destination in the muscles, but passes first to the ganglion, and is thence *reflected* from the sensory surface to the muscles.* This, which at first sight appears a roundabout or cumbrous sort of process, is really the most economic that is available. For we must remember the enormous number and complexity of the stimuli to which all the higher animals are perpetually exposed, and the consequent necessity that arises for there being some system of co-ordination whereby these innumerable stimuli shall be suitably responded to. And such a system of co-ordination is rendered possible, and actually realized, through this principle of reflex action. For the animal body is so arranged that the innumerable nerve-centres, or ganglia, are all more or less in communication one with another, and so receive messages from all parts of the body, to which they respond by sending appropriate messages down the nerve-trunks supplying the particular groups of muscles which under the given circumstances it is desirable to throw into contraction. In other words, when a stimulus falls upon the external surface of an animal, it is not diffused in a general way throughout the whole body of the animal, so causing general and aimless contractions of all the muscles; but it passes at once to a nerve-centre, and is there centralized; the stimulus is dealt with in a manner which leads to an appropriate response of the organism to that stimulus. For the nerve-centres which receive the stimulus only reflect it to those particular muscle-groups which it is desirable for the organism, under the circumstances, to throw into action. Thus, to take an example, when a small foreign body, such as a crumb of bread, lodges in the windpipe, the stimulus which it there causes is immediately conveyed to a nerve-centre in the spinal cord, and this nerve-centre then originates, by reflex action, a highly complicated series of muscular movements which we call coughing, and which clearly have for their very special object the expulsion of the foreign body from a position of danger to the organism. Now it is obvious that so complicated a series of muscular movements could not be performed in the absence of a centralizing mechanism; and this is only one instance among hundreds of others that might be adduced of

* The term, however, is not a happy one, because the process is something more than the reflection of the original stimulus or molecular disturbance; the ganglion adds a new disturbance.

the co-ordinating power which is secured by this principle of reflex action.

Of course we may wonder how it is that the nerve-centres, which preside over reflex action, not being endowed with consciousness, know what to do with the stimuli which they receive. The explanation of this, however, is that the anatomical arrangement of ganglion and nerves in any particular case is such as to leave no choice or alternative of action, if the apparatus is called into action at all. Thus, to begin at the bottom of the series, in the Medusæ the simple ganglia are distributed all round the margin of the animal, and respond by reflex action to the stimuli which are applied at any other part of the surface. This has the effect of increasing the rate and the strength of the swimming-movements, and so of enabling the animal to escape from the source of danger. Now, although this is a true reflex action, and has an obvious purpose to serve, it does not involve any co-ordination of muscular movements. For the anatomical plan of a jelly-fish is so simple, that all the muscular tissue in the body is spread out in the form of one continuous sheet; so that the only function which the marginal ganglia have to perform when they are stimulated into reflex action, is that of throwing into contraction one continuous sheet of muscular tissue.

Hence we may infer that in its earliest stages reflex action is nothing more than a promiscuous discharge of nervous energy by nerve-cells, when they are excited by a stimulus passing into them from their attached nerve-fibres.* But as animals become more highly organized, and distinct muscles are by degrees set apart for the performance of distinct actions, we can readily understand how particular nerve-centres are likewise by degrees set apart to preside over these distinct actions; the nervous centres then perform the part of triggers to the particular muscular mechanisms over which they preside—triggers which can only be loosened by the reception of stimuli along their own particular lines of communication, or nerves. Thus, for instance, in the star-fish—animals which are somewhat higher in the zoological scale than the jelly-fish, and which have a more highly developed neuro-muscular system—the ganglia are arranged in a ring round

* For a full account of reflex action in Medusæ, see *Phil. Trans.*, *Croonian Lecture*, 1875; also *Phil. Trans.*, 1877 and 1880.

the bases of the five rays, into which they send, and from which they receive, nerve-fibres; the ganglia are likewise connected with one another by a pentagonal ring of fibres. Now experiment shows that in this simple, and indeed geometrical plan of a nervous system, the constituent parts are able, when isolated by section, to preside over the movements of their respective muscles; for if a single ray be cut off at its base, it will behave in all respects just like the entire star-fish—crawling away from injury, towards light, up perpendicular surfaces, and righting itself when turned upon its back. That is to say, the single nerve-centre at the base of a single separated ray is able to do for that ray what the entire pentagonal ring, or central nervous system, is able to do for the entire animal; it is for that ray the trigger which, when touched by the advent of a stimulus, throws the muscular mechanism into appropriate action. Thus it is evident that each of the five nerve-centres stands in such anatomical relation to the muscles of its own ray, that when certain stimuli fall upon the ray, the process of reflex action leaves no choice of response. The beauty and delicacy of this mechanism is shown when in the unmutilated animal all the nerve-centres are in communication as one compound nerve-centre. For now, if one ray is irritated, all the rays will co-operate in making the animal crawl away from the source of irritation; if two opposite rays are simultaneously irritated, the star-fish will crawl away in a direction at right angles to an imaginary line joining the two points of irritation. And, more prettily still, in the globular Echinus, or sea-urchin (which is, anatomically considered, a star-fish whose five rays have become doubled over in the form of an orange, soldered together and calcareous so as to make a rigid box), if two equal stimuli be applied simultaneously at any two points of the globe, the direction of escape will be the diagonal between them; if a number of points be simultaneously irritated, one effect neutralizes the other, and the animal rotates upon its vertical axis; if a continuous zone of injury be made all the way round the equator, the same thing happens; but if the zone be made wider at one hemisphere than the other, the animal will crawl away from the *greatest amount of injury*. So that in the Echinoderms the geometrical distribution of the nervous system admits of our making experiments in reflex action with very precise quantitative results; we can, as it were, play upon this beautifully adjusted mechanism, so as to produce at will the balancing or this stimulus against that one—the results, as expressed in the movements of the animal, being so many exemplifications of the mechanical principle of the parallelogram of forces.*

As we proceed through the animal series we find nervous systems becoming more and more integrated; nerve-centres multiply, become larger, and serve to innervate more numerous and more complex groups of muscles. It is, however, needless for me to devote space to describing this advance of structure, because the subject is one belonging to comparative anatomy. It is enough to say that everywhere the nervous machinery is so arranged that, owing to the anatomical plan of a nerve-centre with its attached nerves, there is no alternative of action presented to the nerve-centre other than that of co-ordinating the group of muscles over the combined contraction of which it presides. The next question, therefore, which arises is—How are we to explain the fact that the anatomical plan of a ganglion, with its attached nerves, comes to be that which is needed to direct the nervous tremours into the particular channels required? The following is the theory whereby Mr. Herbert Spencer seeks to answer this question, and in order fully to understand it we must begin by noticing the effects of stimulation upon undifferentiated protoplasm. A stimulus, then, applied to homogeneous protoplasm, which is everywhere contractile and nowhere presents nerves, has the effect of giving rise to a visible wave of contraction, which spreads in all directions from the seat of stimulation as from a centre. A nerve, on the other hand, conducts a stimulus without undergoing any contraction, or change of shape. Nerves, then, are functionally distinguished from undifferentiated protoplasm by the property of conducting invisible or molecular waves of stimulation from one part of an organism to another, so establishing physiological continuity between such parts without the necessary passage of visible waves of contraction.

Now, beginning with the case of undifferentiated protoplasm, Mr. Spencer starts from the fact that every portion of the colloidal mass is equally excitable and equally contrac-

* For a full account of these experiments, see *Croonian Lecture, Phil. Trans.*, 1882.

tile. But soon after protoplasm begins to assume definite shapes, recognized by us as specific forms of life, some of its parts are habitually exposed to the action of forces different from those to which other of its parts are exposed. Consequently, as protoplasm continues to assume more and more varied forms, in some cases it must happen that parts thus peculiarly situated with reference to external forces will be more frequently stimulated to contract than are other parts of the mass. Now in such cases the relative frequency with which waves of stimulation radiate from the more exposed parts, will probably have the effect of creating a sort of polar arrangement of the protoplasmic molecules lying in the line through which these waves pass, and for other reasons also will tend ever more and more to convert these lines into passages offering less and less resistance to the flow of such molecular waves,—*i.e.*, waves of stimulation as distinguished from waves of contraction. And lastly, when lines offering a comparatively low resistance to the passage of molecular impulses have thus been organically established, they must then continue to grow more and more definite by constant use, until eventually they become the habitual channels of communication between the parts of the contractile mass through which they pass. Thus, for instance, if such a line has been established between the points A and B of a contractile mass of protoplasm, when a stimulus falls upon A, a molecular wave of stimulation will course through that line to B, so causing the tissue at B to contract—and this even though no wave of *contraction* has passed through the tissue from A to B. Such is a very meagre epitome of Mr. Spencer's theory, the most vivid conception of which may perhaps be conveyed in a few words by employing his own illustration, viz., that just as water continually widens and deepens the channel through which it flows, so molecular waves of the kind we are considering, by always flowing in the same tissue tracts, tend ever more and more to excavate for themselves functionally differentiated lines of passage. When such a line of passage becomes fully developed, it is a nerve-fibre, distinguishable as such by the histologist; but before it arrives at this its completed stage, *i.e.*, before it is observable as a distinct structure, Mr. Spencer calls it a "line of discharge."*

* A certain amount of experimental verification has been lent to this

Such being the manner in which Mr. Spencer supposes nerve-fibres to be evolved, he further supposes nerve-cells to arise in positions where a crossing or confluence of fibres gives rise to a conflict of molecular disturbances; but it is unnecessary for our present purposes to enter upon this more elaborate and less satisfactory part of his theory.* All I desire now to point out is the *à priori* probability that nervous channels become developed where they are required simply from the fact of their being required—that is by use.

And this *à priori* probability derives so much confirmation from facts that it is scarcely possible to refrain from accepting it as an answer to the question above propounded, namely, How are we to explain the fact that the anatomical plan of a ganglion with its attached nerves comes to be that which is needed to direct the nervous tremours into the particular channels required? It is a matter of daily observation that "practice makes perfect," and this only means that the co-ordinations of muscular movement which are presided over by this or that nerve-centre admit of more ready performance the more frequently they have been previously performed—which, in turn, only means that the discharges taking place in the nerve-centre travel more and more readily through the channels or nerve-fibres which are being rendered more and more permeable by use. So much, indeed, is this the case, that when an associated muscular

theory by my own work on the physiology of nerves in Medusæ. For a full account of this, I may refer to a lecture published in the *Proceedings of the Royal Institution* for 1877, on "Evolution of Nerves." The principal facts are that when physiological continuity of a sheet of neuro-muscular tissue is interrupted by overlapping or spiral sections, so that the passage both of visible or muscular waves of contraction and invisible or molecular waves of stimulation are blocked, after a long succession of contraction waves are allowed to break upon the shore of the physiological interruption, they at last begin to force a passage, and very soon this passage becomes perfectly free, so that neither the waves of contraction nor those of stimulation are any longer hindered. Whether in such a case a definite nerve-fibre is developed, or only a "line of discharge," I cannot say; but most probably the passage is effected through previously existing fibres of the plexus which become more functionally developed by their increase of activity.

* Less satisfactory, not only because more speculative, but because the whole weight of embryological and histological evidence appears to me to be opposed to the speculation. For the whole weight of this evidence goes to show that nerve-cells are the result of the specialization of epithelial or epidermal cells—that is, that they arise, not out of undifferentiated protoplasm, but by way of a further differentiation of a particular kind of already differentiated tissue, where this is exposed to particular kinds of stimulation.

movement takes place with sufficient frequency, it cannot by any effort of the will become again dissociated; as is the case, for instance, with the associated movement of the eyeballs, which does not begin to obtain till some days after birth, but which then soon becomes as closely organized as any of the associated movements in the muscles of the limbs.*

And if this is the case even in the life-time of individuals, we can scarcely wonder that in the life-time of species heredity with natural selection should still more completely adapt the anatomical plan of ganglia, with their attached nerves, to the performance of the most useful—*i.e.*, the most habitual—actions. Thus we may see in a general way how such nervous machinery may at last come to be differentiated into specially distributed anatomical structures, which, on account of their special distribution, are adapted to minister only to particular co-ordinations of muscular movements. That is to say, we are thus able to understand the rise and development of Reflex Action.

* Mr. Darwin called my attention to the following passage in the writings of Lamarck (*Phil. Zool.*, tom. ii, pp. 318–19):—"Dans toute action, le fluide des nerfs qui la provoque, subit un mouvement de déplacement qui y donne lieu. Or, lorsque cette action a été plusieurs fois répétée, il n'est pas douteux que le fluide qui l'a exécutée, ne se soit frayé une route, qui lui devient alors d'autant plus facile à parcourir, qu'il l'a effectivement plus souvant franchie, et qu'il n'ait lui-même une aptitude plus grand à suivre cette route frayée que celles qui le sont moins."

CHAPTER III.

The Physical Basis of Mind.

We have already taken it for granted that Mind has a physical basis in the functions of the nervous system, or that every mental process has a corresponding equivalent in some neural process. I shall next endeavour to show how precise this equivalency is.

We have seen that ganglionic action consists of waves of nervous tremours originating in the cells, coursing along the attached fibres to other cells, and there arousing fresh impulses of the same kind. Moreover, we have seen that this coursing of nervous impulses through nervous arcs is not, as it were, promiscuous, but that, owing to the anatomical plan of a ganglion, it takes place in certain determinate directions, so that the result, when expressed in muscular movement, shows the function of a ganglion to be that of centralizing nervous action, or of directing nervous tremours into definite channels. Lastly, we have seen that this directing or centralizing function of ganglia has probably in all cases been due to the principle of use combined with that of natural selection. Now it is known from experiments on the lower animals, as well as from the effects of cerebral disease in man, that the part of the nervous system in all the Vertebrata which appears to be exclusively concerned in all mental operations, is the so-called "large brain," or cerebral hemispheres. This is the convoluted part of the brain which appears immediately below the skull, and is above all the series of ganglia or nerve-centres which occupy the rest of the cerebro-spinal tract. As some at least of the bewildering multitude of cells and fibres which constitute the cerebral hemispheres are in connection with these lower ganglia, there is no doubt that the hemispheres are able to "play down" upon these ganglia as upon so many mechanisms, whose function it is to throw this and that group of muscles into action. Much light is at present being thrown upon this subject by the researches of Hitzig, Fritsch, Ferrier, Goltz, and others; but we must pass on to consider that function of these great nerve-centres with which we shall henceforth be exclusively concerned, the function, namely, of being associated with the phenomena of Mind.

As the cerebral hemispheres pretty closely resemble in their intimate structure ganglia in general, there can be no reasonable doubt that the mode of their operation is substantially the same; and as such operation is here attended with the phenomena of subjectivity, there can be equally little doubt that such phenomena must constitute a sort of obverse reflection of ganglionic action. Looking, then, upon this obverse reflection, can we detect any fundamental principles of mental operation which may reasonably be taken to correspond with the fundamental principles of ganglionic operation?

The most fundamental principle of mental operation is that of memory, for this is the *conditio sine quâ non* of all mental life. But memory on its obverse side, or the side of physiology, can only mean that a nervous discharge, having once taken place along a certain route, leaves behind it a molecular change, more or less permanent, such that when another discharge afterwards proceeds along the same route, it finds, as it were, the footprints of its predecessor. And this, as we have seen, is no more than we find to be the case with ganglionic action in general. Even long before movements involving muscular co-ordination have been repeated with sufficient frequency to become consolidated into one organized and indissoluble act, they become, in virtue of the principle which I have termed the principle of use, more and more easy to repeat; in all but in the absence of a mental constituent the nerve-centre concerned *remembers* the previous occurrence of its own discharges; these discharges have left behind them an *impress* upon the structure of the ganglion just the same in kind as that which, when it has taken place in the structure of the cerebral hemispheres, we recognize on its obverse side as an *impress* of memory. The analogy is much too close to be attributed to accident, for it extends into all details. Thus, a ganglion may *forget* its previous activity if too long an interval is allowed to elapse between the repetitions of its activity, as every one must know who is in the habit of playing on a musical instrument, or performing any other actions entailing the acquirement of dexterity. It may also be observed that when such is the case the particular activity forgotten by the ganglion may be more easily re-acquired than originally it was acquired, which is just what we find to be the case with mental attainments.

As particular illustrations of these facts I may state two or three cases, which will also serve to show of how little importance (on the objective side) is the occurrence of consciousness to the memory of a ganglion.

Robert Houdin early in life practised the art of juggling with balls in the air, and after a month's practice was able to keep four balls in the air at once. His neuro-muscular machinery was now so well trained, or remembered so well how to perform the series of actions required, that he could afford to withdraw his attention from the performance to the extent of reading a book without hesitation while keeping up the four balls. Thirty years afterwards, on trying the same experiment, having scarcely once handled the balls between times, he found that he could still read with ease while keeping up *three* balls; the ganglia concerned had partly forgotten their work, but on the whole remembered it wonderfully well. Again, Lewes gives the case of a waiter asleep at a coffee-house, with much noise of talking around him, who was instantly aroused by a low cry of "Waiter;" and Dr. Abercrombie gives the case of a man who had long been in the habit of taking down a repeater watch from the head of his bed to make it strike the last hour, and who was observed to do this when otherwise apparently unconscious from a fit of apoplexy. But perhaps the most remarkable of all the cases that can be adduced are the most familiar ones of walking and speaking. When we remember the immense amount of neuro-muscular co-ordination that is required for either of these actions, and the laborious steps by which each of them is first acquired in early childhood, it is indeed astonishing that in after life they come to be performed without thought of their performance; the ganglia concerned have fully *learned* their work.

So much for memory. But memory would be a useless faculty of mind if it did not lay the basis for another, and really the most important principle of subjectivity; I mean the Association of Ideas. This is the root and branch of the whole structure psychological, and therefore, if mind has a physical basis, we should expect to meet with some very general and essential feature of ganglionic action answering to this very general and essential feature of mental action. And this, beyond question, we do find.

For the association of ideas is merely a development of simple memory. A mental impression, image, memory, or idea having once occurred in juxtaposition with another, not only are the two memories remembered, but also the fact of their juxtaposition, so that when one memory or idea is aroused, the other is aroused likewise. Let us, then, look at the matter a little more closely, in order to see how this great principle of psychology may receive its explanation, so far as the collateral principle of physiology is concerned.

There can be no doubt that in the complex structure of the cerebral hemispheres one nervous arc (*i.e.*, fibres, cells, and fibres) is connected with another nervous arc, and this with another almost *ad infinitum;* and there can be equally little doubt that processes of thought are accompanied by nervous discharges taking place, now in this arc, and now in that one, according as the group of nerve-cells in each arc is excited to discharge its influence by receiving a discharge from some of the other nerve-arcs with which it is united. Again, as we have seen, it is practically certain that the more frequently a nervous discharge takes place through a given group of nervous arcs, the more easy will it be for subsequent discharges to take place along the same routes—these routes having been thus rendered more permeable to the passage of subsequent discharges. And now a very little reflection will show that in this physiological principle we no doubt have the objective side of the psychological principle of the association of ideas. For it may be granted that a series of discharges taking place through the same group of nervous arcs will always be attended with the occurrence of the same series of ideas; and it may be further granted that the previous passage of a series of discharges through any group of nervous arcs, by making the route more permeable, will have the effect of making subsequent discharges pursue the same course when started from the same origin. And if these two propositions be granted, it follows that the tendency of ideas to *re*cur in the same order as that in which they have previously *occurred*, is merely a psychological expression of the physiological fact that lines of discharge become more and more permeable by use.

We thus see that the most fundamental of physiological principles—the association of ideas—is merely an obverse expression of the most fundamental of neurological principles —reflex action; and that such, in general terms, is the fact, seems to be proved beyond question by such instances as those above given of the sleeping waiter and Dr. Abercrombie's unconscious patient, &c.; for such cases prove that actions originally due to a conscious association of ideas may, by a sufficiently long course of ganglionic instruction, cease to be conscious actions, and therefore become in no way distinguishable from reflex actions.*

But the proof of the fundamental correlation between ganglionic action and mental action does not end even here. There is another line of evidence which, although perhaps not quite so definite, nevertheless seems to me most cogent, and even more interesting than the considerations already adduced. If we take ideation to be in the same sense an index of the higher or more complex nervous processes, as muscular movement is of the lower or less complex, we shall find evidence to show that the development of ideation, or mental evolution, implies a further and continuous development of the corresponding nervous processes, which is precisely the same in kind as that which on the lower plane (that of muscular movement) has led to the advancing development of muscular co-ordination. In other words, if we consent to change the index from muscles to ideas, we shall find evidence that the method of nervous evolution has throughout been uniform; we shall find that the progressive elaboration of nervous structures—which in the one case has found expression in the growing complexity of the muscular system, and in the other case has been reflected in the advancing phases of mental evolution—we shall find that this progressive elaboration has throughout been pervaded by the same principles of development.

* A good instance of this may be found in the fact that men always bring their knees together in order to catch a small falling object, such as a coin, while women always spread their knees apart. The reason of course is that the difference of dress has led to a difference of organized habit—the habit in each case having been originally due to intelligent adjustment, but now scarcely distinguishable from a reflex.

Disregarding the philosophical question as to how nervous action is associated with subjective ideation, and concerning ourselves only with the scientific fact that it is thus associated, we may most clearly appreciate the parallel which I am about to draw if we regard the objective processes as the causes of the subjective. Whether or not such is really the case matters nothing to the exposition on which I am about to enter; for I throughout take it for granted that the association of neurosis and psychosis is as invariable and precise as it would be were it proved to be due to a relation of causality. Placing therefore neurosis for the purposes of my argument as the cause of psychosis, I desire to show that there is a very exact parallel between the ganglionic action which produces subjective ideation and that which produces muscular co-ordination; I desire to show that if we interpret the phenomena of ideation in terms of the nervous activity which is supposed to produce it, we shall find that this activity is just the same in all its laws and principles as that which produces muscular co-ordination.

No doubt it sounds absurd, and from a philosophical point of view alone it is absurd, to speak of ideas as the psychological equivalents of muscles. So far as subjective analysis could teach us, it certainly does not seem that an idea presents any further kinship to a muscle than it does to a stone, or to the moon; but when we look at the matter from the objective side, we perceive that the kinship is most intimate. Taking it for granted that the same idea is only and always aroused during the activity of the same nervous structure, element, or group of cells and fibres, it follows that any particular mental change resembles any particular muscular contraction in so far as it is the terminal result of the activity of a particular nervous structure. The incongruity of comparing a mental change to a muscular contraction arises, of course, from the emphatic distinction which must always be felt to exist between mental and dynamical processes. Physiology, which is concerned only with the dynamical processes, can take no cognizance of anything that happens in the region of mind. It can trace nervous action leading to combined muscular movements of greater and greater intricacy as we ascend to more and more elaborated mechanisms; but even when we reach the brain of man, physiology can have nothing to do with the mental side of the nervous processes. All that physiology can see in these processes is a greatly improved power of discriminating between stimuli, and of issuing impulses to a correspondingly greater number and variety of adaptive movements; the mental changes which accompany these nervous processes are as wholly without the ken of physiology as these nervous processes are without the ken of subjectivity. Therefore it is that when we speak of an idea as the analogue of a muscle, we feel the incongruity of confusing two things which are separated from one another by the whole interval that divides subject from object. But although in speaking of an idea as the analogue of a muscle, we do and ought to feel the incongruity, let it not be supposed that by thus speaking we are allowing ourselves to be betrayed into any confusion of thought. I speak of a mental change as the analogue of a muscular contraction only with reference to its being the terminal event invariably associated (whether by way of causality or not) with the activity of a nervous structure. And if we do not seek to press the analogy further than this, there is no fear of our confusing ideas which ought always to be kept fundamentally distinct.

So much, then, by way of introduction to the point which I have to make plain. Now it admits of being abundantly proved that throughout the animal kingdom, so long as we regard the muscular system as our index of the structural advances taking place in the nervous system, we find this index to consist in the growing complexity of the muscular system, and the consequent increase in the number and variety of co-ordinated movements which this system is enabled to execute. Therefore the point which I have to prove will be proved if I can make it clear that the process of mental evolution bears some such resemblance to that of muscular evolution as we should expect that it ought to bear, if they are both dependent on a similar process of nervous evolution. In other words, I have to show that the process of mental evolution consists essentially in a progressive co-ordination of progressively developing mental faculties, analogous to that which takes place in muscular movements.

Beginning with the faculties of simple sensation, we know, for instance, that when a note of music is struck, it appears to produce a single vibration, and yet physical analysis shows that the sound is not a single vibration, but a highly complex structure of vibrations or harmonics, and that the ear takes in all these harmonics by as many separate nervous elements (whatever the elements may be which minister to the perception of pitch), although they are all blended into one compound sensation, which is so well compounded that the evidence supplied by it alone would never have led us to suspect that the sensation was other than simple. The same is known to be the case with sensations of colour, taste, and smell; so that Lewes feels justified in going to the length of saying, "Every sensation is a group of sensible components."* And, taking the same view on the psychological side as I take, he further says in general terms, "The main fact on which our exposition rests is indisputable, namely, that sensation, perception, emotions, conceptions, are not simple undecomposable states, but variously compounded."

To avoid being tedious, I shall not pursue the analysis through all the grades of the psychological faculties; but, taking ideation in its widest sense, as including alike the mere memory of a sensation and the most complex process of abstract thought, I shall briefly show that it everywhere displays a grouping and compounding of subjective elements which, if translated into their objective counterparts, display precisely the same method of nervous evolution as that which obtains in the lower ganglia, as expressed by muscular co-ordination.

As Bain observes, "Movements frequently conjoined become associated, or grouped, so as to arise in the aggregate at one bidding. Suppose the power of walking attained, and also the power of rotating the limbs, one may then be taught to combine the walking pace with the turning of the toes outward. Two volitions are at first requisite for this act, but after a time the rotation of the limb is combined with the act of walking, and, unless we wish to dissociate the two, they go together as a matter of course; the one resolution brings on the combined movement. . . . . Articulate speech largely exemplifies the aggregation of muscular movements and positions. A concurrence of the chest, larynx, tongue, and mouth, in a definite group of exertions, is requisite for each alphabetical letter. These groupings, at first impossible,

* *Problems, &c.*, p. 260.

are, after a time, cemented with all the firmness of the strongest instinct."

Precisely analogous to this process of blending many separate muscular movements into one simultaneous and compounded movement, is the process of blending many simple ideas into one complex or compounded idea. Just as muscular co-ordination is dependent on the simultaneous action of a certain group of nerve-centres for the purpose of securing the combined action of a number of muscles, so we must suppose that a general or a composite idea is dependent on the simultaneous activity of several nerve-centres which minister to the several component parts of the blended idea. The psychological side of this process has been so well expressed by James Mill, that I cannot do better than render it in his words:—"Ideas which have been so often conjoined that whenever one exists in the mind the other exists along with it, seem to run into one another, to coalesce, as it were, and out of many to form one idea, which idea, however, in reality complex, appears to be no less simple than any one of those of which it is compounded. . . . . . The word 'gold,' for example, or the word 'iron,' appears to express as simple an idea as the word 'colour,' or the word 'sound.' Yet it is immediately seen that the idea of each of those metals is made up of the separate ideas of several sensations: colour, hardness, extension, weight. Those ideas, however, present themselves in such intimate union, that they are constantly spoken of as one, not many. We say, our idea of iron, our idea of gold; and it is only with an effort that reflecting men perform the decomposition." And similarly, of course, with the most highly complex ideas, except that the more complex they become the greater is the difficulty of securing the needful composition, and the more easily do they undergo disintegration. Thus it is that, in the words of Mr. Spencer, "In the development of mind there is a progressive consolidation of states of consciousness. States of consciousness once separate become indissoluble. Other states that were originally united with difficulty, grow so coherent as to follow one another without difficulty. And thus there arise large aggregations of states, answering to complex external things—animals, men, buildings—which are so welded together as to be practically single states. But this integration, by uniting a large number of related sensations into one state, does not destroy them. Though subordinated as parts of a whole, they still exist."*

Again, just as the principle of association is exhibited in the case of ideas not only with reference to the *simultaneous* blending of simple ideas into one complex idea, but also with reference to the *successive* sequence or concatenation of ideas; so in the case of muscular co-ordinations we acquire, not only the power of a simultaneous co-operation of muscle-groups, but also that of a successive co-operation. For instance, as Professor Bain observes, "In all manual operations there occur successions of movements so firmly associated, that when we will to do the first, the rest follow mechanically and unconsciously. In eating, the action of opening the mouth mechanically follows the raising of the morsel. . . . . Although the learning of successions of movements involves the medium of sensation, in the first instance, yet we must assume that there is a power, in the system, for associating together movements as such." In fact, it might well have been added, there is such a power that manifests itself long before the dawn of any of the powers of the "will"; it is as true of the polyp as of the man that "in eating, the act of opening the mouth mechanically follows the raising of the morsel."

So with the highest or most abstract powers of mind. For abstraction merely means the mental dissociating of qualities from objects, and, in its higher phases, blending these qualities, or conceptions of them, into new ideal combinations.

Lastly, just as innumerable special mechanisms of muscular co-ordinations are found to be inherited, innumerable special associations of ideas are found to be the same; and in one case as in the other, the strength of the organically imposed connection is found to bear a direct proportion to the frequency with which in the history of the species it has occurred. Thus, the simplest, oldest, and most constant ideas relating to time, space, number, sequence, &c., may be compared, in point of organic integrity, with the oldest and most indissolubly associated muscular movements, such as those concerned in breathing, deglutition, and visceral motions. Again, inherited instincts have their counterparts in such inherited muscular co-ordinations as are not abso-

* *Principles of Psychology*, vol. ii, p. 476.

lutely indissoluble. And similarly, of course, associations of ideas acquired only during the life-history of the individual need to be more or less constantly maintained by repetition, just as muscular co-ordinations similarly acquired can only be maintained by practice.

Upon the whole, therefore, it is impossible that there could be a more precise parallelism between these two manifestations of nervous machinery, and it is one which for recognition in a general way does not require scientific analysis; it has been perceived by the common sense of mankind—witness, for instance, the term "gymnastics" having become applicable to mental no less than to muscular co-ordinations. But, for the sake of systematic completeness, I shall conclude this exposition by briefly pointing out that all those pathological derangements which occur in the nervous centres that preside over muscular activities, have their parallels in similar derangements which occur in the nervous centres that are concerned in mental activities. Thus "nervousness," or a disturbance of the normal balance of nerve-centres, has a strikingly analogous effect in confusing the ideas and in perturbing muscular co-ordinations. Idiotcy has its parallel in inability to perform complex muscular movements, with which inability, indeed, idiotcy is itself almost invariably associated. Lunacy has it counterpart in an unbalanced, or badly correlated power of muscular co-ordination, which in its graver manifestations is known as ataxy; while mania is mental convulsion, and unconsciousness mental paralysis.

I must not, however, take leave of this branch of our subject without briefly alluding to a difficulty which may occur to some minds, and which has been well stated by Professor Calderwood in his recently published work.* The difficulty to which I allude arises from there being an absence of such a constant relationship between the size or mass of the brain, and the degree of intelligence displayed by it, as the foregoing teaching would reasonably lead us to expect.

Now, I do not deny that the relation of intelligence to size, mass, or weight of brain is a perplexing matter when we look to the animal kingdom as a whole; for although there is unquestionably a *general* relation of a quantitative

* Pp. 211—216.

kind, it is not a *constant* relation. Even within the limits of the human species this relation is not so precise as is usually supposed; for, neglecting particular cases that might be quoted of men of genius not having particularly large or heavy brains, the converse cases are perhaps in this connection more remarkable—viz., those of feeble-minded persons having large and apparently well-formed brains. I am indebted to Dr. Frederick Bateman of the Eastern Counties' Asylum for directing my attention to the observations of Dr. Mierzejewskis, which were published at the international congress of psychologists held in Paris in 1878. These observations, which appear to have been carefully made, seeing that casts of the brains were exhibited, went to show that idiotcy is compatible with large and apparently well-developed brains—the amount of grey matter in one instance being "enormous."

And, if we turn to the animal kingdom, we find in a still larger measure that the mere amount of cerebral substance furnishes but a very uncertain index of the level of intelligence which is attained by the animal. This is the case even when we eliminate the element of complexity that is introduced by the differences which obtain in different animals between the bulk of the brain and the bulk of the body—small animals requiring a greater proportional bulk of brains than large ones, because the nervous machinery which ministers to muscular movement and co-ordination has in both cases to be accommodated. But this element of complexity may be removed by considering the cases in which small animals exhibit remarkable intelligence; and in this respect no animals are so remarkable as the more intelligent species of ants alluded to in my former work. As Mr. Darwin has observed, the brain of such an insect deserves to be regarded as perhaps the most wonderful piece of matter in the world.

But if this whole question touching the relation between the mass of brain and degree of intelligence is felt to lie as a difficulty in the way of evolutionary theory, I should reply to it by the following considerations.

In the first place, that there is a *general* relation between size of brain and degree of intelligence, both in the case of man and in that of animals, is unquestionable. It is, therefore, only with the more special exceptions that we have to deal. But here we have to remember that besides size or mass, there must certainly be a no less important factor to be taken into account—that, namely, of structure or complexity. Now we really know so little about the relations of intelligence to neural structure, that I do not think we are justified in forming any very strong conclusions *à priori* concerning the relation of intelligence to mere size or mass of brain. Knowing in a general way that mass *plus* structure of brain is necessary for intelligence, we do not know how far the second of these two factors may be increased at the expense of the first. And, as a mere matter of complexity, or of *multum in parvo*, I am not sure that even the brain of an ant deserves to be considered more wonderful than the ovum of a human being. Lastly, in this connection it may be as well to observe that there is as good evidence to show the importance of cerebral structure as a factor in determining the level of mental development, as there is to show the importance of cerebral mass. Throughout the vertebrated series of animals the convolutions of the brain—which are the coarser expressions of more refined complexities of cerebral structure—furnish a wonderfully good general indication of the level of intelligence attained; while in the case of ants Dujardin says that the degree of intelligence exhibited stands in an inverse proportion to the amount of cortical substance, or in direct proportion to the amount of the peduncular bodies and tubercles. In view of these considerations, therefore, I do not feel that the supposed difficulty, which I have thought it desirable to mention, is one of any real solidity.

CHAPTER IV.

THE ROOT-PRINCIPLES OF MIND.

ALTHOUGH the phenomena of Mind, and so of Choice, are both complex, and as to their causation obscure, I think we have now seen that we are justified in believing that they all present a physical basis. That is to say, whatever opinion we may happen to entertain regarding the ultimate nature of these phenomena, in view of the known facts of physiology, we ought all to be agreed concerning the doctrine that the mental processes which we cognize as subjective, are the psychical equivalents of neural processes which we recognize as objective. As already stated, I have elsewhere considered the various hypotheses concerning the nature and the various attempts at an explanation of this equivalency between mental processes and neural processes; but here I desire to consider the fact of this equivalency merely as a fact. It will therefore signify nothing to my discussion whether, with the materialists, we rest in this fact as final, or endeavour, with men of other schools, to seek an explanation of the fact of some more ultimate character. It is enough if we are agreed that every psychical change of which we have any experience is invariably associated with a definite physical change, whatever we may suppose to be the nature and significance of this association.

Looking, then, at the phenomena of Mind as invariably presenting a physical, or, as we may indifferently call it, a physiological side, I shall endeavour to point out what I conceive to be the most ultimate principle of physiology which analysis shows to be common to them all. On the mental side, as we have already seen, we have no difficulty in distinguishing this ultimate principle, or common characteristic, as that which we designate by the term Choice. Now if the power of choice is the distinctive peculiarity of a mental being, and if, as we have taken for granted, every change of Mind is associated with some change of Body, it follows that this distinctive peculiarity ought to admit of being translated into some physiological equivalent. Further, if there is any such physiological equivalent to be found, we should expect to find it much lower down in the scale of physiological development than in the functions of the human brain. For not only do the lower animals manifest, in a long descending scale, powers of choice which gradually fade away into greater and greater simplicity; but we should be led *à priori* to expect, if there is a physiological principle which constitutes the objective basis of the psychological principle, that the former should manifest itself more early in the course of evolution than the latter. For, whatever views we may entertain concerning the relation of Body and Mind, there can be no question, on the basis of the evolution theory which I assume, that, as a matter of historical sequence, the principles of physiology were prior to those of psychology; and therefore, if in accordance with our original agreement we allow that the latter have a physical basis in the former, it follows that the principles of physiology, which now constitute the objective basis of choice, whatever they may be, probably came into operation long before they were sufficiently evolved thus to constitute the foundation of psychology.

Now I think that the *à priori* expectation thus briefly sketched is fully realized in the occurrence of a physiological principle, which first appears very low down in the world of life, and which, in its relation to psychology, has not yet received the attention which it deserves. I may best state the principle by giving an example. I have observed that if a sea-anemone is placed in an aquarium tank, and allowed to fasten upon one side of the tank near the surface of the water, and if a jet of sea water is made to play continuously and forcibly upon the anemone from above, the result of course is that the animal becomes surrounded with a turmoil of water and air-bubbles. Yet, after a short time, it becomes so accustomed to this turmoil that it will expand its tentacles in search of food, just as it does when placed in calm water. If now one of the expanded tentacles is gently touched with a solid body, all the others close around that body, in just the same way as they would were they expanded in calm water. That is to say, the tentacles are able to discriminate between the stimulus which is supplied by the turmoil of the water and that which is supplied by their contact with the solid body, and they respond to the latter stimulus notwithstanding that it is of incomparably less intensity than the former. And it is this power of discriminating between stimuli, *irrespective of their relative mechanical intensities*, that I regard as the objective principle of which we are in search; it constitutes the physiological aspect of Choice.

A similar power of discriminative response has long been known to occur in plants, though the most carefully observed facts with regard to this interesting subject are those which we owe to the later researches of Mr. Darwin and his son. The extraordinary delicacy of discrimination which these researches show the leaves of plants to exercise between darkness and light of the feeblest intensity, is not less wonderful than the delicacy of discrimination which they show the roots of plants to exercise in feeling about for moisture and lines of least resistance in the soil. But in the present connection the most suggestive facts are those which have been brought to light by Mr. Darwin's previous researches on the climbing and insectivorous plants. For, from these researches it appears that the power of discriminating between stimuli, irrespective of relative mechanical intensity or amount of mechanical disturbance, has here proceeded to an extent that rivals the function of nerve-tissue, although the tissues which manifest it have not in structure passed beyond the cellular stage. Thus, the tentacles of Drosera, which close around their prey like the tentacles of a sea-anemone, will not respond to the violent stimulation supplied by rain-drops falling upon their sensitive surfaces or glands, while they will respond to an inconceivably slight stimulus of the kind caused by an exceedingly minute particle of solid matter exerting by gravity a continuous pressure upon the same surfaces. For Mr. Darwin says, "The pressure exerted by a particle of hair, weighing only $\frac{1}{78740}$ of a grain, and supported by a dense fluid, must have been inconceivably slight. We may conjecture that it could hardly have equalled the millionth of a grain; and we shall hereafter see that far less than the millionth of a grain of phosphate of ammonia in solution, when absorbed by a gland, acts on it and induces movement. . . . It is extremely doubtful whether any nerve in the human body, even if in an inflamed condition, would be in any way affected by such a particle supported in a dense fluid, and slowly brought into contact with the nerve. Yet the cells of the glands of Drosera are thus excited to transmit a motor impulse to a distant point, inducing movement. It appears to me that hardly any more remarkable fact than this has been observed in the vegetable kingdom."

But the case does not end here. For in another insectivorous plant, Dionœa, or Venus' Fly-trap, the principle of discriminating between different kinds of stimuli has been developed in a direction exactly the opposite to that which obtains in Drosera. For while Drosera depends for capturing its prey on entangling the latter in a viscid secretion from its glands, Dionœa closes upon its prey with the suddenness of a spring-trap; and in relation to this difference in the mode of capturing prey, the principle of discrimination between stimuli has been correspondingly modified. In Drosera, as we have seen, it is the stimulus supplied by continuous *pressure* that is so delicately perceived, while the stimulus supplied by *impact* is disregarded; but in Dionœa the smallest impact upon the irritable surfaces, or filaments, is immediately responded to, while the stimulus supplied even by comparatively great pressure upon the same surfaces is wholly disregarded. Or, in Mr. Darwin's own words, "Although the filaments are so sensitive to a momentary and delicate touch, they are far less sensitive than the glands of Drosera to prolonged pressure. Several times I succeeded in placing on the tip of a filament, by the aid of a needle moved with extreme slowness, bits of rather thick human hair, and these did not excite movement, although they were more than ten times as long as those which caused the tentacles of Drosera to bend; and although in this latter case they were largely supported by the dense secretion. On the other hand, the glands of Drosera may be struck with a needle, or any hard object, once, twice, or even thrice, with considerable force, and no movement ensues. This singular difference in the nature of the sensitiveness of the filaments of Dionœa and of the glands of Drosera evidently stands in relation to the habits of the two plants. If a minute insect alights with its delicate feet on the glands of Drosera, it is caught by the viscid secretion, and the slight, though prolonged pressure gives notice of the presence of prey, which is secured by the slow bending of the tentacles. On the other hand, the sensitive filaments of Dionœa are not viscid, and the capture of insects can only be assured by their sensitiveness to a momentary touch, followed by the rapid closure of the lobes." So that in these two plants the power of discriminating between these two kinds of stimuli has been developed to an equally astonishing extent, but in opposite directions.

But we find definite evidence of this power of discriminative selection even lower down in the scale of life than the cellular plants; we find it even among the protoplasmic organisms. Thus, to quote an instructive case from Dr. Carpenter:—

"The Deep-Sea researches on which I have recently been engaged have not 'exercised' my mind on any topic so much as on the following:—Certain minute particles of living jelly, having no visible differentiation of organs . . . . build up 'tests' or casings of the most regular geometrical symmetry of form, and of the most artificial construction . . . From *the same sandy bottom*, one species picks up the *coarser* quartz-grains, cements them together with *phosphate of iron* (?), which must be secreted from their own substance; and thus constructs a flask-shaped 'test' having a short neck and a single large orifice. Another picks up the *finer* grains, and puts them together with the same cement into perfectly spherical 'tests' of the most extraordinary finish, perforated with numerous small tubes, disposed at pretty regular intervals. Another selects the *minutest* sand-grain and the terminal points of sponge-spicules, and works these up together—apparently with no cement at all, but by the 'laying' of the spicules—into perfect spheres, like homœopathic globules, each having a single fissured orifice." *

Thus, co-extensive with the phenomena of excitability, that is to say, with the phenomena of life, we find this function of selective discrimination; and, as I have said, it is this function that I regard as the root-principle of Mind. I so regard it because, if we consider all the faculties of mind, we shall observe that the one feature which on their objective side they present as common, is this power of discriminating among stimuli, and responding only to those which, irrespective of relative mechanical intensity, are the stimuli to which

* *Contemporary Review*, April, 1873.

responses are appropriate. In order to see this, let us take the principal faculties of mind in their ascending order, and consider what they are, in their last analysis, upon their physiological side. First we have the organs of special Sensation, the physiological functions of which clearly constitute the basis of the whole structure psychological. Yet no less clearly, these functions in their last analysis are merely so many specially developed aptitudes of response to special modes of stimulation. Thus, for instance, the structure of the eye is specially adapted to respond only to the particular mode of stimulation that is supplied by light, the ear to that which is supplied by sound, and so on. In other words, the organs of special sense are so many structures which have been variously and extremely differentiated in several directions, for the express purpose of attaining a severally extreme sensitiveness to special modes of stimulation without reference to any other mode. And this is merely to say that the function of an organ of special sense is that of sorting out, selecting, or discriminating the particular kind of stimulation to which its responsive action is appropriate.

Again, many of the nervous mechanisms which minister to various Reflex Actions are only thrown into activity by special modes of stimulation. This is notably the case with those highly complicated neuro-muscular mechanisms which are thrown into activity only by the mode of stimulation which we call tickling. Such instances are of special interest in the present connexion from the fact that the distinguishing peculiarity of this mode of stimulation consists in its being a stimulation of low intensity. The comparatively violent stimulation that is caused by the passage of food down the gullet, or by contact of the soles of the feet with the ground, is unproductive of any response on the part of the mechanisms which are thrown into violent activity by the gentlest possible stimulation of the same surfaces. Similarly with regard to Instincts. These, physiologically considered, are the activities of highly differentiated nervous mechanisms which have been slowly elaborated, through successive generations, for the express purpose of responding to some particular stimulus of a highly wrought character, and which, on its psychological side, is a recognition of the circumstances to which the instinctive adjustment is appropriate. And so with the Emotions. For, physiologically considered, the emotions are the activities of highly wrought nervous mechanisms, and these activities are only excited by the very special stimuli which, on their subjective side, we recognize as the particular kind of ideas which are appropriate to call up particular emotions. We do not laugh at a painful sight, nor does a ludicrous sight cause us to weep; and this, physiologically considered, merely means that the nervous machinery whose action is accompanied by one emotion, will only respond to one kind of very specialized and complex stimulation; it will not respond to another and probably in many respects very similar kind of stimulation, which, nevertheless, is competent to evoke responses from another and probably very similar piece of nervous machinery. And thus, also, it is with Reasoning and Judgment. Reasoning, on its physiological side, is merely a series of highly complicated nervous changes, regarding which the only thing we certainly know is, that not one of them can take place without an adequate physical accompaniment, and therefore that on its physiological side a train of reasoning is a series of nervous changes, every one of which must be produced by physical antecedents. And hence on its objective side every step in a train of reasoning consists in a selective discrimination among all those exceedingly delicate stimuli which, on their subjective side, we know as arguments. Similarly regarded, Judgment is likewise nothing more than the final result of the incidence of a vast number of very delicate stimuli; and this final result, like all the intermediate steps of the reasoning which led to it, is nothing more than the exercise of a power to discriminate between the stimulus which on its subjective side we recognize as the right, and that which we similarly recognize as the wrong. Lastly, Volition, subjectively considered, is the faculty of consciously selecting motives; and motives, objectively considered, are nothing more than immensely complex and inconceivably refined stimuli to nervous action.

If we turn from the ascending scale of mental faculties in man, to the ascending scale of mind in the animal kingdom, we shall meet with further and still more definite evidence that the distinguishing property of mind, on its physiological side, consists in this power of discriminating between different kinds of stimuli, irrespective of their degrees of mechanical intensity. But, before giving a brief review of the evidence on this point, I may here meet a difficulty which has already arisen. The difficulty is that I began by showing it necessary to define Mind as the power of exercising Choice, and then proceeded to define the latter as a power belonging only to agents that are able to feel. Yet, on looking at the objective side of the problem, I pointed out that the physiological or objective equivalent of Choice is found to occur in its simplest manifestations at least as low down as the insectivorous plants, which are certainly not agents capable, in any proper sense of the term, of feeling. Therefore it seems that my conception of what constitutes Choice is in antagonism with my view that the essential element of Choice is found to occur among organisms which cannot properly be supposed to feel. And this antagonism, or inherent contradiction, is a real one, though I hold it to be unavoidable. For it arises from the fact that neither Feeling nor Choice appears upon the scene of life suddenly. We cannot say, within extensive limits, where either can properly be said to begin. They both dawn gradually, and therefore in our everyday use of these terms we do not wait to consider where they are first applicable; we only apply them where we see their applicability to be apparent. But when we endeavour to use these same terms in strict psychological analysis, we are at once met with the difficulty of drawing the line where the terms are applicable and where they are not. There are two ways of meeting the difficulty. One is to draw an arbitrary line, and the other is not to draw any line at all; but to carry the terms down through the whole gradation of the things until we arrive at the terminal or root-principles. By the time that we do arrive at these root-principles, it is no doubt true that our terms have lost all their original meaning; so that we might as well call an acorn an oak, or an egg a chicken, as speak of a Dionœa *feeling* a fly, or of a Drosera *choosing* to close upon its prey. Yet this use, or rather let us call it abuse, of terms serves one important purpose if, while duly regarding the change of meaning which during their gradual descent the terms are made gradually to undergo, we thus serve to emphasize the fact that they refer to things which are the product of a gradual evolution—things which came from other things as unlike to them as oaks to acorns or chickens to eggs. And this is my justification for tracing back the root-principles of

Feeling and of Choice into the vegetable kingdom. If it is true that plants manifest so little evidence of Feeling that the term can only be applied to them in a metaphorical sense, it is also true that the power of Choice which they display is of a similarly undeveloped character; it is limited to a single act of discrimination, and therefore no one would think of applying the term to such an act, until analysis reveals that in such a single act of discrimination we have the germ of all volition.

Let it therefore be understood that the difficulty which we are considering arises merely from the gradual manner in which the faculties in question arose. The rudimentary power of discriminative excitability which a plant displays is commensurate with the rudimentary power of selective adjustment which it manifests in its movements; and, just as the one is destined by developmental elaboration to become a self-conscious subjectivity, so the other is destined, by a similar elaboration, to become a deliberative volition.

I shall now briefly glance at the ascending scale of organisms, with the view of showing that this proportional relation between the grade of receptive and that of executive ability is manifested throughout the series. I desire to make it plain that the power of discrimination which in its higher manifestations we recognize as Feeling, and the power of selective adjustment which in its higher manifestations we recognize as Choice, are developed together, and throughout their development are commensurate.

Amœba is able to distinguish between nutritious and non-nutritious particles, and in correspondence with this one act of discrimination it is able to perform one act of adjustment; it is able to enclose and to digest the nutritious particles, while it rejects the non-nutritious. Some protoplasmic and unicellular organisms are able also to distinguish between light and darkness, and to adapt their movements to seek the one and shun the other; while in "Animal Intelligence" some observations are given which seem to show that the discriminative and adjustive powers of these organisms may go farther even than this. The insectivorous plants, as we have already seen, are able to distinguish, not only between nutritious and non-nutritious particles, but also between different kinds of contact; and, in correspondence with this advance in receptive power, we observe a commensurate advance in the mechanism of adaptive movement. Numberless other cases of such simple powers among plants might here be noticed; but none of them rise above the level of distinguishing between one or two alternatives of stimulation, and supplying the correspondingly simple movements of response. Where nerve-structure first appears, we find that the animals which present it—the Medusæ—have organs of special sense wherewith to distinguish with comparative delicacy and rapidity between light and darkness, and probably also between sound and silence. They are also provided with an elaborate tentacular apparatus, wherewith they are able to distinguish quickly and accurately between moving and not moving objects coming upon them from various sides, as well as between nutritious and non-nutritious particles. And in correspondence with this advance of receptive capacity we observe a considerable advance of executive capacity—the animals being highly locomotive, swimming away rapidly from sources of contact which they distinguish as dangerous, and manifesting several other reflex actions of a similarly adaptive kind. Thus, also, the higher organizations of Star-fish, Worms, &c., while serving to supply the neuro-muscular mechanisms with still more detailed information regarding the outer world, serve likewise to supply them with the means of executing a greater variety of adaptive movements. In the Mollusca, again, we observe another advance in both these respects; the animals feel their way with sensitive feelers, select varied kinds of food, choose mates of their own species to pair with, and may even remember a particular *locus* as their home, &c. Among the Articulata the lower forms present co-ordinated movements which are few and simple as compared with the many and varied movements of the higher members of the class; and their powers of distinguishing between stimuli are proportionally small. But in the complicated anatomy of the Crabs and Lobsters there is a large provision for the co-ordination of movements, and the selective actions are correspondingly numerous and varied; while among the Insects and Spiders the power of muscular co-ordination surpasses that of the lower Vertebrata, and the power of intelligent adaptation, assisted by delicate antennæ and highly perfected organs of special sense, is also greater. And the same principles hold throughout the Vertebrated series. It has already been remarked by Mr. Spencer that there is here a general correspondence to be observed between the possession of organs capable of varied actions, and the degree of intelligence to which the animal attains. Thus of Birds the Parrots are the most intelligent, and they, more than any other members of their class, are able to use their feet, beaks, and tongues in the examination of objects. Similarly, the wonderful intelligence of the Elephant may be safely considered as correlated with the no less wonderful instrument of co-ordinated movement which he possesses in his trunk; while the superior intelligence of the Monkey, and the supreme intelligence of Man may no less safely be considered as correlated with the still more wonderful instrument of co-ordinated movement which has attained to almost ideal perfection in the human hand. Again, and more generally, we may say that throughout the animal kingdom the powers of sight and of hearing stand in direct ratio to the powers of locomotion; and the latter are conducive to the growth of intelligence.*

We may now observe that this correlation between muscular and mental evolution—or, more generally, between power of discrimination and variety of adaptive movements—is only what we should expect to find *à priori*. For it is clear that the development of the one function could be of no *use* without that of the other. On the one hand, it would be of no use to an organism that it should be able to discern a stimulus as hurtful or beneficial, if at the same time it lacked the power of co-ordinated movement necessary to adapting itself to the result of its discernment; and, on the other hand, it would be equally useless that an organism should possess the needful power of co-ordinated movement, if at the same time it lacked the power of discernment which alone could render the power of co-ordinated movement useful. Now we know that all the mechanisms of muscular co-ordination are correlated with mechanisms of nervous co-ordination, and, indeed, that the former without the latter would be utterly useless. Yet we know next to nothing of

* The Dog and Cat seem at first sight to constitute an exception to the principle above set forth; but it must be remembered that both these animals, and all their tribe, possess very efficient instruments of touch and movements in their tongues, lips, and jaws, as well as to some extent in the paws. I think the superior intelligence of the Octopus, among mollusks, is to be attributed to the exceptional advantages which are rendered by its large and flexible, sensitive and powerful arms.

the ultimate nervous mechanisms which play down upon the muscular mechanisms; we only see a mazy mexus of cells and fibres, the very function of which, much less their intimate mechanism, could not be guessed, were it not that we have the grosser mechanisms of the muscular system whereby to study the effects of these finer mechanisms.

Muscular co-ordinations, then, are so many indices, "writ large," of corresponding co-ordinations taking place in the nervous system. Now we have seen that mental processes may be regarded as indices in precisely the same way, and indeed that, like muscular movements, they are the only indices we have of the operations of the nervous mechanisms with which they are connected. Moreover, we have seen that when this new set of indices has reached a certain level of development, marking of course a corresponding level of development in the nervous system, it begins unmistakeably to show that the functions of receptive discrimination and of adaptive movement are taking yet another point of departure in the upward course of their development—that the nervous system is beginning to discriminate between novel and enormously complex stimuli, having reference not only to immediate results, but also to remote contingencies; we see in short that the nervous mechanism is beginning to develope those higher functions of discriminative and adaptive ability which on their subjective side we know as rational.

Therefore it is clear that these two faculties not only *do* but *must* proceed together. Every advance in the power of discrimination will be followed, in the life of the individual and in that of the species, by efforts towards the movements of needful adaptation, and in all cases where such movements require an advance on the previous power of co-ordination, such advance will be favoured by natural selection. Thus every advance in the power of discrimination favours an advance of the power of co-ordination. And, conversely, we may now remark that every advance in the power of co-ordination favours an advance of the power of discrimination. For, as a greater power of co-ordinated movement implies the bringing of nerve-centres into new and more varied relations with the outer world, there is thus afforded to the nerve-centres a proportionately increased opportunity of discrimination—an opportunity which will sooner or later be sure to be utilized by natural selection.

Thus the two faculties are, as it were, necessarily bound together. But here another consideration arises. They are thus bound together only up to the point at which the adaptive movements are dependent upon the machinery supplied by nature to the organism itself. As soon as the power of discrimination has advanced far enough to be, not only consciously precipient, but deliberatively rational, a wholly new state of things is inaugurated. For now the organism is no longer dependent for its adjustments upon the immediate results of its own co-ordinated movements. From the time that a stone was first used by a monkey to crack a nut, by a bird to break a shell, or even by a spider to balance its web, the necessary connexion between the advance of mental discrimination and muscular co-ordination was severed. With the use of tools there was given to Mind the means of progressing independently of further progress in muscular co-ordination. And so marvellously has the highest animal availed itself of such means, that now, among the civilized races of mankind, more than a million per cent. of his adjustive movements are performed by mechanisms of his own construction. Wonderful as are the muscular co-ordinations of a tight-rope dancer, they are nothing in point of utility as compared with the co-ordinated movements of a spinning-jenny. Therefore, although man owes a countless debt of gratitude to the long line of his brutal ancestry for bequeathing to him so surpassingly exquisite a mechanism as that of the human body—a mechanism without which it would be impossible for him, with any powers of mind, to construct the artificial mechanisms which he does—still man may justly feel that his charter of superiority over the lower animals is before all else secured by this, that his powers of adjustive movement have been emancipated from their necessary alliance with his powers of muscular co-ordination.

I say, from his powers of *muscular* co-ordination, because it is evident that our powers of adjustive movement, and so of adaptation in general, have never been, *and can never be*, emancipated from a necessary alliance with our powers of *nervous* co-ordination.

I shall now sum up the results of our enquiry so far as it has hitherto gone. First, we found the Criterion of Mind, ejectively considered, to consist in the exhibition of

Choice, and the evidence of Choice we found to consist in the performance of adaptive action suited to meet circumstances which have not been of such frequent or invariable occurrence in the life-history of the race, as to have been specially and antecedently provided for in the individual by the inherited structure of its nervous system. The power of learning by individual experience is therefore the criterion of Mind. But it is not an absolute or infallible criterion; all that can be said for it is that it is the best criterion available, and that it serves to fix the upper limit of non-mental action more precisely than it does the lower limit of mental; for it is probable that the power of feeling is prior to that of consciously learning.

Having thus arrived at the best available criterion of Mind considered as an eject, we next proceeded to consider the objective conditions under which known Mind is invariably found to occur. This led us briefly to inspect the structure and functions of the nervous system, and, while treating of the physiology of reflex action, we found that everywhere the nervous machinery is so arranged that there is no alternative of action presented to the nerve-centres other than that of co-ordinating the group of muscles over the combined contractions of which they severally preside. The question therefore arose—How are we to explain the fact that the anatomical plan of a nerve-centre with its attached nerves comes to be that which is needed thus to direct the nervous stimuli into the channels required? The answer to this question we found to consist in the property which is shown by nervous tissue to grow by use into the directions which are required for further use. This subject is as yet an obscure one—especially where the earliest stages of such adaptive growth are concerned—but in a general way we can understand that hereditary usage, combined with natural selection, may have been alone sufficient to construct the numberless reflex mechanisms which occur in the animal kingdom.

Passing from reflex action to cerebral action, we first noticed that as the cerebral hemispheres pretty closely resemble in their intimate structure ganglia in general, there can be no reasonable doubt that the mode of their operation is substantially the same. Moreover we noted that, as such operation is here unquestionably attended with mental action, a strong presumption arises that the one ought to constitute a kind of obverse reflection of the other. Turning, therefore, to contemplate this presumably obverse reflection, we found that in many respects it is most strikingly true that the fundamental principles of mental operation correspond with the fundamental principles of ganglionic operation. Thus, we found that such is the case with memory and the association of ideas, both of which we found to have their objective counterparts in the powers of non-mental acquisition which are presented by the lower ganglia. For we found that these ganglia unconsciously learn such exercises as they are made frequently to perform, that they forget their exercises if too long an interval is allowed to elapse between the times of practising them, but that even when apparently quite forgotten such exercises are more easily re-acquired than originally they were acquired. More particularly we found that the association of ideas by contiguity presents a remarkably detailed resemblance to the association of muscular movements by contiguity. For, agreeing to take ideas as the objective analogues of muscular movements, we observed when we thus changed the index of nervous operation from muscles to ideas, that the strongest evidence was yielded of the method of nervous evolution being everywhere uniform. Thus we remarked that sensations, perceptions, ideas, and emotions all more or less resemble muscular co-ordinations in that they are usually blended states of consciousness, wherein each constituent part must correspond with the activity of some particular nervous element—a variety of such elements being therefore concerned in the composite state of consciousness, just as a variety of such elements are concerned in a combined movement of muscles. Further, just as the association of ideas is not restricted to a blending of simultaneous ideas into one composite idea, but extends to a linking of one idea with another in serial succession; so we saw that muscular movements exhibit a precisely analogous tendency to recur in the same serial order as that in which they have previously occurred. Lastly, we noted that all the pathological derangements which arise in the nerve-centres that preside over muscular activities, have their parallels in similar derangements which arise in the nerve-centres that are concerned in mental activities.

Having thus dealt with the Physical Basis of Mind, we passed on in the next chapter to consider the Root-principles of Mind. Here the object was to trace the ultimate principles of physiology that might be taken as constituting the objective side of those phenomena which on their subjective and ejective sides we regard as mental. These principles we found to be the power of discriminating between different kinds of stimuli irrespective of their relative degrees of mechanical intensity, coupled with the power of performing adaptive movements suited to the results of such discrimination. These two powers, or faculties, we saw to occur in germ even among the protoplasmic and unicellular organisms, and we saw that from them upwards all organization may be said to consist in supplying the structures necessary to an ever-increasing development of both these faculties, which always advance, and must necessarily advance, together. When their elaboration has proceeded to a certain extent, they begin gradually to become associated with Feeling, and when they are fully so associated, the terms Choice and Purpose become to them respectively appropriate. Continuing in their upward course of evolution, they next become consciously deliberative, and eventually rational. But although when viewed from the subjective or ejective side they thus appear, during the upward course of their development, to become transformed from one entity to another, such is not the case when they are viewed from their objective side. For, when viewed from their objective side, the most elaborate process of reasoning, or the most comprehensive of judgments, is seen to be nothing more than a case of exceedingly refined discrimination, by highly-wrought nervous structures, between stimuli of an enormously complex character; while the most far-sighted of actions, adapted to meet the most remote contingencies of stimulation, is nothing more than a neuro-muscular adjustment to the circumstances presented by the environment.

Thus, if we again take mental operations as indices whereby to study the more refined working of nervous centres, as we take muscular movements to be so many indices, "writ large," of the less refined working of such centres, we again find forced upon us the truth that the method of nervous evolution has everywhere been uniform; it has everywhere consisted in a progressive development of the power of discriminating between stimuli, combined with the complementary power of adaptive response.

CHAPTER V.

EXPLANATION OF THE DIAGRAM.

We have now sufficiently considered the sundry first principles and preliminary questions which lie at the threshold of our subject proper. It seemed to me desirable to dispose of these principles and questions before we enter upon our attempt at tracing the probable history of Mental Evolution. But now that these first principles and preliminary questions have been disposed of, so far as their nature renders possible, the way is as clear as it can be for us to pursue our enquiry concerning the Genesis of Mind. In order to give definition to the somewhat laborious investigation on which we are thus about to embark, I have thought it a good plan to draw a diagram or map of the probable development of Mind from its first beginnings in protoplasmic life up to its culmination in the brain of civilized man. The diagram embodies the results of my analysis throughout, and will therefore be repeatedly alluded to in the course of that analysis—*i.e.*, throughout the present and also my future work. I may therefore begin by explaining the plan of this diagram.

The diagram, as I have just said, is intended to represent in one view the whole course of mental evolution, supposing, in accordance with our original hypothesis, such evolution to have taken place. Being a condensed epitome of the results of my analysis, it is in all its parts carefully drawn to a scale, the ascending grades or levels of which are everywhere determined by the evidence which I shall have to adduce. The diagram is therefore not so much the product of my individual imagination, as it is a summary of all the facts which science has been able so far to furnish upon the subject; and although it is no doubt true that the progress of science may affect the diagram to the extent of altering some of its details, I feel confident that the general structure of our knowledge concerning the evolution of mind is now sufficiently coherent to render it highly improbable that this diagrammatic representation of it will, in the future, be altered in any of its main features by any advances that science may be destined to make.

From the groundwork of Excitability, or the distinguishing peculiarity of living matter, I represent the structure of mind as arising by a double root—Conductility and Discrimination. To what has already been said on these topics it is needless to add more. We have seen that the distinguishing property of nerve-fibre is that of transmitting stimuli by a propagation of molecular disturbance irrespective of the passage of a contraction wave; and this property, laying as it does the basis for all subsequent co-ordination of protoplasmic (muscular) movements, as well as of the physical aspect of all mental operations, deserves to be marked off in our map as a distinct and important principle of development; it is the principle which renders possible the executive faculty of appropriately responding to stimuli. Not less deserving of similar treatment is the cognate principle of Discrimination, which, as we have seen, is destined to become the most important of the functions subsequently distinctive of nerve-cells and ganglia. But we have also seen that both Conductility and Discrimination first appear as manifested by the cellular tissues of plants, if not even in some forms of apparently undifferentiated protoplasm. It is, however, only when these two principles are united within the limits of the same structural elements that we first obtain optical evidence of that differentiation of tissue which the histologist recognizes as nervous; therefore I have represented the function of nerve-tissue in its widest sense, Neurility, as formed by a confluence of these two root-principles. Neurility then passes into Reflex Action and Volition, which I have represented as occupying the axis or stem of the psychological tree. On each side of this tree I have represented the outgrowth of branches, and for the sake of distinctness I have confined the branches which stand for the faculties of Intellect on one side, while placing those which represent the Emotions upon the other. The level to which any branch attains represents the degree of elaboration which the faculty named thereon presents; so that, for instance, when the branch Sensation, taking origin from Neurility, proceeds to a certain level of development, it gives off the commencement of Perception, and then continues in its own line of development to a somewhat higher level. Similarly, Imagination arises out of Perception, and so with all the other branches. Thus, the fifty levels which are drawn across the diagram are intended to represent degrees of elaboration; they are not intended to represent intervals of time. Such being the case, the various products of mental evolution are placed in parallel columns upon these various levels, so as to exhibit the comparative degrees of elaboration, or evolution, which they severally present. One of these columns is devoted to the psychological scale of intellectual faculties, and another to the psychological scale of the emotional. But for the danger of rendering the diagram confused, these faculties might have been represented as secondary branches of the psychological tree; in a model this might well be done, but in a diagram it would not be practicable, and therefore I have restricted the branching structure to represent only the most generic or fundamental of the psychological faculties, and relegated those of more specific or secondary value to the parallel columns on either side of the branching structure. In these two columns I have throughout written the name of the faculty at what I conceive to be the earliest stage, or lowest level of its elaboration; *i.e.*, where it first gives evidence of its existence. In another parallel column I have given the grades of mental evolution which I take to be characteristic of sundry groups in the animal kingdom, and in yet another column I have represented the grades of mental evolution which I take to be characteristic of different ages in the life of an infant.

In my subsequent work I shall fill up all the levels in these vertical columns which are now left blank, on account of the text of the present work being restricted to the mental evolution of animals. At first I intended in this work to truncate the whole diagram at the level where mental evolution in animals ends—*i.e.*, at the level marked 28—and to reserve the continuation of the stem and branches, as well as that of the parallel columns, for my ensuing work. But afterwards I thought it was better to supply the continuation of the stem and branches, in order to show the proportion which I conceive to obtain between the elaboration of the higher faculties as they occur in animals and the same faculties as they occur in man.

Confining, then, our attention to the first twenty-eight

levels with which alone the present essay is to be concerned, if we pitch upon any one of them at random, we shall obtain a certain rough estimate of the grade of mental evolution which is presented by the animals named upon that level.

To avoid misapprehension I may add that in thus rendering a diagrammatic representation of the probable course of mental evolution with the comparisons of psychological development exhibited in the parallel columns, I do not suppose that the representation is more than a rough or general outline of the facts; and, indeed, I have only resorted to the expedient of thus representing the latter for the sake of convenience in my subsequent discussion. Rough as this outline of historical psychology may be, it will serve its purpose if it tends to facilitate the exposition of evidence, and afterwards serves as a dictionary of reference to the more important of the facts which I hope this evidence will be able to substantiate.

Such being the general use to which I intend to put the diagram, I may here most fitly make this general remark in regard to it. In the case alike of the stem, branches, and the two parallel columns on either side—*i.e.*, all the parts of the diagram which serve to denote psychological faculties—we must remember that they are diagrammatic rather than truly representative. For in nature it is as a matter of fact impossible to determine any hard and fast lines between the completed development of one faculty and the first origin of the next succeeding faculty. The passage from one faculty to another is throughout of that gradual kind which is characteristic of evolution in general, and which, while never preventing an eventual distinction of species, always renders it impossible to draw a line and say—Here species A ends and species B begins. Moreover, I cannot too emphatically impress my conviction that any psychological classification of faculties, however serviceable it may be for purposes of analysis and discussion, must necessarily be artificial. It would, in my opinion, be a most erroneous view to take of Mind to regard it as really made up of a certain number of distinct faculties—as erroneous, for example, as it would be to regard the body as made up of the faculties of nutrition, excitability, generation, and so on. All such distinctions are useful only for the purposes of analysis; they are abstractions of our own making for our own convenience, and not naturally distinct parts of the structure which we are examining.

But although it is desirable to keep these caveats in our memory, I do not think that either the artificial nature of psychological classification or the fact that we have to do with a gradual process of evolution, constitutes any serious vitiation of the mode of representation which I have adopted. For, on the one hand, some classification of faculties we must have for the purposes of our inquiry; and, on the other hand, I have as much as possible allowed for the unavoidable defect in the representation which arises from evolution being gradual, by making the branches of the arborescent structure wide at their bases, and by allowing each of them, after giving off the next succeeding branch, to continue on its own course of development; so that both the parent and daughter faculty are represented as occupying for a more or less considerable distance the same levels of development—in each case my estimate of the comparative elaboration which the completed faculty betokens being represented by the vertical height of its apex. Besides, as already stated, faculties named in the two parallel columns are written upon those levels where I have either *à priori* reasons or actual evidence to conclude that they first definitely appear in the growing structure of Mind; in this way the difficult question of assigning the lower limit of evolution at which any particular faculty begins to dawn is as much as possible avoided.

It is almost needless to add that in preparing this diagram I have resorted to speculation in as small a measure as the nature of the subject permits. Nevertheless it is obvious that the nature of the subject is such that, in order to complete the diagram in some of its parts, I have been obliged to resort to speculation pretty largely. I think, however, that as the exposition proceeds, it will be seen that, if the fundamental hypothesis of mental evolution having taken place is granted, my reasoning as to the probable history of the process does not anywhere involve speculation of an extravagant or dangerous kind. In matters of detail—such, for instance, as the comparative elevation of the different branches in the psychological tree—my estimates may, probably enough, be more or less erroneous; but the main facts as to the sequence of the faculties in the order of their comparative degrees of elaboration are mere corollaries from our fundamental hypo-

E 2

thesis; and, as we shall see, these facts, as I have presented them, are sustained or corroborated by many others drawn from observations on the psychology of animals and children. Again, in the columns devoted to the emotions and faculties of intellect, the results of actual observation predominate over those yielded by speculation; while in the remaining columns the results tabulated are for the most part due to observation.

Therefore I submit that if the hypothesis of mental evolution be granted, and if all the matters of observable fact which the diagram serves to express are eliminated, comparatively little in the way of deductive reasoning is left; and of this little most follows as necessary consequence from the original hypothesis of mental evolution having taken place. Of course any one who does not already accept the theory of evolution in its entirety, may object that I am thus escaping from the charge of speculation only by assuming the truth of that which grants me all that I require. To this I answer that as far as the evidence of Mental Evolution, considered as a fact, is open to the charge of being speculative, I must leave the objector to lodge his objection against Mr. Darwin's "Origin of Species" and "Descent of Man." I shall be abundantly satisfied with my own work if, taking the process of Mental Evolution as conceded, I can make it clear that the main outlines of its history may be determined without any considerable amount of speculation, as distinguished from deduction following by way of necessary consequence from the original hypothesis.

Having thus explained the plan and principles of the diagram, I shall now consider the levels from the lowest as far as the rise of the first branch, *i.e.*, from 1 to 14. After what has already been said in the foregoing chapters on the Physical Basis and Root-principles of Mind, our consideration of this part of the diagram need not detain us long.

Levels 1 to 4 are occupied by Excitability, Protoplasmic Movements, Protoplasmic Organisms, and the generative elements which have not yet united to start the Embryo of Man. From 4 to 9 we have the levels filled by the rise and progress of the functions Conductility and Discrimination, which by their subsequent union at 9 lay the basis of Neurility, or the stem of Mind; in these levels occur the

Non-nervous Adjustments, Unicellular Organisms, and part of the Life-history of the Embryo. Between 9 and 14 is represented the development of Neurility and its passage into Reflex Action; the parallel columns within this space are therefore respectively filled with Partly-nervous Adjustments and the beginning of True Nervous Adjustments, Unknown Animals, probably Cœlenterata, perhaps extinct, and another portion of the Life-history of the Embryo. I here speak of "unknown animals" because, so far as investigation has hitherto gone, the animals in which nerve-tissue first began to be differentiated have not yet been found. In the lowest animals where this tissue has been found—the Medusæ—it appears as already well differentiated. The ganglion cells, however, show in a most unmistakeable manner their parentage from epithelium—their structure, in fact, often resembling that of modified epithelium more than that of true nerve-cells.* In these structures, therefore (as in the analogous histological elements met with in the embryonic nerve-tissue of higher animals), we have a link which connects true nerve-tissue with its cellular ancestry, and thus it is comparatively immaterial whether or not the animals which presented the earlier stages of this histological transition are still in existence. Thus we need not wait to discuss Kleinenberg's view on the "neuro-muscular" cells of Hydra.

* See Prof. E. A. Schäfer on *Nervous System of Aurelia Aurita*, *Phil. Trans.*, 1878, and Profs. O. and R. Hertwig on *Das Nervensystem und die Sinnesorgane der Medusen.*

CHAPTER VI.

CONSCIOUSNESS.

HITHERTO in this work I have been considering, as exclusively as the nature of the subject permits, the physical or objective aspect of mental processes, and of the antecedents of these processes in the non-mental activities of living organisms. It now devolves upon us to turn to the subjective side of the matter, and still more closely, I may observe, to the ejective side of it. That is to say, from this point onward my endeavour will be to trace the probable course of Mental Evolution by having regard to truly mental phenomena, so far as these admit of analysis by subjective or ejective methods. I desire, therefore, to draw prominent attention to the fact that from this point in my treatise I take, as it were, a new departure; for if this is not kept in mind, my exposition may appear to resemble two separate essays bound together rather than one continuous whole. In my endeavour to draw a sharp line of demarcation between the physiology and the psychology of my subject, I have found it impossible to discuss the one without numerous allusions to the other—the consequence being that hitherto, while treating as exclusively as I could of the physiology of vital processes, I have been obliged frequently to refer to the psychology of mental processes, a knowledge concerning the main facts of which I have taken for granted on the part of any one who is likely to read this book. Thus it happens that in now turning to investigate the psychology of these processes, it is impossible to avoid a certain amount of overlapping with what has gone before. For example, in my chapter on the Physical Basis of Mind, it was clearly impossible not to allude to such leading principles of psychology as sensation, perception, ideation, and others. Therefore, in now undertaking an investigation of these various principles in the order of their probable evolution, it may often appear that I am, as it were, going back upon, or in part repeating, what I have already said. But this apparent defect in the method of my exposition will, I think, be seen on closer attention to be more than compensated for by the advantage of avoiding confusion between physiology and psychology. It would, for instance, have been easy to have split up the chapter on the Physical Basis of Mind already alluded to, and to have apportioned its various parts to those among the succeeding chapters which treat of the psychological aspects of the physiological principles set forth in those various parts; but the result would have been largely to have obscured the doctrine which I desired to make plain throughout—viz., that all mental processes must be regarded as presenting physical counterparts.*

So much in explanation of my method being understood, I shall begin the psychology of mental evolution by considering that in which the mind-element must be regarded as consisting—namely, Consciousness. Turning to the diagram, it will be observed that I have written the word "Consciousness" in a perpendicular direction, beginning at level 14 and extending to level 19. My reason for doing this is because the rise of Consciousness is probably so gradual, and certainly so undefined to observation, that any attempt to draw the line at which it does arise would be impossible, even on the rough and general scale wherewith I have endeavoured to draw the lines at which the sundry mental faculties may be regarded as taking origin. Therefore I have represented the rise of Consciousness as occupying a considerable area in our representative map, instead of a definite line. This area I make to begin with the first development of "Nervous Adjustments," and to terminate with the earliest appearance of the power of associating ideas.

In now proceeding to justify this assignment of limits between the earliest dawn of Consciousness and the place where Consciousness may first be regarded as truly such, I may best begin by saying that I shall not attempt to define

* It seems almost needless to add that the impossibility of entirely separating psychology from physiology for the purposes of exposition will, *mutatis mutandis*, continue to meet us more or less throughout the following, as it has throughout the preceding chapters; but I shall endeavour always to make it clear when I am speaking of mental processes and when of physical.

what is meant by Consciousness. For, like the word "Mind," "Consciousness" is a term which serves to convey a meaning well and generally understood, but a meaning which, from the peculiar nature of the case, cannot be comprehended in any definition. If we say that a man or an animal is conscious, we mean that the man or animal displays the power of Feeling, and if we ask what we mean by Feeling, we can only, I think, answer—that which distinguishes Non-extended Existence from Extended. Deeper than this we cannot go, because Consciousness, being itself the basis of all thought, and so of all definition, cannot be itself defined except as the antithesis of its logical correlative—No-consciousness.

Let us first regard the phenomena of Consciousness as disclosed in our own or subjective experience. We shall subsequently see that the elementary or undecomposable units of consciousness are what we call sensations. If we interrogate experience we find that an elementary state of consciousness, or sensation, may exist in any degree, from that of an almost unrecognizable affection, up to that of unendurable pain, which monopolizes the entire field of consciousness. More than this, from the lowest limit of perceptible sensation there arises a long and indefinite descent through sensation that is not perceptible, or through sensation that is sub-conscious, before we arrive at nervous action which we feel entitled to regard as unconscious. This is proved by those grades of almost unconscious action, passing at last into wholly unconscious action, which we all know as frequently occurring in the descent, through repetition or habit, of consciously intelligent adjustments to automatic adjustments, or adjustments performed unconsciously. Thus it is evident, not only that consciousness admits of numberless degrees of intensity, but that in its lower degrees its ascent from no-consciousness is so gradual, that even within the range of our own subjective experience we find it impossible to determine within wide limits where consciousness first emerges.*

With this gradual dawn of consciousness as revealed to subjective analysis, we should expect some facts of physiology, or of objective analysis, to correspond; and this we do find.

* Any one who has gradually fainted, or has slowly been put under the influence of an anæsthetic, will remember the peculiar experience of feeling consciousness becoming obliterated by stages.

For in our own organisms we know that reflex actions are not accompanied by consciousness, although the complexity of the neuro-muscular systems concerned in these actions may be very considerable. Clearly, therefore, it is not mere complexity of ganglionic action that determines consciousness. What, then, is the difference between the mode of operation of the cerebral hemispheres and that of the lower ganglia, which may be taken to correspond with the great subjective distinction between the consciousness which may attend the former and the no-consciousness which is invariably characteristic of the latter? I think the only difference that can be pointed to is a difference of rate or time. We know by actual measurement, as we shall subsequently see in more detail, that the cerebral hemispheres work more slowly while undergoing those changes which are accompanied by consciousness than is the case with the activities of the lower centres. In other words, the period between the fall of a stimulus and the occurrence of responsive movement is notably longer if the stimulus has first to be *perceived*, than it is if no perception is required. And this is proved, not only by comparing the latent period (or the time which elapses between the stimulation and the response) in the case of an action involving one of the lower centres and that of an action involving the cerebral hemispheres in perception; but also by comparing the latent period in the case of one and the same cerebral action which from having originally involved perception has through repetition become automatic. An old sportsman will have his gun to the shoulder, by an almost unconscious act, the moment that a bird unexpectedly rises; a novice similarly surprised will spend a valuable second in "taking in" the situation. And any number of similar facts might be given to show that if few things are "as quick as thought," reflex or automatic action is one that is quicker. Further, in a general way it can be shown that the more elaborate a state of consciousness is, the more time is required for its elaboration, as we shall see more in detail when we come to treat of Perception.

Now what does this greater consumption of time imply? It clearly implies that the nervous mechanism concerned has not been fully habituated to the performance of the response required, and therefore that instead of the stimulus merely needing to touch the trigger of a ready-formed apparatus of response (however complex this may be), it has to give rise in the nerve-centre to a play of stimuli before the appropriate response is yielded. In the higher planes of conscious life this play of stimuli in the presence of "difficult circumstances" is known as indecision; but even in a simple act of consciousness—such as that of signalling a perception—more time is required by the cerebral hemispheres in supplying an appropriate response to a non-habitual experience, than is required by the lower nerve-centres for performing the most complicated of reflex actions by way of response to their habitual experience. In the latter case the routes of nervous discharge have been well worn by use; in the former case these routes have to be determined by a complex play of forces amid the cells and fibres of the cerebral hemispheres. And this complex play of forces, which finds its physiological expression in a lengthening of the time of latency, finds also a psychological expression in the rise of consciousness.

The function, then, of the cerebral hemispheres is that of dealing with stimuli which, although possibly and in a comparative sense simple, are yet so varied in character that special reflex mechanisms have not been set aside to deal with them in one particular way; and it is the consequent perturbation of these highest nerve-centres in dealing with such stimuli that is accompanied by the phenomena of consciousness. Or, in the words of Mr. Spencer, "there cannot be co-ordination of many stimuli without some ganglion through which they are all brought into relation. In the process of bringing them into relation, this ganglion must be subject to the influence of each—must undergo many changes. And the quick succession of changes in a ganglion, implying as it does perpetual experiences of differences and likenesses, constitutes the raw material of consciousness."*

Thus we see, so far as we can ever perhaps hope to see, how conscious action gradually arises out of reflex. As the stimuli to be dealt with become more complex and varied (owing to the advancing evolution of organisms bringing

* *Principles of Psychology*, vol. i, p. 435. I think, however, that Mr. Spencer is not sufficiently explicit, either in the above quoted passage or elsewhere, in showing that "the raw material of consciousness" is not necessarily constituted by the mere *complexity* of ganglionic action. Indeed, as I have said, such complexity in itself does not appear to have anything to do with the rise of consciousness, except in so far as it may be conducive to what we may term the ganglionic friction, which is expressed by delay of response.

them into more and more complex and varied relations with their environment), the primitive assignment of a special nervous mechanism to meet the exigencies of this or that special group of stimuli becomes no longer practicable, and the higher nerve-centres have therefore to take on the function of focussing many and more or less varied stimuli, in order to attain to that higher aptitude of discrimination in which we have already seen to consist the distinctive attribute of Mind. And, as Mr. Spencer has observed, "the co-ordination of many stimuli into one stimulus is, so far as it goes, a reduction of diffused simultaneous changes into concentrated serial changes. Whether the combined nervous acts which take place when the fly-catcher seizes an insect are regarded as a series passing through its centre of co-ordination in rapid succession, or as consolidated into two successive states of its centre of co-ordination, it is equally clear that the changes going on in its centre of co-ordination have a much more decided linear arrangement than have the changes going on in the scattered ganglia of a centipede." And this linear character of the change is, of course, one of the most distinctive features of consciousness as known to ourselves subjectively.

It will have been observed that this interpretation of the rise of consciousness is purely empirical. We know by immediate or subjective analysis that consciousness only occurs when a nerve-centre is engaged in such a focussing of varied or comparatively unusual stimuli as have been described, and when as a preliminary to this focussing or act of discriminative adjustment there arises in the nerve-centre a comparative turmoil of stimuli coursing in more or less unaccustomed directions, and therefore giving rise to a comparative delay in the occurrence of the eventual response. But we are totally in the dark as to the causal connection, if any, between such a state of turmoil in a ganglion and the occurrence of consciousness. Whether it is the Angel that descends to trouble the waters, or the troubling of the waters that calls down the Angel, is really the question which divides the Spiritualists from the Materialists; but with this question we have nothing to do. It is enough for all the objects of the present work that we never get the Angel without the troubling, nor the troubling without the Angel; we have an empirical association between the two which is as valid for the purposes of merely historical psychology as would be a full understanding of the causal connection, if there is any such connection to be understood.

So much, then, for the physical conditions under which consciousness is always and only found to occur. It remains briefly to conclude this chapter by showing that these conditions may most reasonably be regarded as first arising within the limits between which I have represented the origin of consciousness.

Remembering what has already been said concerning the gradual or undefined manner in which consciousness probably dawned upon the scene of life, and that I therefore represent its rise as occupying a wide area on the diagram instead of a definite line, I think it least objectionable to place the beginning of this dawn in nervous adjustments or reflex action, and the end of it in the association of ideas. For, on the one hand, it is clear from what has been said that it is impossible to draw any definite line between reflex and conscious action, inasmuch as, considered objectively or as action, the latter differs from the former, not in kind, but only in a gradual advance in the degree of central co-ordination of stimuli. Therefore, where such central co-ordination is first well established, as it is in the mechanism of the simplest reflex act, there I think we may with least impropriety mark the advent of consciousness. On the other hand, where vague memory of past experiences first passes into a power of associating simple ideas, or of remembering the connections between memories, there I think consciousness may most properly be held to have advanced sufficiently far to admit of our regarding it as fairly begun.

In this scheme, therefore—which of course it is needless to say I present as a somewhat arbitrary estimate where no more precise estimate is possible—the Cœlenterata are represented as having what Mr. Spencer calls "the raw material of consciousness," the Echinodermata as having such an amount of consciousness as I think we may reasonably suppose that they possess, if we consider how multifarious and complicated their reflex actions have become, and if we remember that in their spontaneous movements the neuromuscular adjustments which they exhibit almost present the appearance of being due to intelligence.* The Annelida I

* See *Phil. Trans.*, *Croonian Lecture*, 1881.

place upon a still higher level of consciousness, because, both from the facts mentioned in "Animal Intelligence" and from those published by Mr. Darwin,* it seems certain that their actions so closely border on the intelligent that it is difficult to determine whether or not they should be classed as intelligent. Upon this level, also, I represent the period of the embryonic life of Man as coming to a close; for although the new-born child, from the immaturity of its experience, displays no adjustments that can be taken as indicative of intelligence, still, as its nerve-centres are so elaborate (embodying the results of a great mass of hereditary experience, which although more latent in the new-born child than in the new-born of many other mammals and all birds, must still, we should infer from analogy, count of something), that we can scarcely doubt the presence of at least as much consciousness as occurs among the annelids. Moreover, pain appears to be felt by a new-born child, inasmuch as it cries if injured; and although this action may be largely or chiefly reflex, we may from analogy infer that it is also in part due to feeling. The remaining levels occupied by the dawn of consciousness may be considered as assigned to the lower Mollusca—an assignment which I think will be seen to be justified by consulting the evidence given in my former work of actions performed by these animals of a nature which is unquestionably intelligent.

* See his work on *Earthworms*, 1881.

CHAPTER VII.

SENSATION.

BY Sensation I mean simply Feeling aroused by a stimulus. In my usage, therefore, the term is of course exclusive of all the metaphorical meanings which it presents in such applications as "sensitive plates," &c. It is also exclusive, on the one hand, of Reflex Action, as well as of non-nervous adjustments, and on the other, of Perception. Thus, too, it is exclusive of the carefully defined meaning which it bears in the writings of Lewes. He defined Sensation as the reaction of a sense-organ, whether or not accompanied by Feeling, and thus he habitually speaks of unfelt sensations. In his nomenclature, therefore, Sensation is a process of a purely physical kind, with which consciousness may or may not be involved. In my opinion, however, it is most desirable, notwithstanding his elaborate justification of this use of the term, to abide by its original signification, which I have explained. When I have occasion to speak of the physical reaction of a sense-organ, I shall speak of it as a physical reaction, and not as a sensation. The distinction which, in common with other psychologists, I draw between a Sensation and a Perception, I shall explain more fully in the chapter where I shall have to treat of Perception. Meanwhile it is enough to say that the great distinction consists in Perception involving an element of Cognition as well as the element of Feeling.

It is more difficult to draw the distinction between Sensation and non-nervous adjustments, and still more so between Sensation and nervous adjustments which are unfelt (Reflex Action). Here, however, we are but again encountering the difficulty which we have already considered, viz., that of drawing the line where consciousness begins; and, as we have previously seen, this difficulty has nothing to do with the validity of a classification of psychical faculties; it only has to do with the question whether such and such a faculty occurs in such and such an organism. Therefore, so long as the question is one of classifying psychical faculties, we can only say that wherever there is Feeling there is Sensation, and wherever there is no Feeling there is no Sensation.* But where the question is one of classifying organisms with reference to their psychical faculties, it is clear that the difficulty of determining whether or not this and that particular low form of life has the beginnings of Sensation, is one and the same as the question whether it has the beginnings of Consciousness. Now we have already considered this question, and we have found it impossible to answer; we cannot say within broad limits where in the animal kingdom consciousness may first be regarded as present. But for the sake of drawing the line somewhere with reference to Sensation, I draw it at the place in the zoological scale where we first meet with organs of special sense, that is to say, at the Cœlenterata. In doing this, it is needless to observe, I am drawing the line quite arbitrarily. On the one hand, for anything that is known to the contrary, not only the sensitive plant which responds to a mechanical stimulus, but even the protoplasmic organisms which respond to a luminous stimulus by congregating in or avoiding the light, may, while executing their responses, be dimly conscious of feeling; and, on the other hand, the mere presence of an organ of special sense is certainly no evidence that its activities are accompanied by Sensation. What we call an organ of special sense is an organ adapted to respond to a special form of stimulation; but whether or not the process of response is accompanied by a sensation is quite another matter. We infer by a strong analogy that it is so accompanied in the case of organisms like our own (whether of men or of the higher animals); but the validity of such inference clearly diminishes with the diminishing strength of the analogy—*i.e.*, as we recede in the zoological and psychological scales from organisms like our own towards organisms less and less like.

Having thus made it as clear as I can that it is only for the matter of convenience that I have supposed the rise of Sensation to coincide with the rise of organs of special

* Although this sounds like a truism, it is in direct opposition to the classification of Lewes, alluded to above.

sense, I shall next proceed to take a brief survey of the animal kingdom with reference to the powers of special sense. In doing this, however, it is needless, and indeed undesirable, that I should enter with much closeness into the anatomy of the innumerable organs of special sensation which the animal kingdom presents. My object is merely to give a general outline of the powers of special sensation probably enjoyed by different classes of animals; for, as these powers constitute the foundation of all the other powers of mind, it is of importance for us to have a general idea of the grade of their development in the sundry grades of the zoological scale.

In some of his recently published experiments, Engelmann found that many of the protoplasmic and unicellular organisms are affected by light; that is to say, their movements are influenced by light, in some cases causing acceleration, in others slowing, of their movements; in some cases the organisms seeking the light, while in other cases they shun it, &c., &c. He found that all these effects were reducible to one or other of three causes: (1) alteration produced by the light in the interchange of gases, (2) consequent alteration in the conditions of respiration, and (3) specific processes of luminous stimulation. It is with the latter only that we are concerned, and the organism which Engelmann names as exhibiting it typically is *Englena viridis*. After precautions had been taken to eliminate causes 1 and 2, it was still found that this organism sought the light. Moreover, it was found that it would only do so if the light were allowed to fall upon the anterior part of its body. Here there is a pigment-spot, but careful experiment showed that this was not the point most sensitive to light, a colourless and transparent area of protoplasm lying in front of it being found to be so. Hence it is doubtful whether this pigment-spot is or is not to be regarded as an exceedingly primitive organ of special sense. Of the rays of the spectrum, *Englena viridis* prefers the blue.*

The remarkable observation recorded by Mr. H. J. Carter, F.R.S., and quoted from him in my previous work,† seems to display almost incredible powers of special sense among the

* For full account of these experiments, see *Pflüger's Archiv. f. d. ges. Physiologie*, Bd. XXIX, 1882.

† *Animal Intelligence*, pp. 19–21.

Rhizopoda; and Professor Hæckel observes, in his essay on the "Origin and Development of the Sense-Organs," that "already among the microscopic Protista there are some that love light, and some that love darkness rather than light. Many seem also to have smell and taste, for they select their food with great care. . . . Here also we are met by the weighty fact that sense-function is possible without sense-organs, without nerves. In place of these, sensitiveness is resident in that wondrous, structureless, albuminous substance which, under the name of protoplasm, or organic formative material, is known as the general and essential basis of all the phenomena of life."

Again, Engelmann describes a chase of one infusorium by another. The former in its free course happened to cross the route of a free-swarming vorticella. There was no contact, but it immediately gave chase, and for five seconds the two darted about with the utmost activity, the chasing infusorium maintaining a distance of about $\frac{1}{15}$ mm. behind the chased one. Then, owing to a sudden sideward dart of the vorticella, its pursuer lost the object of pursuit. The powers of discrimination shown by certain deep-sea protoplasmic organisms in selecting sand-grains of a particular size wherewith to construct their tests has already been alluded to.

But passing now to animals in which we first meet with nerves, viz., the Medusæ, it is among them also that we first meet with organs of special sense. I have myself observed that several species of Medusæ seek the light, following a lantern if this is moved round a bell-jar containing them in a dark room. The pigmented bodies round the margin of the swimming-disk were proved to be the organs of special sense here concerned, and the rays in the spectrum by which they are affected were shown to be confined to the luminous part. It was further observed that some genera of Medusæ had more highly developed visual sensation than others. The least efficient occurs in *Tiaropsis polydiademata*, as shown by the prolonged interval of delay between the fall of a luminous stimulus and the occurrence of the response. As the case is an interesting one, I shall state the particulars more fully. This Medusa, then, always responds to strong luminous stimulation by going into a spasm or cramp; but it will not respond at all unless the light is allowed to fall upon its sense-organs for a period of more than one second; if a slip-shutter is opened and closed again for a shorter period, no response is made. It therefore seems certain that here we have not to deal with what physiologists call the period of latent stimulation, but with the time during which the light requires to fall in order to constitute an adequate stimulus; just as a photographic plate requires a certain period of exposure in order to admit of the luminous vibrations throwing down the salt, so with the ganglionic material of this sense-organ. How different is the efficiency or development of such a visual apparatus from that of a fully perfected retina, which is able to effect the needful nervous changes in response to a stimulus as instantaneous as that supplied by a flash of lightning.* It is remarkable, looking to the Medusæ as a whole, in what a wonderful degree these primitive sense-organs vary as to their minute structure in different species. Nerve-cells and fibres, wrought up into more or less complex forms, are clearly discernible in all those which have hitherto been carefully examined; but when the particular specific forms are compared with one another, it seems almost as if organs of special sense, where they first undoubtedly occur in the animal kingdom, revel, as it were, in the variety of forms which they are able to present.

It is probable, from the structure of the lithocysts, that the Medusæ are also affected by sonorous vibrations, and it is certain that they are richly supplied with a variety of organs ministering to the sense of touch. For not only are they furnished with numerous long, highly sensitive, and contractile tentacles, but in some species the marginal ganglia are provided with minute hair-like appendages, which must enable the nerve-cells to which they are attached to be exceedingly sensitive to anything touching the hairs. And, in connection with the sense of touch in the Medusæ, I may allude to my own observations on the precision with which the point of contact of a foreign body is localized. A Medusa being an umbrella-shaped animal, in which the whole of the surface of the handle and the whole of the concave surface of the umbrella is sensitive to all kinds of stimulation, if any point in the last-named surface is gently touched with a camel-hair brush or other soft (or hard)

* For a full account of these experiments, see *Phil. Trans.*, vol. 166, Pt. I, *Croonian Lecture*, where it is shown that in other species of Medusæ, the sense-organs of which are more highly developed, there is no such prolonged delay in the response to luminous stimulation.

object, the handle or manubrium is (in the case of many species) immediately moved over to that point, in order to examine or to brush away the foreign body. This is especially the case in a species which for this reason I have called *Tiaropsis indicans;* and here it is of interest to observe that if the nerve-plexus, which is spread all over the concave surface of the umbrella, is divided by means of an incision carried in the form of a short straight line parallel to the margin of the umbrella, and if a point below the line of incision is touched, the manubrium is no longer able to localize the seat of contact. Nevertheless it feels that contact is taking place *somewhere,* for it begins actively to dodge about from side to side of the umbrella, applying its extremity now to one point and now to another of the umbrella surface, as if seeking in vain for the offending body. This of course shows that the stimulus, on reaching the ends of the severed nerve-fibres, spreads through the general nerve-plexus, and so arriving at the manubrium by a number of different routes, conveys a corresponding number of conflicting messages to the manubrium as to the point in the umbrella at which the stimulus is being applied. This irradiation of a stimulus into other nerve-fibres when the stimulus reaches the cut ends of the fibres which constitute the habitual route of a stimulus between two points, is rendered the more interesting from the fact that in the case of the external nervous plexus of the Echinodermata there is no vestige of such a phenomenon.

So much for the senses of sight (at least to the extent of distinguishing light from darkness), hearing, and touch, as localized in organs of special sense among the Medusæ. In the allied Actiniæ Mr. Walter Pollock and myself have obtained conclusive evidence of the sense of smell. For we found that when a morsel of food is dropped into a pool or tank containing sea-anemones in a closed state, the animals quickly expand their tentacles.* It has been said that this may be taken to argue a sense of taste no less than a sense of smell; but I conceive that here no distinction can be drawn between these two senses, any more than we can draw such a distinction in the analogous case of fish. Looking, then, to the Cœlenterata as a whole, we find that where we first meet with unmistakeable organs of special sense, we also first meet

* See *Journal Linnean Society*, 1882.

with unmistakeable evidence of the occurrence of all the five senses—or, more correctly, with unmistakeable evidence of a power of adaptive response to all the five classes of stimuli which respectively affect the five senses of man.

Coming next to the Echinodermata, Professor Ewart and myself have observed that Star-fish and Echini crawl towards and remain in the light, even though this be of such feeble intensity as scarcely to be perceptible to human eyes. Moreover, we proved that this exceedingly delicate power of discrimination between light and darkness is localized in the pigmented ocelli situated at the tips of the rays in Star-fish, and occupying the homologous positions in Echini. The sense of touch we found likewise to be highly delicate, and provided for by a variety of specially modified organs. Lastly, I found that the sense of smell occurs in Star-fish, though it is not localized in any special olfactory organs, being in fact distributed equally over the whole of the ventral surface of the animal, to the exclusion, however, of the dorsal.*

Among the Articulata we meet with numberless grades of visual apparatus, from that of a simple ocellus, capable only of distinguishing light from darkness, up to the greatly elaborated compound eyes of insects and the higher Crustacea. These compound eyes are remarkable from the fact that each one of their possibly many thousand facets forms an image of the corresponding portion of the visual field—the multitude of separate sensory impressions being then combined into a mosaic-like whole by a sensorial operation taking place in the cephalic ganglion. In these compound eyes, moreover, the images are thrown upon the receptive nerve-surface without inversion. In the uncompounded or simple type of eye, on the other hand, the image is inverted, and as in the case of ants both kinds of eyes occur in the same individual, it has been thought a psychological puzzle how to explain the fact that mental confusion in the interpretation of images does not result. A little thought, however, will show that the apparent puzzle is not a real one. Thus it is commonly said that we ourselves really see objects reversed, and that long practice enables us to correct the erroneous impressions. But this statement of the case is

* See *Phil. Trans.*, 1881, Pt. III, *Croonian Lecture;* and, for smell in Star-fish, *Journ. Linn. Soc.*, 1883.

not correct. "We do not really *see* things reversed, for the mind is not a perpendicular object in space, standing behind the retina in the manner that a photographer stands behind his camera. To the mind there is no up or down in the retina, except in so far as the retina is in relation to the external world; and this relation can only be determined, not by sight, but by touch. And if only this relation is constant, it can make no difference to the mind whether the images are direct, reversed, or thrown upon the retina at any angle with reference to the horizon; in any case the correlation between sight and touch would be equally easy to establish, and we should always *see* things, not in the position in which they are *thrown upon* the retina, but in that which they occupy with *reference to* the retina. Thus it really requires no more 'practice' correctly to interpret inverted images than it does similarly to interpret upright images; and therefore the fact that some eyes of an ant are supposed to throw direct images, while others are supposed to throw inverted, is not any real objection to the theory" that they do.*

There is no one group in the animal kingdom where we have so complete a series of gradations in the evolution of an organ of special sense as is presented by the organ of sight in Worms. "In the lowest Vermes,"—I quote from Professor Hæckel†—"the eye is only made up of individual pigment-cells In others, refractive bodies are associated with these, and form a very simple lens. Behind these refractive bodies sensory cells are developed, forming a retina of the simplest order presenting a single layer, the cells of which are in connection with extremely delicate terminal fibres of the optic nerve. Lastly, in the Alcipidæ, which are highly organized Annelidæ that swim on the surface of the sea, adaptation to this mode of life has brought about such perfection of the eye that this organ in these animals is in no way inferior to that of the lower vertebrata. In these creatures we find a large globular eye-ball, enclosing externally a laminated globular lens, internally a vitreous body of large circumference. Immediately investing these are rods of the usual cells sensitive to light, which are separated by a layer of pigment-cells from the outer expansion of the optic nerve or retina. The ex-

* Quoted from an article of my own in *Nature*, June 8, 1882.

† *Essay on Origin and Development of Sense-organs.*

ternal epidermis invests the whole of the prominent eye-ball, and forms in front of it a transparent horny layer, the cornea." Further, from the more recent observations of Mr. Darwin, it is certain that Earthworms, although destitute of eyes, are able to distinguish with much rapidity and precision between light and darkness; and as he found that it is only the anterior extremity of the animal which displays this power, he concludes that the light affects the anterior ganglia immediately, or without the intervention of a sense-organ.* Lastly, Schneider says that Serpulæ will suddenly withdraw their expanded tufts when a shadow falls upon them; but the shadow must be that of an object moving with some rapidity.†

Turning now to the sense of hearing in the Articulata, we find the simplest type of ear among the Vermes, where it occurs as a closed globular vesicle containing fluid in which there is suspended an otolith.‡ In some of the Crustacea, such as the cray-fish and lobster, the organ of hearing is much more complex, and here, "if we give rise, by playing the violin, to notes of varying pitch, and at the same time observe the auditory organ under the microscope, we see that at each note only a particular auditory hair is set in vibration."§ Among Insects organs of hearing certainly occur, at least in some species, although the experiments of Sir John Lubbock appear to show that ants are deaf. The evidence that some insects are able to hear is not only morphological, but also physiological, because it is only on the supposition that they do that the fact of stridulation and other sexual sounds being made by certain insects can be explained; and Brunelli found that when he separated a female grasshopper from the male by a distance of several metres, the male began to stridulate in order to inform her of his position, upon which the female approached him.|| I have myself published observations proving the occurrence of a sense of hearing among the Lepidoptera.¶ Turning to the morphological side of the subject, it is remarkable that in the Articulata the

* See *Earthworms*, pp. 19-45. † *Der thierische Wille*, s. 194.

‡ Earthworms have no ears and are totally deaf, although very sensitive to vibrations communicated through contact with solid bodies. (See Darwin, *loc. cit.*, pp. 26-7.)

§ Hæckel, *loc. cit.*, English translation, *International Library of Science and Freethought*, vol. vi, p. 325.

|| See Houzeau, *Fac. Mém. des Animaux*, t. i, p. 60.

¶ See *Nature*, vol. xv, p. 177.

auditory organs occur among different members of the group in widely different parts of the body. Thus in the lobster and cray-fish they are situated in the head at the base of the antennules, while in some of the crabs (*e.g.*, *Mysis*) they occur in the tail. Among the Orthoptera, again, they are found in the tibiæ of the front legs, or, in other species, upon the sides of the thorax. In other insects, probability points to the organs of hearing being placed in the antennæ. These facts prove that in the Articulata the sundry kinds of auditory organs must have arisen independently, and have not been inherited from a common ancestor of the group; and it is remarkable that this should have been the case even within the limits of so comparatively small a subdivision as that which separates a crab from a crayfish or a lobster.*

There can be no question that the sense of smell is well developed in at least many of the Articulata, although, save in a few cases, we are not yet in a position to determine the olfactory organs. Thus the account which I quoted in "Animal Intelligence" (p. 24), from Sir E. Tennent, concerning the habits of the land leeches of Ceylon, proves that these animals must be accredited with a positively astonishing delicacy of olfactory perception, seeing that they smell the approach of a horse or a man at a long distance. In earthworms the sense of smell is feeble, and seems to be confined to certain odours.† Sir John Lubbock has proved by direct experiment that ants are able to perceive odours, and that they appear to do so by means of their antennæ. The same remark applies to bees, and the general fact that many insects can smell is shown by the general fact that so many species of flowering plants, which depend for their fertilization upon the visits of insects, give out odours to attract them. That the crustacea are able to smell is rendered evident by the rapidity with which they find food. I have recently been able to localize the olfactory organs of crabs and lobsters by a series of experiments which I have not yet published, and which would occupy too much space here to detail. I shall therefore merely say that they are situated in the pair of small antennules, the ends of which are curiously modified in order to perform the olfactory function. That is to say, the

* Analogous facts are to be observed in the case of the Eye among Vermes, and also, as we shall presently see, among Mollusca.

† Darwin, *loc. cit.*, p. 30.

terminal joint works in a vertical plane, and supports the sensory apparatus, which is kept in a perpetual jerking motion up and down, so as to bring that apparatus into sudden contact with any minute odoriferous particles which may be suspended in the water—just on the same principle as we ourselves smell by taking a number of small and sudden sniffs of air. Any one visiting an aquarium can have no difficulty in observing these movements upon any crab or lobster in a healthy condition.

The sense of taste certainly occurs at least among some species of the Articulata (as, *e.g.*, among the honey-feeding insects), and the sense of touch is more or less elaborately provided for in all.

Turning now to the Mollusca, we pass in a tolerably uniform series from the simple eye-spots of certain of the Lamellibranchiata, through the Pteropoda, to the more completely organized eyes of the Gasteropoda and the Heteropoda. But when we arrive at the Cephalopoda, we encounter, as it were, a vast leap of development; for the eye of an octopus, in point of organization, is equal to that of a fish, which it so closely resembles. And, while remembering that the resemblance, striking though it be, is only superficial, we must not fail to note that this enormous development in the organization of the molluscous eye, which brings it so strangely to resemble the eye of a fish, is clearly correlated with the no less enormous development of the neuro-muscular system of the animal, in which respect it more resembles a fish than it does the other Mollusca. This case is therefore analogous to the similarly high development which has been attained by the eye of the swimming worm previously described.

If we look to the Mollusca as a class, we meet with the same kind of variation in the position of the eye which we have already noticed with respect to the ear in the Articulata. Thus, while in the Cephalopoda and Gasteropoda the eyes are situated in the head, in some of the latter group there are supplementary eyes upon the back, which greatly differ in structure from the eyes in the head. In the Lamellibranchiata, again, the eyes occur in large numbers on the margin of the mantle.

The sense of hearing is general to all the Mollusca, and the auditory organs exhibit a progressive elaboration as we ascend from the lower to the higher groups, which is analogous to that already noticed with reference to the organs of sight. Thus, among the lower Mollusca the organs of hearing consist of a pair of small vesicles attached to auditory nerves, and filled with fluid in which an otolith is suspended. In the Cephalopoda, however, while the same general plan of structure is adhered to, we find an approximation to the auditory apparatus of a fish; for the vesicle or sac is now embedded in the cartilage of the head, is of larger size, and in general analogous to the organ of hearing of the Vertebrata. That at all events the majority of the Mollusca are able to smell, is proved by the readiness with which they find food, and the octopus is said to show a strong aversion to certain odours (Marshall). In the Cephalopoda the olfactory organs are probably two small cavities near the back of the eye, and in the other Mollusca they are surmised to be situated in the small tentacles near the mouth. Touch is provided for both by these and by the larger tentacles (as well as by the general soft exterior); but in the Cephalopoda by the long, snake-like arms, which I think must be regarded as giving these animals a greater power of receiving tactile impressions than is enjoyed by any other marine animal.

Among Fish sight is well developed. A trout will distinguish a worm suspended in muddy water; a salmon can avoid obstacles when swimming with immense velocity; and a *Chelmon rostratus* can take unerring aim with its little water projectile at a fly. The blind fish, which live habitually in the dark, have lost their eyes merely from disuse; but in this connection it must be noted that we meet with a curious biological puzzle in the case of many of the deep sea fishes dredged by the *Challenger*. For although living at depths to which no light can be supposed to penetrate, some of these fish have large eyes. It may be suggested that the use of these eyes is that of seeing the many self-luminous forms of life which, as the *Challenger* dredgings also show, inhabit the deep sea. But if this is suggested, the question immediately arises as to why these forms have become luminous; for if thus rendered conspicuous to the fish, their luminosity must so far be a disadvantage to them. In the case of the luminous animals which themselves have eyes, we may suppose that this disadvantage is more than compensated for by the advantage of enabling the sexes to find each other; but this explanation does not apply to the blind forms.

Fish, as we have already observed, are well provided with the organs both of hearing and of smell, Amphioxus being the only member of the class which is destitute of ears, and the olfactory lobes in the case of some species (*e.g.*, the Skate) being of enormous size in relation to the other parts of the brain. The sense of touch is provided for in many species by tentaculæ in the neighbourhood of the mouth. The soft lips, and in some species the pectoral fins, are also tactile in function, and in certain gurnards there are digitate appendages connected with the latter which doubtless serve to increase their efficiency as organs of touch. It is doubtful whether taste, as distinct from smell, occurs in fish; but we must remember, as before observed, that in the case of an aquatic animal there is no true distinction to be drawn between these two senses. For as there is here no gaseous medium (like the air) in question, the only distinction that can be drawn is as to whether the nerve terminations, which are affected by the suspended particles in the water, happen to be distributed over any part of the mouth where the food passes, or over any other part of the animal. I say over any other part of the animal (and not only in the nasal fossæ), because in some species of fish there are embedded in the skin along the sides of the body a number of curiously-formed papillæ, which on morphological grounds may reasonably be regarded as ministering to the sense of smell, or, as we may indifferently call it, of taste. Hæckel, however, speculates upon these organs, and is inclined to think that they minister to some unknown sense.

The sense of sight in Amphibia and Reptiles offers nothing specially worthy of remark, except that the crystalline lens has not so high a refracting power as in Fish. The transition from an eye adapted to see under water and an eye adapted to see in air, appears to be curiously shown by one and the same eye in the case of the Surinam Sprat. This animal has its eyes placed on the top of its head, so that when it comes to the surface of the water part of the eyes come into the air, and "the pupil is partly divided, and the lens is also composed of two portions, so that it is supposed that one part of this curious eye is adapted for aërial, and the other for aquatic, vision."* The senses of hearing, smell, taste, and touch, although all present in the

* Marshall, *Outlines of Physiology*, vol. i, p. 603.

Amphibia and Reptiles, are not much, if at all, in advance of these senses as they occur in Fish.

Among Birds the sense of sight is proverbially keen, and in point of fact the animal kingdom has no parallel to the excellence of the organ of vision as it occurs in some species of this class. Whether we consider the eye of a Hawk, which is able to distinguish from a great height a protectively coloured animal from the surface of the ground which it so closely imitates; or the eye of a Solen Goose, which is able from a height of a hundred feet in the air to see a fish at the depth of many fathoms in the water; or the eye of a Swift, which is able so suddenly to form its adjustments; we must alike conclude that the visual apparatus has attained to its highest perfection among birds. And in this connection it is of interest to note that protective colouring has attained its highest degree of perfection among animals which constitute the prey of birds. So surprising, indeed, is the perfection to which protective colouring has attained in some of these cases, that it has been adduced as a difficulty against the theory of evolution; for it seems incredible that such perfection should have been attained by slow stages through natural selection before the species exhibiting it had been exterminated by the birds. The answer to this difficulty is that the visual organs of the birds cannot be supposed to have been always so perfect as they are now, and therefore that a degree of protective colouring which might have afforded efficient protection at an earlier stage in the evolution of those organs would not supply such protection at the present day. In other words, the evolution of the eyes of birds and of the protective coloration of their prey must be supposed to have progressed *pari passu*, each stage in the one acting as a cause in the succeeding stage of the other. The crystalline lens is flat in birds which are remarkable for long sight, such as the vulture; rounder in owls, which are very nearsighted; and becomes progressively more spherical in aquatic birds, according to their aquatic habits.

All birds are able to hear, and it is in this class that we first meet with definite evidence of an ear capable of appreciating with delicacy differences of pitch. Among many species of birds the delicacy of such appreciation (as well as that of *timbre*) is so remarkable that it may be questioned whether even human ears are more efficient in this respect.

The anatomical difficulty of accounting for this fact I need not wait to consider. I am myself inclined to think that the sense of hearing in birds (at all events of some species) is likewise highly delicate with reference to the *intensity* of sound. My reason for so thinking is that I have observed Curlews dig their long bills up to the base into smooth unbroken surfaces of sea-sand left bare by the tide, in order to draw up the concealed worms. Under such circumstances no indication can be given by the worm of its position to any other sense of the curlew than that of hearing. Similarly, I suspect that the common Thrush is guided to the worm buried beneath the turf by the sense of hearing, and my suspicion is founded on the peculiar habits of feeding shown by the bird, which I have described elsewhere.*

The sense of smell in Birds is in advance of that of Reptiles, but not to be compared with its excellency in Mammals; for the old hypothesis that vultures find their prey by the aid of this sense has been abundantly disproved.† The sense of taste in Birds is likewise very obtuse as compared with this sense in Mammals; and as compared with the same class they are also defective in their organs of touch. Indeed, the parrot tribe is the only one in which this sense is well or specially provided for, except the ducks, snipes, and other mud-feeding species, in which the bill is specially modified for this purpose.

If we regard Mammals as a class we must say that, with the exception of the sense of vision which reaches its greatest supremacy in Birds, all the special senses are more highly developed than in any other class. This is more particularly the case with the senses of smell, taste, and touch.

The sense of smell reaches its highest perfection among the Carnivora and the Ruminants, and, on the other hand, is totally absent in some of the Cetacea. Any one accustomed to deer-stalking must often have been astonished at the precautions which it is needful to take in order to prevent the game from getting the "wind" of the sportsman; indeed to a novice such precautions are apt to be regarded as implying a superstitious exaggeration of the possibilities of the olfac-

* *Nature*, vol. xv, pp. 177 and 292, where also see in more detail my observations on the feeding habits of the curlew.

† See *Animal Intelligence*, pp. 286–7.

tory sense; and it is not until he has himself seen the deer scent him at some almost incredible distance that he lends himself without disguised contempt to the discretion of the keeper. But among the Carnivora the sense of smell is even more extraordinary in its development on account, no doubt, of its being here of so much service in tracking prey. I once tried an experiment with a terrier of my own which shows, better than anything that I have ever read, the almost supernatural capabilities of smell in Dogs. On a Bank holiday, when the broad walk in Regent's Park was swarming with people of all kinds, walking in all directions, I took my terrier (which I knew had a splendid nose, and could track me for miles) along the walk, and, when his attention was diverted by a strange dog, I suddenly made a number of zig-zags across the broad walk, then stood on a seat, and watched the terrier. Finding I had not continued in the direction I was going when he left me, he went to the place where he had last seen me, and there, picking up my scent, tracked my footsteps over all the zig-zags I had made until he found me. Now in order to do this he had to distinguish my trail from at least a hundred others quite as fresh, and many thousands of others not so fresh, crossing it at all angles.

Such being the astonishing perfection of smell in dogs, it has been well observed that the external world must be to these animals quite different from what it is to us; the whole fabric of their ideas concerning it being so largely founded on what is virtually a new sense. But in this connection I may point out that speculation on such a subject is shown to be useless by the fact that the sense of smell in dogs does not appear to be merely our own sense of smell greatly magnified. For if this were the case it seems incredible that highly bred sporting-dogs, which have the finest noses, should be those which take the keenest pleasure in rolling in filth which literally stinks in our nostrils to the degree of being physically painful.

The sense of hearing is acute in Mammals as a class, and it is worthy of remark that this is the only class provided with movable ears. As Paley observes, in beasts of prey the external ear is habitually directed forwards, while in species which they prey upon the ear admits of being directed backwards. With the exception of the singing monkey (*Hylobates agilis*), there is no evidence of any mammal other than man having any delicate perception of pitch. I have, however, heard a terrier, which used to accompany a song by howling, follow the prolonged notes of the human voice with some approximation to unison; and Dr. Huggins, who has a good ear, tells me that his large mastiff "Kepler" used to do the same to prolonged notes sounded from an organ.

The sense of taste is much more highly developed in the Mammalia than in any other class, and the same general statement applies to the sense of touch. Looking to the class as a whole, the principal organs of the latter are the snout, lips, and tongue; the modified hairs, or "whiskers," are also very generally present. Among the Rodents, some of the Mustelidæ and all the Primates, the principal organs of touch are the hands. And it would appear that the extreme modification which these members have undergone in the Cheiroptera has been attended with an extraordinary exaltation of their power of tactile sensation. For in the celebrated experiment of Spallanzani (since repeated and confirmed by several other observers), it was found that when a Bat is deprived of its eyes, and has its ears stopped up with cotton-wool, it is still able to fly about without apparent inconvenience, seeing that it avoids all obstacles in its flight, even though these be but slender strings stretched through the room in which the animal is allowed to fly. The only explanation of this surprising fact is that the membranous expanse of the wing, which is richly supplied with nerves, has developed a sensibility to touch, to temperature, or to both, so extreme as to inform the bat of the proximity of a solid body even before contact—either through the increase in the air-pressure as the wing rapidly approaches the solid body, or through the difference in the exchange of heat between the wing and the solid as compared with such exchange between the wing and the air. When groping our way through a dark room we are ourselves able to feel a large solid body (such as a wall) before we actually touch it, especially, I have observed, with the skin of the face. Probably, therefore, it is a great exaltation of this power which enables these night-flying animals to avoid so slender a solid body as a stretched string. But when we remember the rapidity and accuracy with which the sensation must here be aroused, we may well consider it to equal, if not to surpass, in the domain of touch, the evolutionary development of sense-organs as it occurs in the sight of the vulture or the smell of the dog. Indeed, Hæckel and others have speculated whether the facts in this case do not call for the supposition of some additional and unknown sense, different in kind from any that we ourselves possess. But I think it is safer not to run into any such obscure hypothesis unless actually driven to do so, and therefore I shall not here entertain it. For this reason, also, I shall not follow Hæckel in his view that the "homing" faculty of certain animals is due to some additional and inexplicable sense, and therefore I shall reserve my treatment of this topic for my chapters on Instinct.

After this rapid survey of the powers of Special Sense as they severally occur in different classes of the animal kingdom, I shall conclude the present chapter by briefly considering certain general principles connected with Sensation.

The muscular sense, the sense of hunger, thirst, satiety, and others of the like general kind need not detain us; for although their causation is somewhat obscure, we know at least that they are dependent upon nervous adjustments, and, being of so much importance to animals, we infer that they have been developed under the general principles of neuro-muscular evolution already considered in previous chapters. My object here is rather to consider the mechanisms of certain more special senses from the point of view of those general principles.

First as to the sense of Temperature, there is good evidence that in ourselves and at least in all the higher animals, thermal sensations can only be received by the nerve-endings in the skin and adjacent parts of the mucous membranes; if the nerve-fibres immediately above their terminations in these localities (as in the raw surface of a wound) be stimulated by heat or cold, the sensation produced is merely one of pain. There is strong evidence that not only the nerve-endings, but even the whole of the nerve-tracts of which they are the endings, are specialized for the purpose of receiving thermal impressions. These impressions, when received, are not absolute, but relative to the temperature of the part receiving them—the greater the difference of temperature between the part and the object touching it, the greater being the impression. Moreover, the greater the extent of the receiving surface, the greater is the impression; so that if the whole hand be immersed in water at 102°, the temperature of the water will be erroneously judged to be higher than that of another body of water at 104°, the temperature of which is simultaneously estimated by a single finger of the other hand; and, similarly, smaller differences of temperature can be appreciated by the whole hand than by a single finger. According to Weber, the left hand is considerably more sensitive to temperature than the right; and it is certain that different parts of the body differ greatly in this respect. The more sudden the change of temperature, the greater is the sensory effect. We have no means of testing the truth of any of these statements with reference to any of the Invertebrata, or even with reference to the cold-blooded Vertebrata; but we can scarcely doubt that they apply in a general way to all the warm-blooded. The facts certainly show an elaborate provision for appreciating local changes of temperature occurring upon this and that part of the external surface (the general comfort or discomfort arising from the body being kept at a normal temperature or not is another matter, and one with which the special mechanism we are considering is not concerned); and therefore we have to contemplate the probable cause of its origin and development.

At first sight we appear to encounter a difficulty which I wonder never to have seen adduced by opponents of evolution. For in nature the only differences of temperature which normally occur in objects with which animals have any opportunity of coming into contact, are those between ice and objects heated by a tropical sun; and no one animal ever has the opportunity of experiencing changes of temperature extending through anything like so great a range; for in the arctic regions there is no tropical sun, in the tropics there is no ice, and in the temperate zones the solar heat is moderate. Of course since the introduction of fire by man, the sense of temperature has become of much use to sundry species of animals for the examination of food, &c., and in this connection is of almost indispensable service to man himself; but, looking to the antecedents of these animals and also to the antecedents of man, it may at first sight seem remarkable that such an elaborate provision should have been developed, and, as I have said, I wonder that no opponent of evolution has pointed to the fact. For it might be argued that here we have a complicated piece of special organic machinery constructed in obvious anticipation of the advent of cookery and warm baths. But I think the matter may be explained on evolutionary principles, if we remember that the only use of a sense of temperature is not that of examining food. We know that differences of temperature on the surface of the body (whether local or general) greatly modify the conditions of the circulation in the part or parts affected, and therefore it must always have been of use for animals to be provided with a sensory apparatus upon the surface of their bodies to give them immediate information of such differences. Its development along special lines (so that some parts of the body should be more sensitive to changes of temperature than other parts) is easily to be explained by the effects of habit or use. Thus, for example, the fact that the lips of man, although provided with a skin so delicate and so sensitive to tactile impressions, are nevertheless able to endure a sudden rise of temperature which would be painful to the skin of the face, must be taken to mean that habit has adapted the nerves in the lips to withstand a sudden rise of temperature—and this certainly within the period since the invention of cookery.

Mr. Grant Allen takes a more general view of this subject, and says: "To an animal, cold is death, and warmth is life. Hence it is not astonishing that animals should very early have developed a sense which informed them of changes of temperature taking place in their vicinity; and that this sense should have been equally diffused over the whole organism. . . . . As soon as moving creatures began to feel at all, they probably began to feel heat and cold."* The truth of such a general statement of this must be obvious, and the step between a sense of temperature equally diffused over the whole organism, and the specialization of superficial nerve-endings to minister to this sense alone, is not a large step. Moreover, the step between this and the development of a rudimentary visual organ is likewise not a large one. For the deposition of dark-coloured pigment in particularly exposed parts of the skin must have been of benefit to animals by enabling (in virtue of the increased absorption of heat thus secured) the nerve-endings

* *Colour Sense*, p. 13.

in those parts to be more sensitive to changes of temperature. But the deposition of pigment in such localities constitutes a favouring condition to the origination of an eye, or of an organ whose sense of temperature becomes sufficiently developed to enable it to begin to distinguish between light and darkness. Thus, as Professor Hæckel eloquently remarks: "The ordinary nerves of the skin which pass to these dark pigment-cells of the integument, have already trodden the first steps of that magnificent march, at the end of which they have attained to the highest development of the nerves of sensation—the optic nerves."

Turning next to the sense of Colour, it appears from the experiments of Engelmann already alluded to, that colour-sense of a kind occurs as low down in the zoological scale as the protoplasmic and unicellular organisms, inasmuch as particular species showed particular preferences for certain rays of the spectrum. But as in these organisms there are no organs of special sense, and probably no beginnings of consciousness, I do not think that any true analogy can be drawn between these cases and those in which there is a true sensation of colour. Nor have we any evidence of such a true sensation till we arrive at the Crustacea. Here we have proof, furnished by the direct experiments of Sir John Lubbock, that *Daphnia pulex* prefers certain rays of the spectrum to others,* and the Chameleon Shrimp (*Mysis chameleo*) is known to change its colour in imitation of the surface on which it reposes, provided that it is not blinded or otherwise prevented from seeing that surface. Precisely analogous facts occur among the Cephalopoda (*e.g.*, *octopus*), Batrachia (*e.g.*, Common Frog), Reptilia (*e.g.*, Cameleon), and Pisces (*e.g.*, Flounder); in all these cases, if the animals are blinded, the effects no longer occur. Moreover, Pouchet found that in the Pleuronectidæ the mechanism whereby these imitative changes of colour are produced is bilaterally disposed, so that if only one eye of the animal is stimulated by coloured light, only one side of the animal changes colour. M. Fredericq afterwards found the same thing to be true of the Octopus, and in conjunction with Professors Burdon-

* *Journ. Linn. Soc.*, 1881. These observations have been adversely criticized by Merejkowsky (*Comptes Rendus*, xciii, pp. 160–1), but his criticisms have been fully met by further experiments recently published by Sir John (*Journ. Linn. Soc.*, 1883).

Sanderson, Cossar Ewart, and Mr. W. D. Scott, I have corroborated M. Fredericq's observations by a number of experiments; stimulation of one eye alone by means of light produces immediate unilateral flushing of the whole of the same side of body, but no change of colour beyond the median line.

As further proof that a well-developed sense of colour occurs in some of the Articulata, I may allude to the experiments of Sir John Lubbock on the Hymenoptera; but as these have been already twice published in the International Scientific Series,* I need not here wait to recapitulate them, and shall therefore only remark that it is without any reasonable question to the presence of this sense in insects that we owe the beauty both of floral and of insect coloration. Again, as further proof that a well-developed sense of colour occurs in Fish, I may remark that the elaborate care with which anglers dress their flies, and select this and that combination of tints for this and that locality, time of day, &c., shows that those who are practically acquainted with the habits of trout, salmon, and other fresh-water fish, regard the presence of a colour-sense in them as axiomatic. And, with reference to the sea-water fish in general, we have the highly competent opinion of Professor H. N. Moseley to the effect that the great majority of the colours of marine animals have been acquired either for the protection or the allurement of prey, and that they refer particularly to the eyes of Fish, and also to those of Crustacea.†

The fact that a sense of colour occurs in Birds is unquestionable, and meets with its most general proof in the more or less conspicuous coloration of the fruits on which they feed; for as in the analogous case of conspicuously coloured flowers depending on insects for their fertilization, so conspicuously coloured fruits depend for the dissemination of their seeds upon being eaten by birds or mammals. Again, I have already mentioned the fact that nowhere in the animal kingdom does the protective and imitative colouring of animals attain to such nicety as it does where the eyes of birds are concerned. Lastly, the elaborate coloration of birds themselves, and the pleasure which some species take in the decoration of their nests, constitute supplementary

* Viz., in *Ants, Bees, and Wasps*, and in *Animal Intelligence*.

† *Quarterly Journ. Micro. Science*, New Series, vol. xvii, pp. 19–22.

proof of the high development to which the colour-sense has attained in this class.

All the remarks just made with reference to Birds, apply likewise, though not perhaps in quite so high a degree, to Mammals, considered as a class. And here it becomes needful to consider the speculation of Dr. Magnus and Mr. Gladstone, that the colour-sense of man has undergone a great improvement within the last two thousand years, inasmuch as before that time mankind are supposed by this speculation to have perceived only the lower colours of the spectrum, or red, orange, and yellow, and to have been colourblind to the higher, or green, blue, and violet. Professor Hæckel lends his support to this speculation; but to me it seems a highly improbable one, and this for the following reasons.

In the first place the speculation is based merely on etymological grounds, which in a matter of this kind are exceedingly unsafe. For the absence in a language of words denoting particular colours is, at best, but negative evidence that the men who spoke the language were blind to those colours; the absence of such words may quite as well be due to the imperfection of language as to the imperfection of the visual sense. Thus, for instance, Professor Blackie tells us that the Highlanders call both sky and grass "gorm," and are nevertheless quite able to discriminate between the colours blue and green. In the next place, it is antecedently improbable, upon the general principles of evolution, that a considerable change in the visual apparatus of man should have taken place within so short a period as the speculation in question assigns—especially in view of the fact that other Mammals, Birds, and even some of the Invertebrata unquestionably distinguish the higher as well as the lower colours of the spectrum. Lastly, Mr. Grant Allen has taken the trouble to enquire, by means of a table of questions addressed to educated Europeans in all parts of the world, whether any of the savage races of mankind now living display any inability to distinguish between the colours of the spectrum, and the answers which he has received have been uniformly in the negative.* I think, therefore, we may safely dismiss the speculation of Dr. Magnus and Mr. Gladstone as opposed to all the evidence which is at once trustworthy and available. But in saying this I do not intend to

* *Colour-sense*, Chapter X.

dispute the probability, which indeed amounts almost to a certainty, that as civilization advances and the fine arts become developed, the colour-sense undergoes a progressive improvement in its power of distinguishing between fine shades, and also in its power of ministering to a more and more evolved condition of æsthetic feeling. And this, I believe, is the true explanation of the class of facts alluded to by Professor Hæckel as proof of the speculation which I have now discarded—the fact, namely, that "nowadays we see in the surviving savage races a crudity as to their sense of colour . . . . Our little ones, also, like the savages, love assemblages of glaring hues which grate upon us, and susceptibility to the harmony of delicate shades of colour is the latest product of æsthetic education."

Professor Preyer has published within the last year or two a very interesting theory touching the origin and development of the colour sense, and as it has not, to my knowledge, been noticed in any English publication, I shall here state the main points. The theory is that the colour-sense is a special and highly-exalted development of the sense of temperature. To sustain this theory, Professor Preyer first compares the sensibility of the skin to temperature with that of the retina to light, and points out that the analogy has already been recognized by artists, who speak of colours as "warm" and "cold." "The warm colours arouse sensations of a character antagonistic to those which are aroused by the cold colours, in just the same way as the hot and cold sensations of skin-temperature are antagonistic; and the more this analogy is pursued, the closer is the agreement found to be.' Therefore the suggestion arises, "that the sense of colour has been developed out of the sense of temperature," bespeaking a high refinement of functional activity which has its structural correlative in the extremely differentiated and delicately organized expansion of nerve-endings which we find in the retina.

A further analogy is that of contrasts. A finger that has been warmed or cooled retains its change of temperature for some time after it has ceased to be warmed or cooled; and this is taken to correspond with the phenomena of positive after-images in sensations of colour. Moreover, while the after-effect of warming or cooling a portion of the skin remains, the temperature-sense of that portion is altered in such wise that, if it has been cooled, it over-estimates the temperature of any object it may touch, and *vice versâ*. This is taken to be analogous to the appearance of warm colours in the eyes when closed immediately after having been exposed to intense cold colours, and *vice versâ*. So, too, it is with simultaneous contrasts. It is well known that if a small colourless surface is enclosed between two surfaces of cold or warm colours, the small surface will appear inversely coloured warm or cold, as the case may be; and Professor Preyer has found by experiment, that if a small portion of the skin be enclosed by cold or warm surfaces on either side, the small enclosed area will feel cool if the neighbouring parts are heated, and *vice versâ*.

After showing that in his view illumination is to the sense of colour what contact is to the sense of temperature, and pointing out several subordinate analogies which I have no space to mention, Professor Preyer goes on to remark an important fact in relation to his theory, viz., that different parts of the skin manifest in their estimations of temperature great differences in their estimates of what he calls the "neutral point," *i.e.*, the point at which it cannot be said that a body is felt to be either hot or cold. The retina, then, being supposed to be merely a nerve-expansion having a much higher "neutral point" in the appreciation of temperature (ethereal vibrations) than has any nerve-expansion of the skin, colour-blindness is explained by supposing that the retina of the individual so affected has a neutral point either above or below the normal. "An over-warm eye must be blind to yellow and blue; an over cool one must be blind to red and green." Total colour-blindness, which is a physiological characteristic among certain nocturnal animals, has its parallel in the pathological condition sometimes met with in man, of a total absence of the sense of temperature without impairment in the sense of touch.

Lastly, it is observed that the first condition to the validity of any physiological hypothesis is that it should accord with morphological fact. But this is not the case with the theory of Young and Helmholtz, which ascribes the colour-sense to the functions of three retinal elements; for it has been proved that the number of fibres in the optic nerve immediately before it enters the retina is much smaller than the number of rods and cones in the retina.

In my opinion this theory, in its main outlines, seems a probable, as it certainly is a plausible one. I do not, indeed, quite understand why, in accordance with the theory, the "neutral point" of the colour-blind should not merely be found to be shifted to another part of the spectrum, nor am I quite clear about the explanation of the fact that the warm colours are those having the lowest and not the highest order of vibrations, as analogy would lead us to expect. But the theory has the merit of being antecedently probable, when we remember that in all likelihood the visual sense arose by the progressive elaboration of nerve-endings in particular parts of the skin, which before their special elaboration presumably ministered to the senses of touch and temperature.

And this remark leads me to the last topic that I have to dwell upon in the present chapter. I refer to the body of morphological evidence which we now possess, showing that all the organs of special sense have had their origin in special elaborations of these nerves of the integument. For the uniform result of histological and embryological investigation is to show that all organs of special sense, wherever they occur and whatever degree of elaboration they present in the adult animal, are fundamentally alike in that their receptive surfaces are composed of more or less modified epithelium cells which originally constituted part of the external layer of the animal. Thus, the origin of the olfactory membrane in the embryo of the Vertebrata is found to consist in a pitting of the skin of the fore part of the head—the pits, therefore, being lined by the general layer of epidermic cells. The subsequent growth of the surrounding parts of the face eventually brings this lining to occupy the position which it does in the hollow parts of the nose. Similarly, the organs of hearing first begin as a pair of pits on the side parts of the head, situated somewhat far back, and likewise lined by the cells of the general integument. These pits rapidly deepen, so that their lining is pinched off or separated from the general integument of which it originally formed a part. The deep pit then becomes a closed sac, and the adjacent tissues becoming first cartilaginous and then osseous, this sac is enclosed well within the skull by bony walls. While its structure is undergoing further anatomical and histological changes, the drum, the chain of ear-bones, and the external ear are being formed, and thus eventually the auditory organ is completed. In the case of the eye, again, the earliest sign of commencement consists in a similar pitting of the general integument, but the lining of this pit is not destined, as in the previous cases, to become the receptive surface of the sensory impressions. For, after it has deepened considerably it undergoes sundry changes which result in its forming the cornea, aqueous humour, and crystalline lens; while the retina arises as an offshoot from the brain in the form of a sac growing, as it were, upon a slender stalk towards the crystalline lens. At first the anterior surface of this sac is convex, but the posterior part afterwards becomes pushed into the cavity of the sac; so that the anterior surface eventually becomes strongly concave. Therefore the sac is now, as Professor Huxley graphically describes it, "like a double night-cap, ready for the head, but the place which the head would occupy is taken by the vitreous humour, while the layer of night-cap next it becomes the retina." Thus the rods and cones of the retina are not developed immediately out of the epidermic cells of the integument; but inasmuch as the brain is itself begun as an infolding of the epidermic layer, the rods and cones of the retina are ultimately derived from those epidermic cells. Or, again to quote Professor Huxley, "the rods and cones of the vertebrate eye are modified epidermal cells, as much as the crystalline cones of the insect or crustacean eye are."* Therefore, in the words of Professor Hæckel, "the general conclusion has been reached that in man, and in all other animals, the sense-organs as a whole arise in essentially the same way, viz., as parts of the external integument or epidermis. The external integument is the original general sense-organ. Gradually the higher sense-organs detach themselves from this their primal condition, whilst they withdraw more or less completely into the protecting parts of the body. Nevertheless in many [invertebrate] animals, even at the present hour, they lie in the integument, as *e.g.*, in the Vermes."

I have entered thus fully into this general fact, because it is of importance, not only to the theory of evolution, but also to the philosophy of sensation, to know from such direct historical sources that all the special senses are differentiations of the general sense of touch.

* *Science and Culture*, &c., p. 271.

CHAPTER VIII.

PLEASURES AND PAINS, MEMORY, AND ASSOCIATION OF IDEAS.

IN the diagram I have represented Pleasures and Pains as occupying in their first origin a level not far removed from that at which Sensation takes its rise. I have also represented a short interval between Sensation and the origin of Perception, which is filled up in the lateral column by Memory and Primary Instincts. Therefore, before we pass on to consider the rise of Perception out of Sensation, I shall devote a chapter to a consideration of Pleasures and Pains Memory, and Association of Ideas.

Pleasures and Pains.

On this topic I have little to add to the treatment which it has received at the hands of Mr. Herbert Spencer, and of his disciple, Mr. Grant Allen.* Pains, as Mr. Spencer points out, may be due to the want of action ("craving"), or to an excess of action. These two classes correspond largely, though not entirely, with the division of pains into massive and acute, which is formulated by Professor Bain. It also indicates the doctrine of Sir W. Hamilton and others, that Pain is due to excessive stimulation. But it is important to observe that the statement of Mr. Spencer, while "recognizing at one extreme the positive pain of excessive actions," recognizes also "at the other extreme the negative pains of inactions; the implication is that Pleasures accompany actions lying between these extremes."

Mr. Grant Allen in the course of his able exposition of this subject, shows by many examples that "the Acute Pains, as a class, arise from the action of surrounding

* See *Principles of Psychology* and *Physiological Æsthetics*, in both cases the chapter on "Pleasures and Pains."

destructive agencies; the Massive Pains, as a class, from excessive function or insufficient nutriment:" also that "Massive Pains, when pushed to an extreme, merge into the Acute Class," so that "the two classes are rather indefinite in their limits, being simply a convenient working distinction, not a natural division." Hence it follows that Pains of both classes "are the subjective concomitants of an actual disruption or disruptive tendency in some one (or more) of the bodily tissues, provided the tissue be supplied with afferent cerebro-spinal nerves in unbroken connexion with the brain." Referring the reader to Mr. Allen's own essay for all matters of detail and criticism, I shall merely say that in my opinion he has successfully established this formula as applicable to all cases of Pain. His view concerning the physiology of Pleasure is substantially the same as that of Mr. Spencer already quoted; but it is somewhat more extended and precise. This view is that Pleasure is "the concomitant of a normal amount of activity in any portion or the whole of the organism," supplemented with the important *addendum* that "the strongest Pleasures result from the stimulation of the largest nervous organs, where activities are most intermittent;" so that the *amount* of Pleasure is "in the direct ratio of the number of nerve-fibres involved, and in the inverse ratio of the natural frequency of excitation." Hence "we see wherein the feeling of Pleasure fails to be exactly antithetical to the feeling of Pain, just as their objective antecedents similarly fail. Massive Pleasure can seldom or never attain the intensity of Massive Pain, because the organism can be brought down to almost any point of innutrition or exhaustion; but its efficient working cannot be raised very high above the average. Similarly any special organ or plexus of nerves can undergo any amount of violent disruption or wasting away, giving rise to extremely Acute Pains; but organs are very seldom so highly nurtured and so long deprived of their appropriate stimulant as to give rise to very Acute Pleasure."

Now towards what conclusion do these generalizations point? They clearly point to the conclusion, which I do not think is open to any one valid exception, that Pains are the subjective concomitants of such organic changes as are harmful to the organism, while Pleasures are the subjective concomitants of such organic changes as are beneficial to the organism—or, we must add, to the species. The more this doctrine is pursued in detail, the more unquestionable does its truth become. Thus there is to be perceived, not merely a general qualitative, but also a roughly quantitative relation between the amount of pain and the degree of *hurtfulness*, as well as between the amount of pleasure and the degree of *wholesomeness*.* As Mr. Allen observes, "nothing can more thoroughly militate against the efficiency of the mechanism than the loss of one of its component parts: and we find accordingly that to deprive the body of any one of its members is painful in a degree roughly proportionate to the general value of such member to the organism as a whole. Take, for example, the relative painfulness of severing from the body a leg, an arm, an eye, a finger-nail, a hair, or a piece of skin." Similarly with Pleasures, the least pleasurable are those attending activities of the organism which are least important for its welfare (or for that of its species), while the most pleasurable are those which attend the satisfaction of hunger, thirst, and sexual desire—especially if, in terms of Mr. Allen's formula, the needs to which these cravings minister have been long unsatisfied, so that the organism is either in danger of enfeeblement and death, or in the most fit condition for propagating its kind. Pleasures of the intellectual kind, although subservient to the same general laws of nutrition and exhaustion, have reference to such complex nervous states, involving mental prevision of future contingencies, &c., that for the purposes of clear analysis they had best be here disregarded.

The superficial or apparent objection to the doctrine we are considering which arises from the fact that feelings of Pleasure and Pain are not infallible indices of what is respectively beneficial or injurious to the organism, is easily met by the consideration that in all such exceptional cases it is not the doctrine but its application which is at fault. Thus, again to quote Mr. Allen, who in my opinion has given in brief compass the best analysis of the philosophy of Pleasure and Pain that has hitherto appeared, "every act, so long as it is pleasurable, is in so far a healthy and useful one; and conversely, so long as it is painful, a morbid and destructive one. The fallacy lies in the proleptic employment of the words 'deleterious' and 'useful.' To put it in a simple form,

* I use these antithetical words because their etymology alone suggests forcibly the doctrine in question.

the nervous system is not *prophetic*. It informs us of what is its actual state at the moment, not what the after-effects of that state will be. If we take sugar of lead, we receive at first a pleasant sensation of sweetness, because the immediate effect upon the nerves of taste is that of a healthy stimulation. Later on, when the poison begins to work, we are conscious of a painful sensation of griping, because the nerves of the intestines are then being actually disintegrated by the direct or indirect action of the irritant."

Now if the doctrine before us is found to apply generally to all cases of Pleasure and of Pain, the implication is sufficiently apparent; Pleasures and Pains must have been evolved as the subjective accompaniment of processes which are respectively beneficial or injurious to the organism, and so evolved for the purpose or to the end that the organism should seek the one and shun the other. Or, to quote Mr. Spencer, "if we substitute for the word Pleasure the equivalent phrase—a feeling which we seek to bring into consciousness and retain there, and if we substitute for the word Pain the equivalent phrase—a feeling which we seek to get out of consciousness and to keep out; we see at once that, if the states of consciousness which a creature endeavours to maintain are the correlatives of injurious actions, and if the states of consciousness which it endeavours to expel are the correlatives of beneficial actions, it must quickly disappear through persistence in the injurious and avoidance of the beneficial. In other words, those races of beings only can have survived in which, on the average, agreeable or desired feelings went along with activities conducive to the maintenance of life, while disagreeable and habitually-avoided feelings went along with activities directly or indirectly destructive of life, and there must ever have been, other things equal, the most numerous and long-continued survivals among races in which these adjustments of feelings to actions were the best, tending ever to bring about perfect adjustments.

"If we except the human race and some of the highest allied races, in which foresight of distant consequences introduces a complicating element, it is undeniable that every animal habitually persists in each act which gives pleasure, so long as it does so, and desists from each act which gives pain. It is manifest that, for creatures of low intelligence, there can be no other guidance."

Thus, then, we see that the affixing of painful or disagreeable states of consciousness to deleterious changes of the organism, and the reverse states to reverse changes, has been a necessary function of the survival of the fittest. We may further see that in bringing about this adjustment or correspondence, the zoological principle of the survival of the fittest must have been largely assisted by the physiological principle that Pleasure tends to accompany the normal activity of an organ and Pain to accompany its abnormal. For as organs are invariably of use to the organism, their normal activity must always be beneficial to it; while, conversely, their abnormal activity, tending to cause or being caused by their own disintegration, must always be harmful to the organism. Survival of the fittest is thus provided with a ready-formed condition or tendency of psycho-physiology on which to work—a tendency which survival of the fittest may itself in earlier times have been instrumental in producing; but which, in any case, when once established must greatly assist survival of the fittest in apportioning the appropriate state of consciousness to any particular organic process.

Another principle of pyscho-physiology must likewise have greatly assisted natural selection in its execution of this work. This principle is that which obtains in so-called acquired tastes and distastes. Thus, as Mr. Spencer observes, "Pleasures and Pains may be acquired—may be, as it were, superimposed on certain feelings which did not originally yield them. Smokers, snuff-takers, and those who chew tobacco, furnish familiar instances of the way in which long persistence in a sensation not originally pleasurable, makes it pleasurable—the sensation itself remaining unchanged. The like happens with various foods and drinks, which, at first distasteful, are afterwards relished if frequently taken. Common sayings about the effects of habit imply recognition of this truth as holding with feelings of other orders. That acute pain can be superinduced on feelings originally agreeable or indifferent, we have no proof. But we have proof that the state of consciousness called disgust may be made inseparable from a feeling that once was pleasurable:" so that even in the life-time of the individual the states of consciousness as pleasurable or painful may reverse their character with reference to the same organic changes or sensations, and if this is the case it becomes evident with what plastic material natural selection has had to deal in moulding through numberless generations the form of consciousness which best fits, with reference to the welfare of the organism, the circumstances of stimulation.

Thus we may well believe that survival of the fittest, acting always in co-operation with these principles of psycho-physiology, must have been successful in accomplishing the adjustments here assigned to its agency—the adjustments, that is, between states of consciousness as agreeable or disagreeable and circumstances of stimulation as beneficial or deleterious. And thus it is that in the process of evolution organisms "have gone on establishing a consensus between the various organs of the body, so that at last, for the most part, whatever will prove deleterious to any organ proves deleterious also to the first nerves of the organism which it affects," and therefore disagreeable to consciousness, although of course, as we should from these principles expect, this is only the case "when the deleterious object is found sufficiently often in the environment to give an additional point of advantage to any species which is so adapted as to discriminate and reject it."*

Thus then, it seems to me, we have as full a *rationalé* of Pleasures and Pains as we can expect or need desire. The only difficulty is to understand the connection between the objective fact of injuriousness or the reverse, and the corresponding subjective state of consciousness; how is it that injuriousness or the reverse comes to be, as it were, translated into the language of Pleasure and Pain. But this is only the old difficulty of understanding the connection of Mind with Body, and has no reference to historical psychology, which takes the fact of this connection as granted. Possibly, however—and as a mere matter of speculation, the possibility is worth stating—in whatever way the inconceivable connection between Body and Mind came to be established, the primary cause of its establishment, or of the dawn of subjectivity,

* Grant Allen, *loc. cit.*, p. 27. The latter consideration disarms any criticism which might be advanced against our doctrine on account of the agreeable taste of certain poisons, both to ourselves and to the lower animals. But it is astonishing even here how rapidly the appropriate distaste arises after experience of the injurious effects: witness the dislike of wine which may frequently be caused, even in those who are addicted to excess, by surreptitiously mixing it with nux-vomica.

may have been this very need of inducing organisms to avoid the deleterious, and to seek the beneficial; the *raison d'être* of Consciousness may have been that of supplying the condition to the feeling of Pleasure and Pain. Be this as it may, however, it seems certain, as a matter of observable fact, that the association of Pleasure and Pain with organic states and processes which are respectively beneficial and deleterious to the organism, is the most important function of Consciousness in the scheme of Evolution. And for this reason I have placed the origin of Pleasures and Pains very low down in the scale of conscious life. Indeed, if we contemplate the subject, we shall find it difficult or impossible to imagine a form of consciousness, however dim, which does not present, in a correspondingly undeveloped condition, the capacity of preferring some of its states to others—that is, of feeling a distinction between quiescence and vague discomfort, which, with a larger accession of the mind-element, grows into the vivid contrast between a Pleasure and a Pain. I think, therefore, it is needless to say more in justification of the level on the diagram at which I have written these words.

Memory and Association of Ideas.

It is obvious that Memory must be, and is, a faculty which appears very early in the development of Mind. *A priori*, this must be so, because consciousness without memory would be useless to the animal possessing it, and *à posteriori* we find that this is so whether we contemplate the scale of mental evolution in the animal kingdom or in the growing child. I have therefore assigned the rise of Memory to the level immediately succeeding that which is occupied by the rise of Pleasures and Pains.

In a previous chapter* I have endeavoured to show that, even before the dawn of Consciousness, nervous actions of adjustment when frequently repeated present conclusive evidence that the nervous machinery concerned in them becomes more or less organically adapted to perform them, and so exhibits the objective aspect of memory. This objective aspect I spoke of as the memory of a ganglion. Since that chapter was written, M. Ribot has published his excel-

* On "the Physical Basis of Mind."

lent work on the "Diseases of Memory," which has now been translated, and forms a member of the International Scientific Series. In this work M. Ribot deals fully with the complete analogy that obtains on the objective side between ganglionic memory—or, as he calls it, organic memory—and the physical changes in the cerebral hemispheres which are concerned in true or conscious memory. I should like to express my satisfaction at finding so singularly close a correspondence between the views of M. Ribot and myself upon these matters, which extends into so many details that I have left my chapter already referred to *verbatim* as it was originally written; for it speaks in favour of the truth of one's results when they have been independently arrived at by another worker in the same field.*

And here I may observe that I also agree with M. Ribot in his view that the phenomena of memory, whether "organic" or "psychological," present no point of true analogy with any such purely physical phenomena as the permanent effects upon a photographic plate of a short exposure to light, or any other phenomena where living organisms are not concerned. I further agree with him in his view that the earliest analogy we can find to memory is to be sought in living tissues other than nervous, and that it occurs in protoplasm. Thus he quotes Hering to the effect that muscular fibre "becomes stronger in proportion to its use." To this it may, I think, be objected that there is no evidence of *individual* muscular fibres thus gaining in strength by use. I think a better, because a more unexceptionable, parallel is afforded by the fact that when a constant galvanic current is allowed to pass for a short time through a bundle of muscular fibres, in the direction of their length, and is then opened, a change is found to have been produced in the excitability of the fibres such that they are less excitable than before to a stimulus supplied by again passing the current in the same direction, and more excitable to the stimulus supplied by passing the current in the opposite direction. This memory of a muscle touching the direction in which a galvanic stimulus has passed endures for a minute or two after the current has ceased to pass (Frog). I have found this

* Any one who cares to trace the correspondence may do so by comparing my chapter above alluded to with the first chapter of M. Ribot's work.

curious fact to hold in the case of muscular tissues of various animals, from the Medusæ upwards.*

Again, I concur with M. Ribot in his opinion that the physical basis of memory consists partly in a more or less permanent molecular change or "impress" produced upon the nervous element affected by the stimulus which is remembered, and partly upon "the establishment of stable connections between different groups of nervous elements." I do not think that the view can be too strongly reprobated which crudely supposes that the first of these physical conditions is alone sufficient to explain all the facts of memory, and therefore that a given remembrance is, as it were, stored up in a particular cell, as a particular "impression" upon the substance of that cell. On the contrary, as M. Ribot shows, "Each of these supposed unities (memories) is composed of numberless and heterogeneous elements; it is an association, a group, a fusion, a complexus, a *multiplicity;* . . . . Memory supposes not only a modification of nervous elements, *but the formation among them of determinate associations for each particular act.* We must not, however, forget that this is pure hypothesis—the best available one, no doubt, but still not to be taken as implying that we really know anything definitely concerning the physical substratum of memory."

Profound, however, as our ignorance unquestionably is concerning the physical substratum of memory, I think we are at least justified in regarding this substratum as the same both in ganglionic or organic, and in conscious or psychological memory—seeing that the analogies between the two are so numerous and precise. Consciousness is but an adjunct which arises when the physical processes—owing to infrequency of repetition, complexity of operation, or other causes—involve what I have before called ganglionic friction. And this view is confirmed by the large and general fact noted in

* See "Concluding Observations on the Locomotor System of Medusæ," *Phil. Trans.*, Pt. I, 1880; and on "Modification of Excitability," &c., *Proc. Roy. Soc.*, Nos. 171 and 211. Also, *Journal of Anatomy and Physiology*, vol. x. Another equally good instance of what may be termed protoplasmic memory is to be found in the facts of the so-called "summation of stimuli," which occur more or less in all excitable tissues, *i.e.*, wherever living protoplasm is concerned. These facts are that if a succession of stimuli are applied to the excitable tissue, the latter becomes progressively more and more quick, as well as more and more energetic, in its response; each stimulus leaves behind it an organic memory of its occurrence.

our chapter on the Physical Basis of Mind, that conscious memory may become degraded into unconscious memory by repetition; associations originally mental lapsing into associations that are automatic.

Thus much being premised touching the physical basis of memory, we may next pass on to consider the evolution of memory on its psychological side.

The earliest stage of true or conscious memory may, I think, be regarded as consisting in the after-effect produced upon a sensory nerve by a stimulus, which after-effect, so long as it endures, is continuously carried up to the sensorium. Such, for instance, is the case with after-images on the retina, the after-pain of a blow, &c.*

The next stage of memory that it appears to me possible to distinguish by any definite interval from the first-named, is that of feeling a present sensation to be like a past sensation. In order to do this there may be no memory of the sensation between the two successive occasions of its occurrence, and neither need there be any association of ideas. Only this takes place; when the sensation recurs the second, third, or fourth time, &c., it is recognized as like the sensation when it occurred the first time—as like a sensation which is not unfamiliar. Thus, for example, according to Sigismund, who has devoted much careful attention to the psychogenesis of infants, it appears that the sweet taste of milk being remembered by newly-born infants, causes them to prefer sweet tastes in general to tastes of any other kind. This preference of course endures long after the time of weaning is past, and generally continues through childhood; but the interesting point in the present connection is that it occurs too early in the life of the child to admit of our supposing that any association of ideas can take part in the process. For Sigismund says that the memory of the taste of milk becomes attached to the perception "immediately," and Preyer states, from independent observations of his own, that the preference shown for sweet tastes over tastes of all other kinds may be clearly seen as early as the first day.

The next distinguishable stage of memory is reached when, still without any association of ideas, a present sensation is perceived as *unlike* a past one. Thus, again turning to the observations of Sigismund and Preyer, it appears from

* Compare Wundt, *Grundzüge der philosophischen Psychologie*, p. 791.

them that after the accustomed taste of milk has become well fastened in the memory by several successive acts of sucking, the child when a few days old is able to distinguish a change of milk. Similarly, I find among Mr. Darwin's MSS the following note:—

" It is asserted (by Sir B. Brodie) that if a calf or infant has never been suckled by its mother, it is very much easier to bring it up by hand than if it has sucked only once. So again, Kirby and Spence state (from Réaumur, 'Entomology,' vol. i, p. 391) that larvæ after having 'fed for a time on one plant will die rather than eat another, which would have been perfectly acceptable to them if accustomed to it from the first.'"

It will be observed that in dealing with these stages of memory in very young infants, where as yet no association of ideas can either be supposed to be present or is needed to explain the facts, we at once encounter the question whether the memory is to be considered as really due to individual experience, or as an hereditary endowment, *i.e.*, an instinct. And here it becomes apposite to refer to the old and highly interesting experiment of Galen, which definitely answers this question with reference to animals. For soon after its birth, and before it had ever sucked, Galen took a kid and placed before it a row of similar basins, filled respectively with milk, wine, oil, honey, and flour. The kid, after examining the basins by smell, selected the one which was filled with milk. This unquestionably proves the fact of hereditary memory, or instinct, in the case of the kid; and therefore it is probable that the same, at all events in part, applies to the case of the child. In proof of which I may allude to the experiments of Professor Kuszmaul, who found that even prior to individual experience derived from sucking milk, newly-born children show a preference for sweet tastes over all others. For, on their tongues being wetted with sugar or salt solutions, vinegar, quinine, &c., the new-born infants made all manner of grimaces, being pleased with the sugar solution, but with the others showing displeasure by a " sour face," a " bitter face," and so on.

But although we freely admit that the memory of milk is, at all events in large part, hereditary, it is none the less memory of a kind, and occurs without the association of ideas. In other words, hereditary memory, or instinct, belongs to

H 2

what I have marked off as the second and third stages of conscious memory in the largest acceptation of the term—the stages, that is, where, without any association of ideas, a present sensation is perceived as like or unlike a past one. It makes no essential difference whether the past sensation was actually experienced by the individual itself, or bequeathed to it, so to speak, by its ancestors. For it makes no essential difference whether the nervous changes which constitute the obverse aspect of the perceptive aptitude were occasioned during the life-time of the individual, or during that of the species and afterwards impressed by heredity on the individual. In either case the psychological as well as the physiological result is the same; a present sensation is alike perceived by the individual as like or unlike a past sensation. It is not easy at first to grasp the truth of this statement; but the source of the difficulty is in not clearly distinguishing between memory and the association of ideas. Memory in its lower stages which we are now considering has, in my opinion, nothing to do with the association of ideas. It only has to do with perceiving a present sensation as like or unlike a past sensation, which never can have formed the object of an idea between times, and which does not even arise as an ideal remembrance when the sensation again occurs. In other words, there is no act of conscious comparison between the two sensations; there is not even any act of ideation; but the past sensation has left its record in the nervous tissues of the animal in such wise that when it again occurs it emerges into consciousness as a feeling that is familiar—or if another unlike sensation takes its place, this emerges into consciousness as a feeling that is not familiar. And whether such feelings of familiarity or unfamiliarity arise in the experience of the individual or in that of the species, makes, as I have said, no essential difference either in the physiological or in the psychological aspect of the case.

As showing how close is the connection between hereditary memory, or instinct, and memory individually acquired, I shall briefly state some very interesting experiments which were made by Professor Preyer on newly-hatched chickens. He laid before a newly-hatched chicken some cooked white of egg, some cooked yolk of egg, and some millet seed. The chick pecked at all three, but no more frequently at the two latter than it did at pieces of egg-shell, grains of sand, or the spots and cracks of a wooden floor on which it was placed. But at the yellow yolk it pecked often and earnestly. He then removed all three substances, and after the lapse of an hour replaced them. The chick instantly recognized them all, as proved by its immediately beginning to devour them while showing a complete disregard of all other and inedible objects. Yet in the first experiment the chick only once tasted the white of egg, and only took a single millet seed. The experiment therefore shows how apt a young chicken is to learn by its own individual experience, while in the opinion of Professor Preyer the original preference shown to the yolk of egg proves an inherited faculty of taste-discrimination.

These experiments serve to introduce us to the stage of Memory at which the Association of Ideas is first concerned—a principle which throughout all subsequent stages constitutes what may be termed the vital principle of Memory—for the chickens which first pecked at inedible objects in the presence of edible ones, and an hour later were able to distinguish between the two classes of objects, must have established a definite association of ideas between each of the particular objects of its former experience with reference to their edible or inedible character. But it is noteworthy that, as these definite associations were established so quickly and as the result of only a single individual experience in each case, we can scarcely avoid concluding that heredity must have had a large, if not the largest, part in the process—just as in the case of distinguishing from the first the boiled yolk of egg, we must suppose that heredity had the exclusive part.* And this shows how closely the phenomena of hereditary memory are related to those of individual memory; at this stage in the evolution of mnemonics, where the simple association of ideas first occurs in very young animals, it is practically impossible to disentangle the effects of hereditary memory from those of individual.

Association of Ideas.

I shall reserve for my chapter on Imagination a full

* It seems to me doubtful, however, whether heredity here had reference to taste-discrimination, as Preyer supposes, seeing that in nature a young chicken can never have had an opportunity of tasting boiled yolk of egg. Probably the bright yellow colour had something to do with the selection, as many seeds are more or less yellow in tint.

analysis of Ideation. But in connection with Memory it is necessary to touch upon the Association of Ideas, and therefore I shall do so now, notwithstanding the disadvantage which arises from considering the property that ideas present of becoming associated before we consider the ideas themselves. The truth is that here as elsewhere one labours under a difficulty in dealing with the faculties of Mind in the probable order of their evolution, from the fact that these faculties require to be treated separately, although they have not arisen separately, or in historical sequence. Therefore one has to meet the difficulty by occasionally forestalling in earlier chapters general and well-known principles, the detailed consideration of which forms the subject-matter of later chapters. Such a difficulty arises now, and necessitates a somewhat premature treatment of what I may call the elements of ideation.

Throughout the present work I shall use the word Idea in its widest sense. As few terms have been used with a greater variety of meanings, I think it is better to state here at the outset what I take to be its most general meaning, and therefore the one which, as I have said, I shall always attach to it.

If after looking at a tree I close my eyes and then arouse a mental picture of what I have just seen, I may say indifferently that I remember or imagine the tree, or that I have an idea of it. The idea in this case would be simple or concrete—the mere memory of a previous sensuous perception. Now between this and the highest product of ideation there is all the interval between the lowest and the highest development of Mind. The range of meaning over which the term Idea thus extends has seemed to many writers inconveniently large, and they have therefore imposed upon it various limitations. But as all such limitations are of a purely artificial kind, I shall nowhere limit the term in itself, but whenever I have occasion to specify one or other class of ideas, I shall do so by employing the convenient adjectives, Concrete, Abstract, and General, in the senses which I shall have to explain further on. Meanwhile it is enough to say that whenever I employ the term Idea alone, I mean it to be a generic term.

We have already seen, while treating of the obverse or physiological side of ideation (in the chapter on the Physical

Basis of Mind) that ideas have a strong tendency to cohere together in groups, so as to constitute one compound idea out of many simpler or more elementary ideas; and also that they show no less strong a tendency to cohere together in concatenated series, such that the arousing of the first member determines the successive arousing of the other members. On its physiological side, as we saw, this is precisely analogous on the one hand to the co-ordination of muscular movements in space (*i.e.*, the grouping of such movements to form a simultaneous act, such as striking), and on the other hand to the co-ordination of muscular movements in time (*i.e.*, the grouping of such movements to perform a serial act, such as vomiting). Now it is found by observation that this cohesion of ideas is determined either by contiguity or by similarity. This fact is too well and generally known to call for more than a bare statement.

Association by contiguity is more primitive than association by the similarity, for in order that there should be association by similarity, the similarity must be *perceived;* and this implies a higher level of mental evolution than is required to establish an association by contiguity—which, as we have seen, may be established even in non-mental nervous processes, while there is nothing truly analogous to association by similarity observable in such processes.*

But it will be observed that even association of ideas by contiguity of the simplest possible kind, implies a higher development of the powers of memory than any of the three stages of memory which I have already indicated. For now there is not merely the memory of a past sensation (which is dormant till aroused by another like or unlike sensation); but there is the memory of at least two things, and also the memory of a previous relation of sequence between them.

* The nearest approach to such an analogy is perhaps to be found in the curious fact, which I find to hold true in most persons, that if a pencil is taken in each hand, and while the habitual signature is being written with the right hand, moving from left to right, the movements are imitated by the left hand moving in the opposite direction, the signature will be found to have been written backwards by the left hand, and even the hand-writing can be recognized on holding the paper before a mirror. As the left hand may never have performed this feat before, and cannot perform it unless the right hand is working simultaneously, the case looks like one of association by similarity. But I think it is really due to association by contiguity; and the same applies to the extreme difficulty of moving the two hands simultaneously as if carding wool in opposite directions.

This, therefore, we may mark off as another distinct stage in the evolution of mnemonics.

After this stage has advanced to a considerable extent, so that numerous concrete and compound ideas are associated in a great many chains of more or less length or number of links, a sufficient body of psychological data has been furnished to admit of the next stage of memory being reached, or that of association by similarity. Professor Bain remarks: "The force of contiguity strings together in the mind words that have been uttered together; the force of similarity brings forward recollections from different times and circumstances and connexions, and makes a new train out of many old ones."* And as in these higher planes of human memory, so in the lower ones of animal memory; association by similarity implies a better development of ideation than does association by contiguity.

The next and final stage of Memory is attained when reflection enables the mind to localize in the past the time when an event remembered took place. This is the stage of memory which is called Recollection, and occurs in all cases where the mind knows that some particular association of ideas has previously been formed, and is therefore able deliberately to search the memory until the particular association required is brought into the light of consciousness.

I have now given a sketch of the successive stages in the evolution of Memory, drawing a line to mark off a stage wherever I have been able to distinguish a place where a line could be drawn. It is needless to say that here, as in all similar cases, I deem these lines to be of a purely arbitrary character, and introduce them only to give a general idea of the upward growth of a continuously developing faculty. I shall now conclude this chapter by briefly glancing at the animal kingdom and the growing child with reference to the evolution of Memory.

Taking first the case of the child, I have assigned the seventh week as the appropriate age at which to mark the first evidence of memory in the association of ideas. I do so because I have observed that this is the age at which hand-fed children first recognize the feeding-bottle, *i.e.*, an artificial object without smell or other quality that can arouse any ancestral instincts, and one which young infants always

* *Senses and Intellect*, p. 469.

appear to recognize earlier than any other object. Locke, indeed, mentions recognition of the feeding-bottle as contemporaneous with that of the *rod;* but as our ideas on matters of education have undergone some improvement since his time, this statement would now be difficult to verify. In my own child I observed that the power of associating ideas extended in the ninth week from the feeding-bottle to the bib, which was always and only put on before feeding; for as soon as this was put on the child used to cease to cry for the bottle. At this age, also, I observed that when I put her woollen shoe upon her hand she gazed at it intently, as if perceiving that some curious change had come over the habitual appearance of the hand. At ten weeks she knew her bottle so well that she would place the nipple of it in her own mouth, and, when allowed to do so, would hold the bottle herself while sucking. Generally, however, she would fail in her attempts at introducing the nipple into her mouth, clearly from a lack of co-ordinating power in her muscles—the nipple striking various parts of her face. She would then cry for the nurse to help. Preyer says* that at eight months old his child was able to classify all glass bottles as resembling, or belonging to the same order of objects, as a feeding-bottle. I may add that at seven weeks old my child used to cry when left alone in a silent room for a few minutes —a fact which also seems to show a rudimentary power of associating ideas, with the consequent perception of a change in the habitual environment.

Turning now to the animal kingdom, the first evidence of memory that I have found in the psychological scale is in the Gasteropoda, and consists in the Limpet returning to its groove in the rock after having been crawling about upon a browsing excursion.† This fact, I think, clearly proves the power of remembering locality, and as such a grade of memory can scarcely be regarded as the earliest, we may reasonably suppose that the faculty really occurs still lower in the psychological scale of animals, although we have not as yet any observations to prove the fact. Moreover, as Oysters learn by individual experience, acquired in the "Oyster-schools," to keep their shells closed for a much longer time than is natural to uneducated individuals,‡ we must conclude that a dim power

* *Loc. cit.*, p. 42.

† *Animal Intelligence*, pp. 28-9.

‡ *Ibid.*, p. 25.

of memory is also present in the division of the Mollusca. The Razor-fish, likewise, shows memory, and this in a high degree, inasmuch as if only once alarmed upon coming to the surface of its burrow, it cannot be again induced to approach the surface for a long time, even by the application of irritants.* Still more remarkable is the level of development to which memory has attained in the Snail, if the observation of Mr. Lonsdale is to be accepted of the *Helix pomatia*, which, after leaving its sickly mate and crawling over a garden wall, returned next day to the place where it had left its mate.† But the highest level to which the development of memory attains in the Mollusca is unquestionably in the Cephalopoda, for according to Hollmann an Octopus remembered its encounter with a lobster in a remarkable manner,† while according to Schneider these animals learn to know their keepers.‡

Seeing that memory in various stages of development thus unquestionably occurs among the Mollusca, I thought it worth while to try some experiments in this connection with the Echinodermata, but they all yielded negative results. It has, however, been alleged that if a star-fish be removed from its eggs, it will crawl back again to the place where they were; and if this statement were confirmed, it would of course prove memory in the Echinodermata. Hitherto I have myself had no opportunity of testing it, and therefore my experiments were confined to endeavouring to teach star-fish a few simple lessons, which, as I have already implied, they would not learn. I am more surprised with my failure in this respect with the higher Crustacea; for although I have tried similar experiments with them, I have never been able to teach them the simplest things. Thus, for instance, I have taken a hermit crab, put it into a tank filled with water, and when he had protruded his head from the shell of the whelk in which he was residing, I gently moved towards him a pair of open scissors, and gave him plenty of time to see the glistening object. Then, slowly including the tip of one of his tentacles between the open blades, I suddenly cut off the tip. Of course the animal immediately drew back into the shell, and remained there for a considerable time. When he again came out I repeated the operation as before, and so on

* *Animal Intelligence*, p. 26. † *Ibid.*, p. 27. ‡ *Ibid.*, p. 30.

for a great number of times, till all the tentacles had been progressively cut away little by little. Yet the animal never learnt to associate the appearance of the scissors with the effect which always followed it, and so never drew in until the snip had been given. Nevertheless, that memory does occur among the higher Crustacea is proved by an observation quoted in "Animal Intelligence" (p. 233), concerning a lobster mounting guard upon a heap of shingle beneath which it had previously hidden some food.

In another class of the Articulata, however, the faculty of memory has been developed to an extraordinary degree, and far surpasses that which has been attained by any other class of Invertebrata. The class of the Articulata to which I allude are the Insects, and, more particularly, the Hymenoptera. Without quoting *in extenso* the evidence on this head which has already been given in my previous work, it is enough to say in general terms that ants and bees are unquestionably able to remember the places where many months before they have obtained honey or sugar, &c.; and will also, when occasion requires, return to nests and hives which they have deserted the year before. Many interesting observations have also been made on the rate of acquisition and the length or duration of particular memories in these animals, which, however, it is needless for me again to quote.* Perhaps the most interesting of these are the observations of Sir John Lubbock on bees gradually learning to know the difference between an open and a closed window, and the observations of Messrs. Bates and Belt on the sand-wasps carefully teaching themselves (by taking mental notes of landmarks) the localities to which they intend to return in order to secure the prey which they have temporarily concealed. Incidental evidence of memory in other orders of Insects will also be found on referring to my previous work—viz., for Beetles, pp. 227—9, for Earwigs, p. 229, and for the common House-fly, p. 230.

Turning now to the Vertebrata, we find that in Fish memory is certainly present, although it never reaches more than such a degree of development as is implied by remembering in successive years the locality for spawning, learning to avoid baits, removing young from a nest which has been

* For a full account of all these observations, see *Animal Intelligence*, under the heading "Memory," of Chap. III and IV.

disturbed, and associating the sound of a bell with the arrival of food.*

Batrachians and Reptiles are able to remember localities, and also to identify persons.† The annual migration of Turtles further proves the duration of memory for at least a year.

In Birds the power of memory has advanced considerably beyond that of remembering, as in the case of the swallow, the precise locality of their nests from season to season, and even beyond that of identifying persons from year to year.‡ For the facts which I have previously detailed at length touching the acquisition by talking birds of tones, words, and phrases, show not only an exceedingly high development of the powers of special association, but even the power of genuine recollection to the extent of knowing that there is a missing link in the train of a previously formed association, and of purposely endeavouring to recover it. Quotations from Dr. Wilks, Mr. Venn, and Mr. Walter Pollock were also given, in order to show from direct and careful observation that the process of forming special associations is in such cases identical with that which occurs in man.§

Among Mammals the highest development of memory is presented by the Horse, Dog, and Elephant. Thus there is unexceptionable evidence of a horse remembering a road and a stable after an interval of eight years;|| of a dog remembering the sound of his master's voice after an interval of five years,¶ and the sound of a clinking collar after an interval of three years;¶ and of an elephant remembering his keeper after having run wild for an interval of fifteen years.** It is probable, also, that if observations were made, the memory of Monkeys would be found to be very retentive, as it certainly is most minute, and largely assisted by the intentional efforts of the animals themselves.††

* See *Animal Intelligence*, pp. 248–51. † *Ibid.*, pp. 254–62.
‡ *Ibid.*, p. 266. § For all these facts, see *ibid.*, pp. 266–70.
|| *Ibid.*, p. 330. ¶ *Ibid.*, p. 438. ** *Ibid.*, p. 387.
†† *Ibid.*, pp. 485–98.

CHAPTER IX.

PERCEPTION.

AT the level marked 18 I represent the rise from Sensation to Perception. By this term I mean, in accordance with general usage, the faculty of cognition. "The contrast between Sensation and Perception is the contrast between the sensitive and the cognitive, intellectual, or knowledge-giving functions." (Bain.) "Perception is an establishment of specific relations among states of consciousness; and this is distinguished from the establishment of these states of consciousness themselves," which constitutes Sensation. (Spencer.) "In Perception the material of Sensation is acted on by the mind, which embodies in its present attitude all the results of its past growth." (Sully.)

Sensation, then, does not involve any of the powers of the intellect as distinguished from consciousness, but Perception implies the necessary occurrence of an intellectual or cognitive process, even though it be a process of the simplest possible kind. The term Perception, therefore, may be applied to all cases where a process of cognition occurs, whether such process arises directly or indirectly out of sensation; thus it is equally correct to say that we perceive the colour or the scent of a rose, and that we perceive the truth or the probability of a proposition.

Otherwise phrased we may state the distinction between Sensation and Perception thus. A sensation is an elementary or undecomposable state of consciousness, but a perception involves a process of mentally interpreting the sensation in terms of past experience. For instance, there is a closed book lying on the table before me; my eyes have been resting on its cover for a considerable time while I have been thinking how I should arrange the material of the present chapter. All that time I have been receiving a visual sensation of a particular kind; but, as I did not attend to it, the sensation did not involve any element of cognition, and therefore did not minister to any act of perception. All at once, however, I became conscious that I was looking at a book, and in cognizing that the particular object of sensation was a book, I performed an act of perception. In other words, I mentally interpreted the sensation in terms of past experience; I made a mental synthesis of the qualities of the object, and assigned it to the class of objects which had previously produced a like sensation.

Perception, then, is a mental classifying of sensations in terms of past experience, whether ancestral or individual; it is sensation *plus* the mental ingredient of interpretation. Now, as a condition to the possibility of this ingredient, it is clearly essential that there should be present the power of memory; for only by a memory of past experience can the process be conducted of identifying present sensations or experiences as resembling past ones. Therefore in the diagram I have placed the dawn of Memory on the level, just below that at which the faculty of Perception takes its rise. Both Sensation and Perception are represented as attaining a considerable vertical elevation from base to apex, *i.e.*, from their first origin to their completed evolution. That this ought to be so represented is evident if we reflect on the difference in the sensuous faculties of a medusa and an eagle, or between the perceptive faculties of a limpet and a man. It may, indeed, be thought that in my representative diagram I have not allowed enough for such differences, and therefore have made the vertical elevation of these branches too low. But here we must remember that in the case of Sensation, as already shown, the advance of the faculty from its earliest to its latest stages consists essentially, on its morphological aspect, of a greater and greater degree of specialization of end-organs of nerves; and I think that the degree of such advance is sufficiently expressed by the vertical elevation which I have given to the branch in question, seeing how much more intricate must be the morphological development of the nerve-tissues which are concerned in ministering to the next and to all succeeding faculties. And, as regards Perception, we must remember that in its more highly elaborated phases this faculty shades off into the higher representative branches marked "Imagination," &c.; so that the branch marked "Perception" is not intended to include all that might possibly be included by the term if we did not separately name the higher faculties to which I allude.

Now concerning the development of Perception, I may here make a general remark, which is first applicable at this stage of mental evolution, and which continues to be applicable to the development of all the faculties which we have subsequently to consider. This remark is that we have ceased to possess any data of a morphological kind—such as we had in the case of Sensation and the pre-mental faculties of adjustment—to guide us in our estimate of the degree of elaboration to which the faculty has attained. That morphological evolution has here, as in the coarser instance of Sensation, always gone hand in hand with psychical evolution, is amply proved in a general way by the advancing complexity of the central nerve-organs; but just because this complexity is so great, and the steps in morphological evolution which it represents so refined, we are totally at a loss to follow the process on its morphological side; we are unable even dimly to understand the mechanisms which we see. Therefore, in order to estimate the ascending grades of excellence which these mechanisms present, we require to look to what we may most conveniently regard as the products of their operation; we have to use the mental equivalents as indices of the morphological facts.

We have seen that Perception is essentially a process of mentally interpreting Sensation in terms of past experience, ancestral or individual. The successive steps in the elaboration which this process undergoes in the course of its evolution must now be considered.

The first stage of Perception consists merely in perceiving an external object as an external object, whether by the sense of touch, taste, smell, hearing, or sight. But confining ourselves, for the sake of brevity, to the sense of sight, in this stage Perception simply amounts to a cognition of an object in space, having certain space relations with other objects of perception, and especially with the percipient organism.

The next stage of Perception is reached when the simplest qualities of an object are re-cognized as like or unlike the qualities presented by such an object in past experience. The most universal of such qualities in objects pertain to size form, colour, light, shade, rest, and motion; less universally such qualities pertain to temperature, hardness, softness, roughness, smoothness, and other qualities appealing to the sense of touch, as well as qualities appealing to the senses of smell, taste, and hearing. In the case of these more universal qualities, the part which the mind takes in the process of cognizing them as belonging to the objects is immediate and automatic, and, as Mr. Sully observes, "may be supposed to answer to the most constant and therefore the most deeply organized connections of experience."

The third step in the advance of Perception consists in the mental grouping of objects with reference to their qualities, as when we associate the coolness, taste, &c., of a particular fruit with its size, form, and colour. Here the more frequently a certain class of qualities has been conjoined with another class in past experience, the more readily or automatically is the perceptive association established; but in cases where the conjunction of qualities has not been so frequently or so constantly met with in past experience, we are able by reflection to recognize the perceptive association "as a kind of intellectual working up of the materials supplied us by the past."

A further development of the perceptive faculty is required to meet cases in which the qualities of objects have become too numerous or complex to be all perceived simultaneously. In meeting such cases the faculty in question, while perceiving some of the qualities through sensation, supplements the immediate information so derived with information derived from previously formed knowledge; the qualities which are not recognized immediately through sensation are inferred. Thus, in my perception of a closed book I have no doubt that the covers are filled with a number of printed pages, although none of these pages are actually objects of present sensation. Or, if I hear a savage growl, I immediately infer the presence of an object presenting so complex a group of unseen qualities as are collectively comprised in a dangerous dog. In a later chapter I shall have to dwell more minutely on this, which I may term the inferential stage of perception, and I shall therefore not deal more with it at present.

It will be evident that the various stages which I have named in the development of Perception shade into one another, so as not really to be distinguishable as separate stages; they constitute rather one uniform growth on which, as in the case of Memory, I have arbitrarily marked these several grades of evolution. Moreover, it will be evident that the term "Perception" is really a very wide one, and may be said to cover the whole area of psychology, from the confines of an almost unfelt sensation up to the recognition of an obscure truth in science or philosophy. On this account the term has been condemned by some psychologists as too extensive in its application to be distinctive of any particular faculty; but nevertheless it is clearly impossible to do without it, and if we are careful to remember the sense in which we employ it—whether with reference to the lower or to the higher faculties of mind—no harm can arise from its use.

I have just said that in the highest stage of its development Perception involves Inference; and I have previously said that in its lowest stages it involves Memory. I must now point out more particularly that in its ascending stages Perception involves Memory of ascending stages. Thus the perception shown by a new-born infant of sweet tastes as distinguished from sour tastes and the rest, implies the presence of that lowest stage of memory which we have seen to consist in cognizing a present sensation as like a past sensation. Again, the power of discerning a change of milk implies the power of cognizing a present sensation as unlike a past sensation. Next, when memory advances to the stage of associating ideas by contiguity, perception also advances to the stage of re-cognizing objects with their qualities and relations of coexistence and sequence. This in turn leads to the power of recognizing objects, qualities, and relations by similarity—the power on which we have seen the next phase of memory to depend. And, lastly, from this point onwards perception throughout depends exclusively upon the association of ideas, no matter how elaborate or refined such association may become.

The fact that perception is thus everywhere and indissolubly bound up with memory, is an important fact to be clear about; for when memory becomes so habitual as to be virtually automatic or unconscious, we are apt to lose sight of the connection between it and perception. Thus, as Mr. Spencer observes, we do not speak of remembering that the sun shines; yet we speak of perceiving that the sun shines. As a matter of fact, however, we do remember that the sun shines, and in all the habitual phenomena of experience such memories as this become so blended with our perceptions of the phenomena that the memories may be said to form integral parts of the perceptions. Suppose, for instance, we see a man whose face we know, but cannot remember who the man is. Here the perception that the object which we see is a man, and not any other of the innumerable objects in Nature, is so intimately bound up with a well organized association of ideas, that we do not think of the perception thus far as really depending on memory. It is only when we turn to the incompletely organized association of ideas between the particular face and the particular individual, that we recognize the incompleteness of this part of the perception to depend upon the incompleteness of memory.

Now these considerations, obvious though they appear, constitute the first stage in a disagreement on an important matter of principle, which will become more pronounced when I have to deal with the higher faculties of mind, and which, I regret to say, has reference to the writings of Mr. Spencer. In his chapter on Memory Mr. Spencer takes the view that, so long as "psychical changes are completely automatic, memory, as we understand it, cannot exist—there cannot exist these irregular psychical changes seen in the association of ideas." Now, I have already given my reasons for not restricting the term Memory to the association of ideas; but, passing over this point, I cannot agree that if *psychical* changes (as distinguished from physiological changes) are completely automatic, they are on this account precluded from being regarded as mnemonic. Because I have so often seen the sun shine, that my memory of it, as shining, has become automatic, I see no reason why my memory of this fact, simply on account of its perfection, should be called no-memory. And similarly with all those well-organized memories which constitute integral parts of perceptions. In so far as they involve true "psychological changes," and therefore imply the presence of *conscious recognition* as distinguished from *reflex action*, so far, I think, no line of demarcation should be drawn between them and any less perfect memories. I shall recur to this point when I come to consider Mr. Spencer's views on Instinct and Reason.

Another point which we have here to consider is the part which heredity has played in forming the perceptive faculty of the individual prior to its own experience. We have already seen that heredity plays an important part in forming memory of ancestral experiences, and thus it is that many animals come into the world with their powers of perception already largely developed. This is shown not only by such cases as those of Galen's kid, and Preyer's chickens before mentioned, but by all the host of instincts displayed by newly-born or newly-hatched animals, both Vertebrate and Invertebrate. This subject will be fully considered when I come to treat of Instinct, and then it will be found that the wealth of ready-formed information, and therefore of ready-made powers of perception, with which many newly-born or newly-hatched animals are provided, is so great and so precise, that it scarcely requires to be supplemented by the subsequent experience of the individual. In different classes of animals these hereditary endowments vary much both in kind and in degree. Thus, with mammals as a class, hereditary perception has reference in its earliest stages to the senses of smell and of taste; for while many mammals are born blind, some probably deaf, and all certainly very deficient in powers of locomotion, they invariably show more or less perceptive powers of taste, and very frequently well-advanced perceptive powers of smell. This we have already seen in the case of Galen's kid, and in the case of the dog (whose ancestors have depended so largely upon the perfection of smell) the same thing occurs in so high a degree, that so special an olfactory impression as is produced by the odour of a cat will cause a litter of newly-born puppies to "puff and spit."*

Birds come into the world with better endowments of perception than animals of any other class. For they are in full possession of every sense almost immediately after they are hatched, and, as we shall see later on, they are then able to use their senses nearly as well as they are ever able to use them.

Reptiles are likewise hatched with their powers of perception almost as highly developed as they are ever destined to become,† and the same as a rule is true of invertebrated animals.

I must now say a few words on the physiology of Percep-

* See p. 164. † See *Animal Intelligence*, pp. 256–7.

tion—or, more correctly, on what is known touching the physiological processes which accompany Perception.

In earlier chapters I have already stated that the only distinction which is known on the physiological side between a nervous activity which is accompanied by consciousness, and a nervous activity which is not so accompanied, consists in a difference of time. I shall now give the experimental data on which the statement rests.

Professor Exner has determined the time which is occupied by a nerve-centre of man in executing its part in the performance of a reflex action. That is to say, the rate of transmission of a stimulus along a nerve being known, and the length both of the afferent and efferent nerves concerned in a particular reflex act being known, as also is the "period of latency" of a muscle; the time occupied by the nerve-centre in conducting its operations was determined by subtracting the time occupied by the passage of the stimulus along the afferent and efferent nerves, *plus* the period of latency of a muscle, from the total time between the fall of the stimulus and the occurrence of the muscular contraction. This time was found in the case of the reflex closure of the eye-lid to vary between 0·0471 and 0·0555 of a second according to the strength of the stimulus.* By a similar process Exner has estimated the time required for the central nervous operations which are together comprised in having a simple sensation, perceiving the sensation, and the volitional act of signalling the perception. That is to say, an electrical shock being administered to one hand, and as quickly as possible signalled by the other, the time occupied by the nerve-centre in performing its part of the process was estimated as in the previous case. This time in the case of this experiment was found to be 0·0828″, which is nearly twice as long as that which, as we have just seen, is required for a nerve-centre to perform its part in a reflex action.†

Acts of perception in which different senses are concerned occupy different times. This interesting topic has been investigated by a number of physiologists.‡ According to Donders the total "reaction-time" (*i.e.*, between stimulus and response) is, roughly speaking, for touch $\frac{1}{7}$, for hearing $\frac{1}{6}$,

* *Arch. f. d. ges. Physiol.*, xliii, 526 (1874).
† *Ibid.*, vii, p. 610.
‡ See Herman, *Handb. d. Physiol.*, Bd. II, Th. 2, s. 264.

and for sight $\frac{1}{6}$ of a second.* The observations of Von Wittich†, Vintschgau, and Hönig-Schnied‡ show that the reaction-time for taste varies between 0·1598″ to 0·2351″ according to the kind of taste; being least for salt, more for sugar, and most for quinine. A constant electrical current applied to the tongue gives a reaction-time for the resulting gustatory impression of 0·167″. I am not aware that any experiments have been made with regard to smell. Exner has more minutely determined on himself the reaction-time for touch, sound, and sight, with the results which are embodied in the following table. The signal was in all cases given by the right hand depressing an electrical key:—

Direct electrical stimulation of retina	0·1139″
Electrical shock on left hand	0·1276
Sudden sound	0·1360
Electric shock on forehead..	0·1370
Electric shock on right hand	0·1390
Visual impression from electric spark	0·1506
Electric shock on toe of left foot	0·1749§

It is thus noticeable that although the sensation of light produced by vision of an electric spark is much greater than that produced by electrical stimulation of the optic nerve, the interval between the stimulation and the perception is much longer in the former case. Seeing that the optic nerve is so short, this difference cannot be attributed to the time lost in transmission along the nerve, and must therefore be supposed due to the time required for the nerve-endings in the retina to complete all the changes (whatever they may be) in which their response to luminous stimulation consists. Thus in the case of hearing, as the above table shows, somewhat less time is consumed in the whole act of perception than is consumed in the case of sight by the peripheral changes taking place in the retina.

According to Helmholtz and Baxt, the more complex an object of visual perception is, the greater must be the duration of its image upon the retina, in order that the perception may be made; while, within certain limits, the *intensity* of the image does not affect the time required to make the per-

* *Arch. f. Anat. und Physiol.*, 1868, p. 657.
† *Qt. Ret. Med.* (3), xxxi, p. 113.
‡ *Arch. f. Anat. und Physiol.*, x, p. 1.
§ *Pflüger's Archiv.*, Bd. VII, p. 620.

ception.* The last-named author found that an exposure of $\frac{1}{20}$ second is required for the perception of a row of six or seven letters.

Other experiments prove that the more complex an act of perception, the more time is required for its performance. Thus Donders has shown that when an experiment in reaction-time is made to consist, not merely in signalling a perception, but in signalling one of two or more perceptions, the reaction-time is lengthened, owing to the greater time required for performing the more complex psychical process of distinguishing which of the expected stimuli is perceived, and in determining to make or to withhold the response accordingly. The state of matters thus presented to the mind is called by Donders a "Dilemma," and the following is his table of results :—

Dilemma between two spots of the skin, right or left foot stimulated by an electric shock; signal to be made in one case only	0·066″
Dilemma of visual perceptions between two colours, suddenly exhibited; signal to be made on seeing the one but not on seeing the other	0·184
Dilemma between two letters; signal to be made on seeing one only	0·166
Dilemma between five letters; signal as before	0·170
Dilemma of hearing; two vowels suddenly called; signal to be made on hearing one only	0·056
Dilemma between five vowels; signal as before	0·088

The above table gives, in each case, not the whole period between the occurrence of the stimulus and the occurrence of the response, but the *difference* between the time required for this whole period when a single stimulus has to be answered, and when only one of two or more possible stimuli has to be answered. It will thus be seen that the time required for the act of meeting a dilemma is from $\frac{1}{5}$ to $\frac{1}{20}$ of a second longer than that which is required to signal a simpler perception.†

This "Dilemma-time" has also been estimated where the other senses are concerned by Kries and Auerbach, with the following results :—‡

* *Archiv. f. d. ges. Physiol.*, Bd. IV, p. 329; *Monatsber. d. Ber. Acad.*, June, 1871.

† For Donders' investigations, see *Archiv. f. Anat. und Physiol.*, 1868, pp. 657–81.

‡ *Archiv. f. d. ges. Physiol.*, 1877, pp. 298–380.

Localization by sight	0·011
Distinguishing colour	0·012
Localization by hearing (least interval) ..	0·015
Distinguishing pitch (high notes)	0·019
Localization by touch	0·021
Distinguishing pitch (low notes)	0·034
Localization by hearing (greatest interval) ..	0·062

If a greater number of alternatives are allowed by the preconcerted arrangement, a still longer interval is required for the response.

The time required for perception in the case of all the senses varies with different persons, and, under the name of "personal equation," has to be carefully determined by astronomers. It is increased by old age, sundry kinds of sickness, and sundry kinds of drugs. But it is not necessarily less in young people full of vitality than it is in young people of less vigorous or lively temperaments. According to Exner, persons who are accustomed to allow their ideas to run slipshod are relatively slow in forming their perceptions, or, at least, have a long reaction-time between receiving and responding to a stimulus. He gives the following table to show the difference in the reaction-time of seven individuals:—*

Age.	Reaction-time.	Remarks.
26	0·1337	Rough, lively labouring-man.
23	0·3311	Lively in movements, but rather slow in apprehension.
76	0·9952	Infirm and not intelligent.
24	0·1751	Slow and deliberate in movements.
20	0·2562	Slow and somewhat uncertain in movements.
22	0·1295	Slow and very precise in movements.
35	0·1381	Accustomed to manual work.

Concerning the effects of drugs it is enough to say that Exner found two bottles of Rhine-wine increased his reaction-time from 0·1904″ to 0·2269″;† and I have myself observed while shooting that an amount of alcohol not sufficient to produce any consciously psychical effects, is apt to make one shoot behind one's birds. And here, with reference to the personal equation, I may briefly allude to some

* *Loc. cit.*, p. 612. † *Loc. cit.*, p. 628.

hitherto unpublished observations of my own, which has served to display a positively astonishing difference between different individuals with respect to the rate at which they are able to read. Of course reading implies enormously intricate processes of perception both of the sensuous and of the intellectual order; but if we choose for these observations persons who have been accustomed to read much, we may consider that they are all very much on a par with respect to the amount of practice which they have had, so that the differences in their rates of reading may fairly be attributed to real differences in their rates of forming complex perceptions in rapid succession, and not to any merely accidental differences arising from greater or less facility acquired by special practice.

My experiments consisted in marking a brief printed paragraph in a book which had never been read by any of the persons to whom it was to be presented. The paragraph, which contained simple statements of simple facts, was marked on the margin with pencil. The book was then placed before the reader open, the page however being covered with a sheet of paper. Having pointed out to the reader upon this sheet of paper what part of the underlying page the marked paragraph occupied, I suddenly removed the sheet of paper with one hand, while I started a chronograph with the other. Twenty seconds being allowed for reading the paragraph (ten lines octavo), as soon as the time was up I again suddenly placed the sheet of paper over the printed page, passed the book on to the next reader, and repeated the experiment as before. Meanwhile the first reader, the moment after the book had been removed, wrote down all that he or she could remember having read. And so on with all the other readers.

Now the results of a number of experiments conducted on this method were to show, as I have said, astonishing differences in the *maximum* rate of reading which is possible to different individuals, all of whom have been accustomed to extensive reading. That is to say, the difference may amount to 4 to 1; or, otherwise stated, in a given time one individual may be able to read four times as much as another. Moreover, it appeared that there was no relationship between slowness of reading and power of assimilation; on the contrary, when all the efforts are directed to assimilating as much as possible in a given time, the rapid readers (as shown by their written notes) usually give a better account of the portions of the paragraph which has been compassed by the slow readers than the latter are able to give; and the most rapid reader whom I have found is also the best at assimilating. I should further say that there is no relationship between rapidity of perception as thus tested and intellectual activity as tested by the general results of intellectual work; for I have tried the experiment with several highly distinguished men in science and literature, most of whom I found to be slow readers. Lastly, it is worth observing that every one who tries this experiment finds that it is impossible, with any amount of effort at recollection, to remember, immediately after reading the paragraph, all the ideas which have been communicated to the mind by the paragraph. But as soon as the paragraph is read a second time, the forgotten ideas are instantly recognized as having been present to the mind while reading. This shows that the memory of a full perception may, as it were, be immediately crowded out by rapidly succeeding perceptions, to the extent of being rendered latent, although it may be instantly recalled by the recurrence of the same perception.

So much, then, to show that the personal equation in different individuals varies the more the greater the number and the higher the intricacy of the perceptions which are to be made in a given time. I must now say a few words to show that the personal equation in the same individual admits of being greatly reduced by practice in making particular perceptions. This is well known to astronomers so far as simple acts of perception are concerned, and in all the researches above mentioned touching the time-measurements of simple perceptions, the experimenters found that practice had the effect of reducing the reaction-time. The degree of reduction which might thus be produced was itself made the subject of experiment by Exner, who chose the old man already mentioned in one of the above quoted tables as having the unusually long reaction-time of 0·9952″. After a little more than six months' practice at the rapid signalling of an electric shock, the old man's reaction-time was reduced to 0·1866″.

This universal fact of repetition serving greatly to reduce the physiological time required for the performance of physical processes even of the simplest kind, is a fact of great significance. And, that the same applies to perceptions of the most multitudinous and complex kind, is proved in every-day life by the acquired rapidity with which bankers' clerks are able to add up figures, musicians to read a complicated score at sight, &c. But perhaps one of the best cases to quote in this connection is the celebrated one of the result of a systematic course of training to which the conjuror Houdin submitted his son. The training consisted in making the boy walk rapidly before a shop window, and perceive as many objects in the window as possible. After several months the boy was able to devour so many objects at a glance, that his father advertised him as "gifted with a marvellous second sight; after his eyes have been covered with a thick bandage he will designate every object presented to him by the audience."* That is to say, the boy, before his eyes were bandaged, was able to perceive all the objects in the room which were likely to be presented to him. It is of interest to note that Houdin, who thus paid special attention to the development of rapidity of perception, observes that women as a rule have a greater rapidity than men, and says that he has known ladies who were able while seeing another lady "pass at full speed in a carriage, have time to analyze her toilette from her bonnet to her shoes, and be able to describe not only the fashion and quality of the stuffs, but also say if the lace were real or only machine made."† I mention this opinion of Houdin because in my own observations on rapid reading I have been struck with the fact that ladies nearly always carry off the palm.

Dr. G. Buccola has shown in a recently published essay that the reaction-time is, as a general rule, less among educated than it is among uneducated persons, and greatest among idiots.‡ I may also direct attention to an interesting paper published a few months ago by Mr. G. Stanley Hall,§ "On the lengthening of the Reaction-time under the Influence of Hypnotism:" the lengthening is not so considerable as might have been anticipated.

I have dwelt thus at length on all the main facts which

* *Memories of Robert Houdin*, vol. ii, p. 9. Professor Preyer has also published some observations on this subject. † *Ibid.*, p. 7.

‡ La durata del discernimento e della determinazione volition, *Rivisti di Filos. Scientif.*, I, p. 2. § *Mind*, No. XXX.

are at present known concerning the time-relations observable in Perception, because with reference to the theory of the rise of consciousness, and also of the physiological side of mental evolution in general, these facts are of the highest importance. They prove by actual measurement that the simplest psychical actions are slow as compared with reflex actions, that they can be rendered more rapid by practice, but that they can never be brought to be so rapid as reflex actions. We have a further exemplification of the effects of practice in thus quickening the act of perception in the higher stages of the process. For universally the effect of previous acts of perception is that of placing the mind in readiness, as it were, for performing acts of the same kind. The mental attitude as regards these particular acts of perception is then the attitude of what Lewes appropriately called pre-perception.* When the pre-perceptive stage is well established, the memory, or the memory and inference as the case may be, arise in or together with the act of perception, so forming an integral part of the act. It is owing to the want of special experiences that young children are so slow in forming perceptions of more than the lowest degree of complexity; as Mr. Spencer observes, they take a long time to "integrate" a strange face or other unfamiliar object; and this, otherwise stated, means that their mental attitude of pre-perception has not yet been fully attained for such and such classes of objects; the processes of memory, classification, and inference do not occur immediately in the act of perception, and therefore the full mental interpretation of the object perceived is only arrived at by degrees. Similarly, in adult life the powers of perception may be trained to a marvellous extent in special lines by practice, as we have already seen in the example of Houdin's son, and as we may also see in the fact that an "artist sees details where to other eyes there is a vague or confused mass." The influence of persistent attention is the most important of all influences in developing the rapidity and accuracy of the perceptive powers in which their highest excellence consists.

We have now to consider the important question whether

* *Problems of Life and Mind*, 3rd ser., p. 107. See also Dr. J. Hughlings Jackson in *Brain*, Nos. III and IV; and Mr. Sully, in *Illusions*, pp. 27–30.

Perception arises out of Reflex Action, Reflex Action out of Perception, or whether there is any genetic continuity between the two at all. This is a most difficult question, and one which I do not think we are as yet entitled to answer with any kind of scientific confidence.

According to Mr. Spencer the perceptive faculties arise out of the reflex when these attain a certain level of intricacy in their structure, or a certain degree of rarity in their occurrence. Thus he says, "When, as a consequence of advancing complexity and decreasing frequency in the groups of external relations responded to, there arise groups of internal relations which are imperfectly organized and fall short of automatic regularity; then, what we call Memory, becomes nascent."* But as a matter of fact it seems, I think, very questionable whether the only factors which lead to the differentiating of psychical nervous processes from reflex nervous processes are thus complexity of operation combined with infrequency of occurrence. For it is obvious that in ourselves certain truly reflex actions are of immense intricacy and of exceedingly rare occurrence—such, for example, as vomiting and parturition. The truth is that, so far as definite knowledge entitles us to say anything, the only constant physiological difference between a nervous process accompanied by consciousness and a nervous process not so accompanied, is that of time. In very many cases, no doubt, this difference may be caused by the intricacy or by the novelty of the nervous process which is accompanied by consciousness; but, for the reason which I have given, I do not think we are justified in concluding that these are the only factors, although I have no doubt that they are highly important factors. For all that we know to the contrary, natural selection or other causes may have determined the physiological conditions necessary to the rise of consciousness (and so to the perception of pleasure and pain), without any question as to intricacy or infrequency being concerned; in which case the time-relations needed to meet these conditions would have become evolved together with them. And I think it speaks in favour of some such view as this that the structure of the cerebral hemispheres is in some respects strikingly unlike the structure of the reflex centres.

Be the factors what they may, however, it is a great

* *Principles of Psychology*, vol. i, p. 446.

matter to have the sure ground of experiment on which to rest the fact that universally psychical processes represent comparative delay of ganglionic action. For from this fact the obvious deduction is, as stated in a previous chapter, that psychical processes constitute the subjective expression of objective turmoil among molecular forces; reflex action may be regarded as the rapid movement of a well-oiled machine, consciousness is the heat evolved by the internal friction of some other machine, and psychical processes as the light which is given out when such heat rises to redness. Presumably, therefore, psychical processes arise with a vividness and intricacy proportional to the amount of ganglionic friction—as, indeed, appears to be experimentally proved by the observations of Donders before described. Now it is certain that by frequency of repetition,—*i.e.*, by practice in the performance of any particular psychical act—the amount of this ganglionic friction admits of being lessened (as shown by the time required for the ganglionic action being reduced), and that concurrently with this change on the objective side of matters, a change takes place on the subjective, in that the action which was previously conscious tends to become automatic.

Now from these considerations I think the inference would appear to be, that reflex action and perception probably advance together—each stage in the development of the one serving as the groundwork for the next stage in the development of the other. And in corroboration of this view is the general fact, that throughout the animal kingdom there is a pretty constant correspondence between the complexity of the reflex actions presented by an organism and the level of its psychical development.

CHAPTER X.

IMAGINATION.

We have already considered the psychology of Ideation to the extent of defining the sense in which I employ the word "Idea" or "Image," and also to the extent of tracing, both on the side of physiology and on that of psychology, the principle of the association of ideas.* We have now to analyze the psychology of Ideation somewhat more in detail.

The simplest case of an idea is the memory of a sensation. That a sensation may be remembered even when there has been no perception is proved, not only by the fact before mentioned that an infant only a day or two old can distinguish a change of milk, but also by the fact, which must be familiar to all, that several minutes after an unperceived sensation is past, we are able by reflection to remember that we have had the sensation. For example, a man reading a book may hear a clock strike from one to five strokes (or perhaps more) without perceiving the sound, yet a minute or two afterwards he can recall the past sensation and tell the number of strokes which have occurred. And in simpler instances the memory of a sensation may extend over a much longer time.

The simplest case of an idea, then, being the memory of a past sensation (as distinguished from the memory of a past perception), it follows that the earliest stages of ideation must be held to correspond with those earlier stages of memory which we have already described, wherein as yet there is no association of ideas, but merely a perception of a present sensation as like or unlike a past one. Hence in its most elementary form an idea may be said to consist in the faint revival of a sensation. This view has already been advanced with much clearness by Mr. Spencer, Professor

* See Chapters II and III.

Bain, and others, who also maintain, with considerable probability, that the cerebral change accompanying the idea of a past sensation is the same in kind and place, though not in degree of intensity, as was the cerebral change which accompanied the original sensation.*

In its next stage of development Ideation may be regarded as the memory of a simple perception, and immediately after this the principle of association by contiguity comes in. Later on there arises association by similarity, and from this point onwards Ideation advances by abstraction, generalization, and symbolic construction, in ways and degrees which will constitute one of the topics to be considered in my next work.

From this brief sketch, then, it will be seen that we have already considered the lowest stages of Ideation while treating of Memory and the Association of Ideas. Resuming, therefore, the analysis at the point where we there left it, I shall devote this chapter to a consideration of those higher phases of the idea-forming powers which we may conveniently include under the general term Imagination.

Now, under this general term we include a variety of mental states, which although all bearing kinship to one another, are so diverse in the degree of mental development which they betoken that we must begin by analyzing them.

As used in popular phraseology, the word Imagination is

* Thus, Mr. Spencer says, "The idea is an imperfect and feeble repetition of the original impression . . . There is first a presented manifestation of the vivid order, and then, afterwards, there may come a represented manifestation that is like it except in being much less distinct." (*First Principles*, p. 145.) And Professor Bain says, "What is the manner of occupation of the brain with a resuscitated feeling of resistance, a smell, or a sound? There is only one answer that seems admissible. *The renewed feeling occupies the very same parts, and in the same manner, as the original feeling, and no other parts, nor in any other assignable manner.*" (*Senses and Intellect*, p. 338.) While quite assenting to this view of ideation, so far as the psychology of the subject is concerned, I think we are much too ignorant of the physiology of cerebration to indulge in any such confident assertions respecting the precise seat and manner of the formation of ideas. Again, with reference to Mr. Spencer's views, it is needless to repeat the point in which I disagree with him touching the earliest stages of memory—or those before the advent of the association of ideas. Only I may point out that as the simplest possible idea is held to consist in a faint revival of a sensation (as distinguished from a perception), it follows that the occurrence of the simplest possible idea precedes the occurrence of its association with any other idea; and if so, the *memory* of the sensation, or the faint revival of the sensation in which the idea is held to consist, must also precede any association with other faint revivals of the same kind.

taken to mean the highest development of the faculty in the intentional imaging of past impressions. In this sense we speak of the imaginations of the poet, imaginations of the heart, scientific use of the imagination, and so on; in all of which cases we presuppose the powers of high abstraction as well as those of intentional ideal combinations of former actual impressions. It is needless to say that even in man, long before the faculty in question attains to this degree of development, it occurs in lower degrees. Indeed, this highest degree may be said to bear the same relation to the lower degrees that recollection bears to memory; it implies the introspective searching of the mind with the conscious purpose of forming an ideal structure. But just as recollection is preceded by memory, or the power of intentional association by that of sensuous association, so is imagination of the intentional kind preceded by imagination of the sensuous.

After considering the subject I think we may, for the purposes of analysis, conveniently divide the grades of Imagination into four classes:—

1. On seeing any object, such as an orange, we are at once *re-minded* of the taste of an orange—have an imagination of that taste; and this is called up by the force of mere sensuous association. This is the lowest stage of mental imagery.

2. Next we have the stage in which we form a mental picture of an absent object suggested to us by some other object, as when water may suggest to us the idea of wine.

3. At a still higher stage we may form an idea independently of any obvious suggestion from without, as when a lover thinks of his mistress even in spite of external distractions; the course of ideation is here self-sustained, and no longer dependent for its mind-pictures (ideas) upon the suggestions of immediate sense-perceptions. At this stage we have dreaming in sleep, where the course of ideation runs on in a continuous stream when all the channels of sense are closed.

4. Lastly we have the stage of intentionally forming mind-pictures with the set purpose of obtaining new ideal combinations.

Such being the great differences in the degrees to which the faculty of Imagination may attain, I have made the branch in the diagram which represents the faculty a very long one, reaching from level 19 to level 38. The top of the branch therefore reaches as high as the top of Abstraction, about as high as two-thirds of Generalization, and beyond the origin of Reflection. Of course these comparative estimates are intended here, as elsewhere, to indicate merely with some rough approximation to the probable truth the relative amount of elaboration presented by each of the mental species which we denominate faculties. I consider, indeed, as I have said before, that these species are themselves of an artificial or conventional character—that what we call faculties are abstractions of our own making rather than objective or independent actualities, and therefore that the classification of these faculties by psychologists only deserves in some remote sense to be regarded as a natural one. Still it is the best classification available for the purpose of comparing one grade of mental evolution with another, and there can be no harm in adopting it if we remember, what I desire always to be remembered, that my representative tree is designed only to show the general relation between the faculties of mind as these have been formulated by psychologists.

But even on this rough and general plan it may seem to require explanation why I represent the apex of Imagination as attaining to the same level as the apex of Abstraction, for psychologists might naturally infer from my doing so that I am inadvertently endorsing the doctrine of Conceptualism. Such, however, is not the case. For, although it is true that, if we were able to imagine every abstraction, Conceptualism would become the only rational theory, I do not intend the diagram to favour any so absurd a notion. In my next work, when I shall have occasion to explain the higher branches of the representative tree, it will become apparent that, as I do not intend Abstraction to include Generalization or Reflection, I am careful to keep well within the lines of Nominalism.

Turning now to the lateral columns, it will be seen that I place upon a level with the rise of imagination the classes Mollusca, Insecta, Arachnida, Crustacea, Cephalopoda, and the cold-blooded Vertebrata. My justification for assigning to these animals the first manifestation of this faculty will be found, as in other cases, in "Animal Intelligence." Thus the octopus which followed a lobster with which it had been fighting into an adjacent tank, by laboriously climbing up the perpendicular partition between the two tanks, must have been actuated by an abiding mental image, or memory, of its antagonist; the spiders which attach stones to their webs to hold them steady during gales must similarly be actuated by a faculty of Imagination; and the same is no less true of the crab which, when a stone was rolled into its burrow, removed other stones near its margin lest they should roll in likewise. The limpet which returns to its home after a browsing excursion, must have some dim memory or mental image of the place.

So much, then, for proof of Imagination of the first degree. Imagination of the second degree—or that wherein one object or set of circumstances suggests another and similar object or set of circumstances, occurs first, so far as my evidence goes, among the Hymenoptera. But here the cases of an association of ideas leading to the establishment of a mental imagery more or less remote from the immediate circumstances of perception are much too numerous to quote. I shall therefore merely refer to the headings "General Intelligence" in the chapters on Ants, Bees, and Wasps.* Among the higher animals imagination of this grade is of frequent occurrence and strong force. Thus, to supply only one example, Thompson, in his "Passions of Animals" (p. 59), gives the case of a dog "which refused dry bread, and was in the habit of receiving from his master little morsels dipped in gravy of the meat remaining in the plate, snapped eagerly after dry bread if he saw it rubbed round the plate, and as, by way of experiment, this was repeatedly done till its hunger was satisfied, it is evident that the imagination of the animal conquered for the time its faculties of smell and taste."

To this order of imagination also belongs the wariness of wild animals. Thus Leroy, who in his capacity of Ranger had a large experience, says, "In the first hours of the night, when the countenance of darkness is in itself a fertile source of hope to the fox, the distant yelping of a dog will check him in the midst of his career. All the dangers which he has on various occasions passed through rise before him; but at dawn this extreme timidity is overborne by the calls of

* *Animal Intelligence*, pp. 122–40, and 181–19.

appetite; the animal then becomes bold by necessity. He even runs to meet danger, knowing [*i.e.*, forecasting by imagination] that it will be redoubled by return of light." And again, speaking of the wolf where rendered timid by the hostility of man, he says that it "becomes subject to illusions and to false judgments, which are the fruit of the imagination; and if these false judgments become extended to a sufficient number of objects, he becomes the sport of an illusory system, which may lead him into infinite mistakes, although perfectly consistent with the principles which have taken root in his mind. He will see snares where there are none; his imagination, distorted by fear, will invert the order of his various sensations, and thus produce deceptive shapes, to which he will attach an abstract notion of danger," &c.*

I shall only give one other fact to prove the existence of Imagination of the second order in animals, but I think it is a good one, because showing that this faculty exists in this degree in an animal not having a very high grade of intelligence—I mean the wild rabbit. Every one who has ferreted wild rabbits must have noticed that if the warren has been ferreted before, the rabbits are very unwilling to "bolt," allowing themselves to be seriously injured by the ferrets rather than face the dangers awaiting them outside. This shows that the rabbits associate (owing to past experience) the presence of a ferret in their burrows with the presence of a sportsman outside them (for it does not signify how careful the sportsman may be to keep *silent*), and so vivid is

* *Intelligence of Animals*, pp. 24, 120–1 (English translation). The well-known cunning of the fox and wolf in eluding the hounds is also evidence of a vivid imagination. In addition to the cases of this given in *Animal Intelligence* (pp. 426–30), I may now publish the following, which has recently been communicated to me by Dr. C. M. Fenn, of San Diego:—"Near the south coast of San Francisco a farmer had been much annoyed by the loss of his chickens. His hounds had succeeded in capturing several of the marauding coyotes (a kind of small wolf), but one of the number constantly eluded the pursuers by making for the coast or beach, where all traces of him would be lost. On one occasion, therefore, the farmer divided his pack of hounds, and with two or three of the dogs took a position near the shore. The wolf soon approached the ocean with the other detachment of hounds in close pursuit. It was observed that as the waves receded from the shore he would follow them as closely as possible, and in no instance made foot-prints in the sand that were not quickly obliterated by the swell. When, finally, he had gone far enough, as he supposed, to destroy the scent, he turned inland."

K 2

the mental picture of this outside enemy, that the animal will for a long time suffer the immediate pain and terror at the teeth and claws of the ferret before venturing to expose itself to the more remote but still more deadly pain which it fears at the hands of the man.

Coming now to Imagination of the third degree, or that which implies the power of forming ideas independently of any obvious suggestions from without, we have first to consider how this kind of imagination, even if present in animals, could be expressed. Now, apart from articulate expression or intelligent gesture, it is evident that the objective indices of imagination in this degree are so limited in number as to be well-nigh absent. Even, therefore, if we assume such imagination as present in any given animal, we might find it difficult to suggest the kind of action to which it might give rise, and which might be taken as unequivocal proof of such faculty. What we require, it will be observed, is some class or classes of actions which must be due to imagination of this degree and can be due to nothing else. I only know of three such classes, which, however, are conclusive as establishing the fact of such imagination being present in the animals which display them. It is almost needless to add that imagination, even of this level of development, may well be present among animals lower in the scale, which yet is not apparent on account of being developed in lines which do not express themselves in either of the three classes of actions on which I rely in the case of the higher animals.

The first of these actions is Dreaming. This, wherever it is found to occur, constitutes certain proof of imagination belonging to what I have called the third degree.

The fact that Dogs dream is proverbial, and was long ago remarked by Seneca and Lucretius. According to Dr. Lauder Lindsay the Horse also dreams, as shown by its "shuddering, shivering, quivering, quaking, or trembling. These phenomena are concomitants or results in the waking state of excitement, fear, ardour, impetuosity, or impatience. Hence it is quite legitimately inferred by Montaigne and others that the same feelings or mental conditions are developed during sleep and dreaming, and are likely to be associated in the racehorse with imaginary races, as in the sporting dog with imaginary coursing."*

* *Mind in the Lower Animals*, vol. ii, pp. 95-6.

The authorities which I have been able to find who assert that dreaming occurs in Birds are Cuvier, Jerdon, Houzeau, Bechstein, Bennet, Thompson, Lindsay, and Darwin.* Thompson also says that Crocodiles dream, but as he gives no references to substantiate the statement, I have ignored it, and in the diagram placed dreaming on a level with Birds, as the lowest animals which I feel there is adequate evidence to accredit with this faculty. According to the writer last named, who is generally accurate, "Among Birds the stork, the canary, the eagle, and the parrot; and among the Mammalia the elephant, the horse, and the dog, are incited in their dreams." Bennet noticed that water-birds moved their legs in their sleep, as if in the act of swimming; and Hennabe heard the hyrax utter a faint cry. Bechstein has described dreaming in a bullfinch, and the dreams appeared to be of the character of nightmares, for "the terror begotten during sleep was such that it required its mistress's interference to prevent bad effects. It frequently fell from its perch, but became immediately tranquillised and reassured by the voice of its mistress." Lastly, Houzeau asserts that parrots sometimes talk in their sleep."†

The second class of facts on which I rely as proof of Imagination of the third degree in animals is that of Delusions.

Dr. Lauder Lindsay writes with truth:—"Delusions of sight in animals take the form, as in man, of phantoms or phantasms. . . . of imaginary persons, animals, or things. And, moreover, it would appear to be the same kind of spectral images that occur in other animals as in man, in canine rabies, for instance, as in human hydrophobia."‡ On this subject Fleming writes:—"It (*i.e.*, a rabid dog) appeared as if it was haunted by some horrid phantoms. . . . At times it would seem to be watching the movements of something on the floor, and would dart suddenly forward and bite

* See, for original passages or references, *Birds of India*, vol. i, p. xxi; *Facultes-Mentales des Animaux, &c.*, tome ii, p. 183; *Mind in the Lower Animals*, vol. ii, p. 96; *Passions of Animals*, p. 60; and *Descent of Man*, p. 74.

† According to Pierguin, Guer, Elam, and Lindsay, dreaming in animals may be so vivid as to lead to somnambulism (see Lindsay, *loc. cit.*, p. 97, *et seq.*). Thus Guer asserts that "the somnambulistic watch-dog prowls in search of imaginary strangers or foes, and exhibits towards them a whole series of pantomimic actions," including barking.

‡ *Loc. cit.*, p. 103.

at the vacant air, as if pursuing something against which it had an enmity." And, indeed, this peculiarity of being liable to optical delusions is so usual and well marked a feature in rabid dogs, that it generally constitutes the earliest and most certain symptom of disease.* My friend Mr. Walter Pollock sends me the following account of a Scotch terrier bitch which he possessed:—" She had a curious hatred or horror of anything abnormal—for instance, it was long before she could tolerate the striking of a spring bell, which when I first knew her was a new experience to her. She expressed her dislike and seeming fear by a series of growls and barks, accompanied by setting her hair up on end. She used from time to time to go through exactly the same performance after gazing fixedly into what seemed to be vacancy. This attracted my attention, and I used to be on the look out for it, but carefully avoided in any way tempting her to make any display of this peculiarity. I simply watched her whenever I was alone with her. The constant repetition in these circumstances of her seeming to see some enemy or portent unseen by me, and giving vent to her feelings in the way already described, led me to the conclusion that at these times she was the victim of optical illusion of some kind. I could, as I have already hinted, produce the same effect upon her by doing some unexpected and irrational thing, until she had become accustomed to this kind of experiment. But after this the seeing, as it seemed to be, of some sort of phantom remained unabated. I had no opportunity of discerning whether the phenomena occurred at any regular intervals, or whether they were more frequent after sleep than at other times."

Pierquin describes a female ape which had a sun-stroke, and afterwards use to become terror-struck by delusions of some kind. She also used to snap at imaginary objects, and " acted as if she had been watching and catching at insects on the wing."†

It seems needless for our present purpose to give more evidence on the fact of animals being subject to delusions, and so I shall pass on to the third class of facts on which I rely as evidence that animals present Imagination of what I have called the third order. This class of facts consists of

* See Youat, *On the Dog*, under *Rabies*.
† *Traité de la Folie des Animaux, &c.*, tome i, p. 93.

animals showing by their actions that they have in their "mind's eye" a picture or representation of absent objects.

Every one must have observed, for instance, the greater spirit with which jaded horses return on their homeward journey, as compared with the sluggishness and lack of energy on their out-going journey. This can only be explained by supposing that the animals have a mental picture of their stables, with its ideal accompaniments of food and repose. Again, the desire which many animals show to return to their habitual haunts when removed from them can only be explained by supposing them to retain a mental picture, or imagination, of their previously happy experience. The promptings of this imagination are frequently so strong as to induce the animals to brave the dangers and fatigues of hundreds of miles of travel for the sole purpose of returning to the scenes which occupy their imaginations. "Pigeons, dogs, cats, and horses, when removed from their former homes, give repeated and daily instances of the fact. It crushes and overwhelms the faculties of the mind, and prostrates the energies of the body. Thus many birds, when encaged, become so utterly spirit-broken, that they refuse all nourishment, pine for a few days, and die. This is particularly the case with song-birds. . . . If the Howling Monkey is caught when full-grown, it become melancholy, refuses all food, and dies in a few weeks; it is also the same with the Puma; and Burdach states that death sometimes ensues so immediately, that it can only arise from a sudden and violent pressure on the mind."*

Although it may be objected to this interpretation of pining under confinement that the fact may be due to the mere absence of liberty or changed condition of life, without any mental and contrasted picture of previous experience, I think that this objection is precluded in other and analogous cases to which I shall next refer, and which serve in larger if not in full measure to disarm this criticism as applied to such cases as the above. I allude to all those cases so frequently observed among domestic animals where similar pining occurs without there being any change in the conditions of life, except the sudden withdrawal of a master or companion to which the animal is strongly attached. I have myself known a case in which a terrier of my own household, on the

* Thompson, *Passions of Animals*, pp. 64–5.

sudden removal of his mistress, refused all food for a number of days, so that it was thought he must certainly die, and his life was only saved by forcing him to eat raw eggs. Yet all his surroundings remained unchanged, and every one was as kind to him as they always had been. And that the cause of his pining was wholly due to the absence of his beloved mistress, was proved by the fact that he remained permanently outside her bedroom door (although he knew she was not inside), and could only be induced to go to sleep by giving him a dress of hers to lie upon. No one could have seen this dog without being persuaded that he had a constant mental picture of his mistress in his imagination, and suffered the keenest mental anguish from her continued absence. Similarly there are numberless anecdotes on record, most of which are probably true, of dogs actually dying under such circumstances.

All these facts, then, taken together—viz., dreaming, delusions, "home sickness," and pining for friends—clearly prove the presence among higher animals of Imagination in what I have called the third order. A question may here arise as to whether I have not in the diagram placed the rise of Imagination too low. I place the first origin of this faculty on level 19, which corresponds with that of the Mollusca and an infant seven weeks old. This question, like all others of line-drawing among the psychological faculties, is confessedly a difficult one; but the reasons why I have placed the dawn of Imagination so low in the psychological scale are as follows:—

It will be remembered that the kind of Imagination which we have recently been considering belongs to what I consider a high level of development. That is to say, I consider the power of dreaming to occupy a place about one third of the distance between the first dawn of the imaginative faculty and its maximum development in a Shakespeare or a Faraday. I so consider it because I believe that to pass through what I have called the first three stages, so as to arrive at the power of forming mental pictures independently of sensuous suggestions from without, the imaginative faculty has made so enormous a progress from its earliest beginnings, that the rest of its development along the same lines is really nothing more than a function of the faculty of Abstraction. Superimpose upon the psychology of a terrier which pines for its absent mistress an elaborate structure of abstract ideation, and the terrier's imaginative faculty would begin to rival that of man. Of course it will be said that abstraction presupposes imagination, and so undoubtedly it does; still the two are not identical, as is proved by the fact that for the building up of abstraction to any exalted height, language, or mental symbolism of some kind, is indispensable; and mental symbols are so many artifices for the saving of imagination.

Now if at first sight it seems absurd to accredit a mollusk with imagination, we must remember exactly what we mean by imagination in the lowest possible phase of its development. We mean merely the power of forming a definite mental picture, or of retaining a memory, no matter of how rudimentary a kind; provided that the memory implies some dim idea of an absent object or experience, and not, as in the case of an infant disliking the taste of strange milk, merely an immediate perception of contrast between an habitual and a present sensation. And that we find such a level of mental development as low down in the zoological scale as the Gasteropoda, would seem to be proved by the fact already alluded to of limpets returning to their homes in the rocks after feeding. Of course the mental image which a limpet forms of its home in a rock cannot be supposed to be comparable in point of vividness or complexity with the mental image that a horse retains of its stall, or a dog of its kennel; still, such as it is, it is a mental image, and therefore betokens imagination. More vivid, and therefore more definite, is the mental image that a spider forms of her lair, who when dislodged and carried away to a short distance again returns to her old home. (Level 20.) With a still further advance in the power of mental imagery (level 21) we find supplied the psychological conditions for the ideation of cold-blooded Vertebrata, such as the determination displayed by migratory Fishes (notably the salmon) to visit particular localities in the spawning season. On the next level (22) we reach the higher Crustacea, which, as we have already seen, are able to imagine in a high degree. Next we come to Reptiles, concerning which I may quote the following anecdote from Lord Monboddo: "I am well informed of a tame serpent in the East Indies, which belonged to the late Dr. Vigot, once kept by him in the suburbs of Madras. This serpent was taken by the French, when they invested Madras, in the late war, and was carried to Pondicherry in a close carriage. But from thence he found his way back again to his old quarters, though Madras is over one hundred miles distant from Pondicherry." If we substitute yards for miles, similar cases are on record with regard to frogs and toads—which from being so numerous can scarcely all be false. And that some reptiles have an imagination passing into what I have called the third stage is proved by the case of the python mentioned in "Animal Intelligence," which, when sent to the Zoological Gardens, pined for its previous master and mistress. The Cephalopoda and Hymenoptera have already been alluded to. Lastly, on the next level (25) we attain in Birds to imagination proved to be unquestionably of the third degree by the phenomenon of dreaming. Above this level it is not of so much interest to trace the improvement of the faculty. Such improvement throughout the subsequent levels till man, probably consists only in a progressive advance through imagination of the third degree—it being I think highly improbable, and certainly not betokened by any evidence, that imagination in any animal attains to what I have called the fourth degree, which I therefore consider distinctive of man.

> "For know that in the soul
> Are many lesser faculties that serve
> Reason as chief. Among these, Fancy next
> Her office holds. Of all external things,
> Which the five watchful senses represent,
> He forms imaginations, airy shapes;
> Which Reason joining or disjoining, forms
> All that we affirm, or what deny,
> And call our knowledge."—MILTON.

Before taking leave of Imagination there are two branches of the subject which I should like briefly to consider. One is the opinion held by Comte that the higher animals present ideas of Fetishism. On this topic I cannot more briefly convey the material which I have to render than by quoting a previous publication of my own from "Nature."* "Mr. Herbert Spencer in his recently published work on the 'Principles of Sociology' treats of the above subject. He says, 'I believe M. Comte expressed the opinion that fetichistic conceptions are formed by the higher animals. Holding, as I

* Vol. xvii, p. 168, *et seq.*

have given reasons for doing, that fetichism is not original but derived, I cannot, of course, coincide in this view. Nevertheless I think the behaviour of intelligent animals elucidates the genesis of it. I have myself witnessed in dogs two illustrative cases.' One of these consisted in a large dog, which, while playing with a stick accidentally thrust one end of it against his palate, when 'giving a yelp, he dropped the stick, rushed to a distance from it, and betrayed a consternation which was particularly laughable in so ferocious-looking a creature. Only after cautious approaches and much hesitation was he induced again to lay hold of the stick. This behaviour showed very clearly the fact that the stick, while displaying none but the properties he was familiar with, was not regarded by him as an active agent; but that when it suddenly inflicted a pain in a way never before experienced from an inanimate object, he was led for a moment to class it with animate objects, and to regard it as capable of again doing him injury. Similarly, in the mind of the primitive man, knowing scarcely more of natural causation than a dog, the anomalous behaviour of an object previously classed as inanimate suggests animation. The idea of voluntary action is made nascent; and there arises a tendency to regard the object with alarm, lest it should act in some other unexpected and perhaps mischievous way. The vague notion of animation thus aroused will obviously become a more definite notion, as fast as the development of the ghost-theory furnishes a special agency to which the anomalous behaviour can be ascribed.'

"The other case observed by Mr. Spencer was that of an intelligent retriever. Being by her duties as a retriever led to associate the fetching of game with the pleasure of the person to whom she brought it, this had become in her mind an act of propitiation; and so, 'after wagging her tail and grinning, she would perform this act of propitiation as nearly as practicable in the absence of a dead bird. Seeking about, she would pick up a dead leaf or other small object, and would bring it with renewed manifestations of friendliness. Some kindred state of mind it is which, I believe, prompts the savage to certain fetichistic observances of an anomalous kind.'

"These observations remind me of several experiments I made some years ago on this subject, and which are perhaps worth publishing. I was led to make the experiments by reading the instance given by Mr. Darwin in the 'Descent of Man' of the large dog which he observed to bark at a parasol as it was moved along a lawn by the wind, so presenting the appearance of animation. The dog on which I experimented was a Skye terrier—a remarkably intelligent animal, whose psychological faculties have already formed the subject of several communications to this and other periodicals. As all my experiments yielded the same results, I will only mention one. The terrier in question, like many other dogs, used to play with dry bones, by tossing them in the air, throwing them to a distance, and generally giving them the appearance of animation in order to give himself the ideal pleasure of worrying them. On one occasion, therefore, I tied a long and fine thread to a dry bone, and gave him the latter to play with. After he had tossed it about for a short time, I took the opportunity, when it had fallen at a distance from him and while he was following it up, of gently drawing it away from him by means of the long invisible thread. Instantly his whole demeanour changed. The bone which he had previously pretended to be alive now began to look as if it were really alive, and his astonishment knew no bounds. He first approached it with nervous caution, as Mr. Spencer describes; but as the slow receding motion continued, and he became quite certain that the movement could not be accounted for by any residuum of the force which he had himself communicated, his astonishment developed into dread, and he ran to conceal himself under some articles of furniture, there to behold at a distance the 'uncanny' spectacle of a dry bone coming to life.

"Now in this and all my other experiments I have no doubt that the behaviour of the terrier arose from his *sense of the mysterious*, for he was of a highly pugnacious disposition, and never hesitated to fight any animal of any size or ferocity; but apparent symptoms of spontaneity in an inanimate object which he knew so well, gave rise to feelings of awe and horror, which quite enervated him. And that there was nothing fetichistic in these feelings may safely be concluded if we reflect, with Mr. Spencer, that the dog's knowledge of causation for all immediate purposes being quite as correct and no less stereotyped than is that of 'primitive man,' when an object of a class which he knew from uniform past experience to be inanimate suddenly began to move, he must have felt the same oppressive and alarming sense of the mysterious which uncultured persons feel under similar circumstances. But further, in the case of this terrier, we are not left with *à priori* inferences alone to settle this point, for another experiment proved that the sense of the mysterious in this animal was sufficiently strong in itself to account for his behaviour. Taking him into a carpeted room, I blew a soap-bubble, and by means of a fitful draught made it intermittently glide along the floor. He became at once intensely interested, but seemed unable to decide whether or not the fitful object was alive. At first he was very cautious, and followed it only at a distance; but as I encouraged him to examine the bubble more closely, he approached it with ears erect and tail down, evidently with much misgiving, and the moment it happened to move he again retreated. After a time, however, during which I always kept at least one bubble on the carpet, he began to gain more courage, and the scientific spirit overcoming his sense of the mysterious, he eventually became bold enough slowly to approach one of the bubbles, and nervously to touch it with his paw. The bubble, of course, immediately burst, and I certainly never saw astonishment more strongly depicted. On then blowing another bubble, I could not persuade him to approach it for a good while; but at last he came, and carefully extended his paw as before, with the same result. But after this second trial nothing would induce him again to approach a bubble, and on pressing him he ran out of the room, which no coaxing would persuade him to re-enter.

"One other example will suffice to show how strongly developed was the sense of the mysterious in this animal. When alone with him in a room I once purposely tried the effect on him of making a series of hideous grimaces. At first he thought I was only making fun; but as I persistently disregarded his caresses and whining while I continued unnaturally to disturb my features, he became alarmed; slunk away under some furniture, shivering like a frightened child. He remained in this condition till some other member of the family happened to enter the room, when he emerged from his hiding place in great joy at seeing me again in my right mind. In this experiment, of course, I refrained from making any sounds or gesticulations, that might lead him to think I was angry. His actions therefore can only be explained by his horrified surprise at any apparently irrational behaviour, *i.e.*, by the violation of his ideas of uniformity in matters psychological. It must be added, however, that I have tried the same experiment on less intelligent and less sensitive terriers with no other effect than causing them to bark at me. I will only add that I believe the sense of the mysterious to be the cause of the dread which many animals show of *thunder*. I am led to think this, because I once had a setter which never heard thunder till he was eighteen months old, and on then hearing it I thought he was about to die of fright, as I have seen other animals do under various circumstances. And so strong was the impression which his extreme terror left behind, that whenever afterwards he heard the boom of distant artillery practice, mistaking it for thunder, he became a pitiable object to look at, and, if out shooting, would endeavour to bury himself or bolt home. After having heard real thunder on two or three subsequent occasions, his dread of the distant cannon became greater than ever; so that eventually, though he keenly enjoyed sport, nothing would induce him to leave his kennel, lest the practice might begin when he was at a distance from home. But the keeper, who had a large experience in the training of dogs, assured me if I allowed this one to be taken to the battery in order that he might learn the true *cause* of the thunder-like noise, he would again become serviceable in the field. The animal, however, died before the experiment was made."*

Thus I think we may safely set down the sense of the mysterious as thus undoubtedly displayed by intelligent dogs —and also, I may add, by many horses when going along a dark road, hearing strange sounds, or seeing unaccustomed sights—to the effects of imagination in suggesting vague possibilities in circumstances perceived to be unusual; just as with children under similar circumstances the undefined imagination of possible harm springing out of such circumstances in some unthought-of manner, engenders that feeling of unreasonable dread which we may in both cases call a sense of the mysterious.

* That such would have been the case, however, I have little doubt, for on one occasion when a number of apples were being shot out of bags upon the wooden floor of an apple-room, the sound in the house as each bag was shot closely resembled that of distant thunder. The setter, therefore, became terribly alarmed; but when I took him to the apple-room and showed him the real *cause* of the noise, his dread entirely left him, and on again returning to the house he listened to the rumbling with all cheerfulness.

CHAPTER XI.

INSTINCT.

Definition.

I SHALL begin this important and extensive part of my subject by repeating the definition of Instinct which I laid down in my former work. It will be remembered that for the sake of precision I there limited the term Instinct as follows:—

"Instinct is reflex action into which there is imported the element of consciousness. The term is therefore a generic one, comprising all those faculties of mind which are concerned in conscious and adaptive action, antecedent to individual experience, without necessary knowledge of the relation between means employed and ends attained, but similarly performed under similar and frequently recurring circumstances by all the individuals of the same species."

Referring the reader to the context for my justification of this definition,* I shall here only further make this general statement. It follows from the above definition of Instinct, that a stimulus which evokes a reflex action is, at most, a sensation;† but a stimulus which evokes an instinctive action is a perception. After what I have already said in Chapter IX concerning the distinction between a sensation and a perception, my meaning now will be clearly understood. For if a perception differs from a sensation in that it presents a mental element, and if an instinctive action differs from a reflex action in that it presents a mental element, it is easy to see that a stimulus supplied by a sensation is to a reflex action what a stimulus supplied by a perception is to an instinctive action; because if a sensation could act as a

* *Animal Intelligence*, pp. 10–17.

† I say "at most," because such a stimulus may be *less* than a sensation, in that it may never cross the field of consciousness.

stimulus to an action apparently instinctive, *ex hypothesi* the action could not be (according to my definition) really instinctive; and conversely, if a perception could act as a stimulus to an action apparently reflex, the action could not be (according to my definition) a true reflex. Therefore, if we agree to limit the term Instinct to nervous processes involving a mental element, it follows that this element is perception, and that it is always involved in every stimulus leading to instinctive action.

With reference to general principles of classification it is only needful for me further to quote the following extract from my previous work:—

"The most important point to observe in the first instance is that instinct involves *mental* operations; for this is the only point that serves to distinguish instinctive from reflex action. Reflex action, as already explained, is non-mental neuro-muscular adaptation to appropriate stimuli; but instinctive action is this and something more; there is in it the element of mind. No doubt it is often difficult, or even impossible, to decide whether or not a given action implies the presence of the mind-element—*i.e.*, conscious as distinguished from unconscious adaptation; but this is altogether a separate matter, and has nothing to do with the question of defining instinct in a manner which shall be formally exclusive, on the one hand of reflex action, and on the other of reason. As Virchow truly observes, 'it is difficult or impossible to draw the line between instinctive and reflex action;' but at least the difficulty may be narrowed down to deciding in particular cases whether or not an action falls into this or that category of definition; there is no reason why the difficulty should arise on account of any ambiguity of the definitions themselves. Therefore I endeavour to draw as sharply as possible the line which *in theory* should be taken to separate instinctive from reflex action; and this line, as I have already said, is constituted by the boundary of non-mental or unconscious adjustment, with adjustment in which there is concerned consciousness or mind."

I shall now proceed to show, by a few selected examples, what has been called the Perfection of Instinct; next I shall similarly illustrate the Imperfection of Instinct; and lastly, I shall discuss the important question of the Origin and Development of Instinct.

Perfection of Instinct.

An instinct may be said to be perfect when it is perfectly adapted to meet those circumstances in the life of an animal for the meeting of which the instinct exists; and if it is an instinct this perfection must be exhibited as independent of the animal's individual experience. We may therefore best illustrate the perfection of instinct by considering the wonderful accuracy of many among the highly refined and complex adjustments which are manifested by the newly-born young of the higher animals.

The late Mr. Douglas Spalding in his brilliant researches on this subject has not only placed beyond question the falsity of the view "that all the supposed examples of instinct may be nothing more than cases of rapid learning, imitation, or instruction,"* but also proved that a young bird or mammal comes into the world with an amount and a nicety of ancestral knowledge that is highly astonishing. Thus, speaking of chickens which he liberated from the egg and hooded before their eyes had been able to perform any act of vision, he says that on removing the hood after a period varying from one to three days, "almost invariably they seemed a little stunned by the light, remained motionless for several minutes, and continued for some time less active than before they were unhooded. Their behaviour, however, was in every case conclusive against the theory that the perceptions of distance and direction by the eye are the result of experience, or of associations formed in the history of each individual life. Often at the end of two minutes they followed with their eyes the movements of crawling insects, turning their heads with all the precision of an old fowl. In from two to fifteen minutes they pecked at some speck or insect, showing not merely an instinctive perception of distance, but an original ability to judge, to measure distance, with something like infallible accuracy. They did not attempt to seize things beyond their reach, as babies are said to grasp at the moon; and they may be said to have invariably hit the objects at

* Quoted from his article in *Macmillan's Magazine*, February, 1873, from which likewise all the subsequent quotations are made. We are nowadays so ready to assimilate scientific truth, that in reading this article—not yet ten years old—it seems difficult to realize that so recently there was such a considerable clinging of competent opinion to the non-evolutionary view of instinct as the quotations in the article show.

which they struck—they never missed by more than a hair's breadth, and that, too, when the specks at which they aimed were no bigger, and less visible, than the smallest dot of an *i*. To seize between the points of the mandibles at the very instant of striking seemed a more difficult operation. I have seen a chicken seize and swallow an insect at the first attempt; most frequently, however, they struck five or six times, lifting once or twice before they succeeded in swallowing their first food. The unacquired power of following by sight was very plainly exemplified in the case of a chicken that, after being unhooded, sat complaining and motionless for six minutes, when I placed my hand on it for a few seconds. On removing my hand the chicken immediately followed it by sight backward and forward, and all round the table. To take, by way of example, the observations in a single case a little in detail:—A chicken that had been made the subject of experiments on hearing, was unhooded when nearly three days old. For six minutes it sat chirping and looking about it; at the end of that time it followed with its head and eyes the movements of a fly twelve inches distant; at ten minutes it made a peck at its own toes, and the next instant it made a vigorous dart at the fly, which had come within reach of its neck, and seized and swallowed it at the first stroke; for seven minutes more it sat calling and looking about it, when a hive-bee coming sufficiently near was seized at a dart and thrown some distance, much disabled. For twenty minutes it sat on the spot where its eyes had been unveiled without attempting to walk a step. It was then placed on rough ground within sight and call of a hen with a brood of its own age. After standing chirping for about a minute, it started off towards the hen, displaying as keen a perception of the qualities of the outer world as it was ever likely to possess in after life. It never required to knock its head against a stone to discover that there was 'no road that way.' It leaped over the smaller obstacles that lay in its path and ran round the larger, reaching the mother in as nearly a straight line as the nature of the ground would permit. This, let it be remembered, was the first time it had ever walked by sight."

Further, "When twelve days old one of my little *protégés*, while running about beside me, gave the peculiar chirr whereby they announce the approach of danger. I looked up, and behold a sparrow-hawk was hovering at a great height over head. Equally striking was the effect of the hawk's voice when heard for the first time. A young turkey, which I had adopted when chirping within the uncracked shell, was on the morning of the tenth day of its life eating a comfortable breakfast from my hand, when the young hawk, in a cupboard just beside us, gave a shrill chip, chip, chip. Like an arrow the poor turkey shot to the other side of the room, and stood there motionless and dumb with fear, until the hawk gave a second cry, when it darted out at the open door right to the extreme end of the passage, and there, silent and crouched in a corner, remained for ten minutes. Several times during the course of that day it again heard these alarming sounds, and in every instance with similar manifestations of fear."

Again referring to young chickens, Mr. Spalding continues,—"Scores of times I have seen them attempt to dress their wings when only a few hours old—indeed as soon as they could hold up their heads, and even when denied the use of their eyes. The art of scraping in search of food, which, if anything, might be acquired by imitation—for a hen with chickens spends the half of her time in scratching for them—is nevertheless another indisputable case of instinct. Without any opportunities of imitation, when kept quite isolated from their kind, chickens began to scrape when from two to six days old. Generally, the condition of the ground was suggestive; but I have several times seen the first attempt, which consists of a sort of nervous dance, made on a smooth table."

In this connection I may here insert an interesting observation which has been communicated to me by Dr. Allen Thomson, F.R.S. He hatched out some chickens on a carpet, where he kept them for several days. They showed no inclination to scrape, because the stimulus supplied by the carpet to the soles of their feet was of too novel a character to call into action the hereditary instinct; but when Dr. Thomson sprinkled a little gravel on the carpet, and so supplied the appropriate or customary stimulus, the chickens immediately began their scraping movements.

But to return to Mr. Spalding's experiments, he says:—

"As an example of unacquired dexterity, I may mention that on placing four ducklings a day old in the open air for the first time, one of them almost immediately snapped at and caught a fly on the wing. More interesting, however, is the deliberate art of catching flies practised by the turkey. When not a day and a half old I observed the young turkey already spoken of slowly pointing its beak at flies and other small insects without actually pecking at them. In doing this, its head could be seen to shake like a hand that is attempted to be held steady by a visible effort. This I observed and recorded when I did not understand its meaning. For it was not until after, that I found it to be the invariable habit of the turkey, when it sees a fly settled on any object, to steal on the unwary insect with slow and measured step until sufficiently near, when it advances its head very slowly and steadily till within an inch or so of its prey, which is then seized by a sudden dart."

Mr. Spalding subsequently tried similar experiments, with similar results, on newly born mammals. He found, for instance, that new-born pigs seek to suck almost immediately after birth. If removed twenty feet from the mother, they wriggle straight back to her guided apparently by her grunting. He put a pig into a bag immediately it was born, and kept it in the dark till seven hours old, and then placed it outside the sty ten feet from its mother. It went straight to her, although it had to struggle for five minutes to squeeze under a bar. A pig blindfolded at birth went about freely, though tumbling against things. It had the blinder taken off next day, and then " went round and round as if it had had sight, and had suddenly lost it. In ten minutes it was scarcely distinguishable from one that had had sight all along. When placed on a chair, it knew the height to require considering, went down on its knees, and leaped down.
One day last month, after fondling my dog, I put my hand into a basket containing four blind kittens three days old. The smell my hand had carried with it sent them puffing and spitting in a most comical fashion."*

Here I may quote an observation of my own from the succeeding issue of " Nature."

" *Apropos* to what Mr. Spalding says about the early age at which the instinctive antipathy of the cat to the dog becomes apparent, I may state that some months ago I tried an experiment with rabbits and ferrets somewhat similar to that which he describes with cats and dogs. Into an out-

* *Nature*, vol. xi, p. 507.

house which contained a doe rabbit with a very young family, I turned loose a ferret. The doe rabbit left her young ones, and the latter, as soon as they smelled the ferret, began to crawl about in so energetic a manner as to leave no doubt that the cause of the commotion was fear, and not merely the discomfort arising from the temporary absence of the mother."*

With reference to the instinctive endowments of this kind in kittens, I may also quote the following, which I find among Mr. Darwin's MSS:—

"The many cases of inborn fear or ferocity in young animals directed towards particular objects, as well as the loss of these individualized passions, seems to me extremely curious. Let any one who doubts their existence give a mouse to a kitten taken early from its mother, and which has never before seen one, and observe how soon the kitten growls with hair erect, in a manner wholly different from when at play or when fed with ordinary food. We cannot suppose that the kitten has an inborn picture of a mouse graven in its mind. But, as when an old hunter snorts with eagerness at the very first sound of the horn, we must suppose the old associations excite him almost as instantly as when a sudden noise makes him start, so I imagine, with the difference that the imagination has become hereditary instead of being only fixed by habit, the kitten without any definite anticipation trembles with excitement at the smell of the mouse."

The only other observations made by Mr. Spalding which it is desirable to quote are those by which he proved experimentally that young birds do not require, as was ordinarily supposed, to be taught to fly, but fly instinctively. This fact was proved by keeping young swallows caged until they were fledged, and then allowing them to escape. When we consider the complicated muscular co-ordination required for flight, the fact that young birds when fledged should be able to fly at the first attempt constitutes another remarkable instance of the perfection of instinct. Of course it is true that under ordinary circumstances the parent birds encourage their progeny to fly, but the experiments in question show that such encouragement, or tuition, is not necessary to enable the young birds to practise the art.

But it is among insects that we meet with the most re-

* *Nature*, vol. xi, p. 554.

markable cases. Thus, to give only a few. Réaumur and Swanderdam assert that a young Bee, as soon as its wings are dry, will collect honey and construct a cell as efficiently as the oldest inhabitant of the hive.* Numberless insects, also, can never have seen their parents, and yet they perform instinctive actions perfectly, though it may be only once in their life-times—such, for instance, as the Ichneumon, which deposits its eggs in the body of a larva hidden between the scales of a fir-cone, which it can never have seen, and yet knows where to seek.†

A kind of insect called the Bembex conveys food to its young which are shut up in a cell, and it has recently been made the subject of some interesting experiments by M. Fabre. Of these the following is an epitome:—

"The insect brings from time to time fresh food to her young, and it is remarkable how the Bembex remembers the entrance to her cell, covered as it is with sand, exactly to our eyes like that all round. Yet she never makes a mistake or loses her way. On the other hand M. Fabre found that if he removed the surface of the earth and the passage, thus exposing the cell and the larva, the Bembex was quite at a loss, and did not even recognize her own offspring. It seems as if she knew the doors, nursery, and the passage, but not her child. Another ingenious experiment of M. Fabre's was made with Chalicodoma. This genus is enclosed in an earthen cell through which at maturity the young insect eats its way. M. Fabre found that if he pasted a piece of paper round the cell the insect had no difficulty in eating through it, but if he enclosed the cell in a paper case, so that there was a space even of only a few lines between the cell and the paper, in that case the paper formed an effectual prison. The instinct of the insect taught it to bite through one enclosure, but it had not wit enough to do so a second time."‡

But I think that perhaps the most remarkable instance of all that can be quoted from the insect world to show the extraordinary perfection of early-formed instincts, is one which is apt to be overlooked—and indeed, so far as I know, has been overlooked—on account of its frequency. I refer to the enormous body of instincts, all having reference to a totally different environment and habits of life, which those insects that undergo a complete metamorphosis present fully-

* Kirby and Spence, *loc. cit.*, vol. ii, p. 470. † *Ibid.*, i, p. 357.

‡ Sir J. Lubbock, *Address to Entomol. Soc.*, 1882.

formed and ready for complete action as soon as the imago escapes from its pupa stage. The difference between its previous life as a larva and its new life as an imago, is as great as the difference between the lives of two animals belonging to two different sub-kingdoms; and the complete adaptation which all the new class of instincts exhibit to the requirements of this new life, is quite as remarkable as is the adaptation of the new structures to the same requirements.

Imperfection of Instinct.

I shall first give a few cases to show that instinct is not an infallible guide to action, and for this purpose shall choose aberrations of those instincts which we should expect to be most fixed, because of most importance to the well-being of the animals or their progeny—I mean the instincts of propagation and the procuring of food.

The flesh-fly (*Musca carnaria*) deposits its eggs in the flowers of the "carrion plant" (*Stapelia hirsuta*), the smell of which resembles that of putrid meat, and so deceives the fly.* Similarly, the house-fly has been observed to deposit eggs in snuff.†

Again, the Rev. Mr. Bevan and Miss C. Shuttleworth, write me independently that they have seen wasps and bees visiting representations of flowers upon the wall-paper of rooms; and Trevellian saw the same mistake made by the sphinx-moth.‡ Swainson in his "Zoological Illustrations," gives an analogous case in a vertebrated animal; an Australian parrot, whose food is taken from the flowers of the Eucalyptus, was observed endeavouring to feed on the representation of flowers on a cotton-print dress. Likewise, Professor Moseley, F.R.S., informs me that he has noticed honey-seeking insects mistake for flowers the bright coloured salmon flies stuck in his hat while fishing; and Mr. F. M. Burton, writing to "Nature" (xvii, p. 162), says that he has observed the humming bird hawk-moth (*Macroglossa stellatarum*) mistake artificial flowers in a lady's bonnet for real ones. Still more curiously, the naturalist Couch observed a

* E. Darwin, *Zoonomia*, i, § 16, art. 11. Also Kirby and Spence, *loc. cit.*, ii, 469, who state the fact on the authority of Dr. Zinken.

† Zinken, in *Germar. Mag. der Ento.*, Bd. I, abth. 4, § 189.

‡ See Houzeau, *loc. cit.*, I, 210.

bee mistake a sea-anemone (*Tealia crassicornis*), which was "covered merely by a rim of water," for a flower—darting into the centre of the disk, "and though it struggled a good deal to get free, was retained till it was drowned, and was then swallowed.* The fact, alluded to by Mr. Darwin in the Appendix, that the workers of the humble-bee attempt to devour the eggs laid by their own queen, appears to constitute a remarkable case of imperfect instinct. Again, Huber saw a bee begin a cell in a wrong direction, and other bees tear it to pieces. Bees have also been observed to collect rye-flower when damp instead of pollen.† "Pollen-getting, according to Gebien, is the weak point in the character of bees;" for this author observes (p. 74) that they "lay up useless hoards of it, which they go on augmenting every year, and this is the only point on which they can be accused of want of prudence."

Mr. Darwin's MS notes contain a brief record of a number of observations on ants (*F. rufa*) carrying pupa skins, with a great and apparently useless expenditure of labour, far away from the nest, and even up trees. He tried taking away the skins from some of the carriers, and replacing them near the nest; the first ants that happened to fall in with them again carried them off. This, as the notes observe, appears to be a case of "blundering instinct;" and the same epithet may be applied to mistakes made by the harvesting ants observed by Mr. Moggridge, which carefully stored in their granaries the gall-apples of a small species of *Cynips*, clearly imagining that they were nuts; and also, under a similar delusion, stored small beads which Moggridge, in order to test their instinct, scattered in their harvesting fields.‡

Among Birds we find mistaken instinct exhibited by the cuckoo when it lays two eggs in the same nest, with the inevitable result that one of the young birds will afterwards eject the other. In the same category we may place the promiscuous dropping of her eggs on the part of the rhea; small birds frequently mistaking a larger and unfamiliar bird for a hawk, as shown by their mobbing it; and numberless special cases could be given of mistaken instinct in the matter of nest-building—in the selection of unsuitable sites, unsuitable materials, and so on.

* *Critic*, March 24, 1860. † *Cottage Gardener*, April, 1860, p. 48.

‡ *Harvesting Ants and Trap-door Spiders*, p. 37, *et seq.*

Among Mammals it must be deemed a mistaken instinct which leads the Norwegian lemming to swim out to sea in its migrations, and perish by millions in consequence. Under existing circumstances it is an imperfect instinct which leads the quadrupeds in South Africa, mentioned by Mr. Darwin in the Appendix, to migrate, seeing that by so doing they expose themselves to persecution. The shrewmouse, also mentioned by Mr. Darwin in the Appendix, which "continually betrays itself by screaming out when approached," is another, and perhaps a better instance. The instincts of rabbits with regard to the attacks of weasels appear to me to be imperfect, or not completely formed. For, as I observe in "Animal Intelligence" (p. 359), I have witnessed the mode of capture practised by weasels in the open field, and it consists merely in the rabbit "toddling along, with the weasel toddling behind, until tamely allowing itself to be overtaken . . . There seems to have been here a remarkable failure of natural selection in doing duty to the instincts of these swift-footed animals"—a failure, however, which time would doubtless remedy, if weasels were sufficiently numerous in relation to the breeding power of the rabbit to give natural selection the opportunity of perfecting the instinct of escape from this particular enemy.

Many other instances of the imperfection of instinct might be quoted, but enough have now been given to render unquestionable the only point with which we are concerned, viz., that although well established instincts are, as a rule, adjusted with astonishing nicety to certain definite and frequently recurring circumstances, the adjustment is made *only* with reference to these, so that a very small variation in them is sufficient to lead the instinct astray. It is also of interest here to note what seems to be a complementary truth, viz., that small variations taking place in the organism itself when not in normal converse with its environment, are sufficient to throw the delicate mechanism of instinct out of gear when this is afterwards brought into such converse. This fact, for instance, is familiar enough in the case of tamed animals (which when again "turned down" in their native haunts are not at first at home in them), but is brought out in a much more striking manner by the experiments of Mr. Spalding. Thus he says:—

"Before passing to the theory of instinct, it may be

worthy of remark that, unlooked for, I met with in the course of experiments some very suggestive, but not yet sufficiently observed, phenomena; which, however, have led me to the opinion that not only do the animals learn, but they can also forget—and very soon—that which they never practised. Further, it would seem that any early interference with the established course of their lives may completely derange their mental constitution, and give rise to an order of manifestations, perhaps totally and unaccountably different from what would have appeared under normal conditions. Hence I am inclined to think that students of animal psychology should endeavour to observe the unfolding of the powers of their subjects in as nearly as possible the ordinary circumstances of their lives. And perhaps it may be because they have not all been sufficiently on their guard in this matter, that some experiments have seemed to tell against the reality of instinct. Without attempting to prove the above propositions, one or two facts may be mentioned. Untaught, the new-born babe can suck—a reflex action; and Mr. Herbert Spencer describes all instinct as 'compound reflex action;' but it seems to be well known that if spoon-fed, and not put to the breast, it soon loses the power of drawing milk. Similarly, a chicken that has not heard the call of the mother until eight or ten days old then hears it as if it heard it not. I regret to find that on this point my notes are not so full as I could wish, or as they might have been. There is, however, an account of one chicken that could not be returned to the mother when (? until) ten days old. The hen followed it, and tried to entice it in every way; still it continually left her and ran to the house or to any person of whom it caught sight. This it persisted in doing, though beaten back with a small branch dozens of times, and indeed cruelly maltreated. It was also placed under the mother at night, but it again left her in the morning. Something more curious, and of a different kind, came to light in the case of three chickens that I kept hooded until nearly four days old—a longer time than any I have yet spoken of. Each of these on being unhooded evinced the greatest terror of me, dashing off in the opposite direction whenever I sought to approach it. The table on which they were unhooded stood before a window, and each in its turn beat against the glass like a wild bird. One of them darted behind some books, and, squeezing itself into a corner, remained cowering for a length of time. We might guess at the meaning of this strange and exceptional wildness; but the odd fact is enough for my present purpose. Whatever might have been the meaning of this marked change in their mental constitution—had they been unhooded on the previous day they would have run to me instead of from me—it could not have been the effect of experience; it must have resulted wholly from changes in their own organization."

Subsequently Mr. Spalding tried the experiment of keeping young ducklings away from the water for several days after they were hatched; on then bringing them to a pond they showed as much dislike to the water as young chickens would have done. (See Lewes, article *Instinct*, "Problems of Life and Mind.")

The change produced in the instincts of male animals by castration may also be mentioned in the present connection, and particularly the tendency which is thus induced among cock birds to adopt the incubating and other habits of the hen. I quote the following from a recently published article by Dr. J. W. Stroud of Port Elizabeth, who has devoted a good deal of attention to the subject of caponizing:—

"Aristotle, more than two thousand years ago, tells us of a cock that performed all the duties of a hen. ('Hist. An. Lib.' ix, 42.) Pliny, too, speaks of the motherly care bestowed by a cock on chickens. 'He did everything for them,' says he, 'like to the very hen that hatched them, and ceased to crow.' ('Pliny Trans.' i, 299.) Albertus Magnus witnessed the same thing; and Ælian ('Hist.' iv, 29) mentions a cock which on the death of the hen while hatching, took to the eggs, sat on them, and brought out chicks.' Says Willoughby (in 'Ray's Willoughby's Natural History'), 'We have beheld more than once, not without pleasure and admiration, a Capon bringing up a brood of chickens, like a hen clucking over them, feeding them, and brooding them under his wings with as much care and tenderness as their dams are wont to do.' 'Once accustomed to this office,' says Baptista Rosa ('Magia Naturalis' iv, 26), 'a Capon will never abandon it, but when one brood is grown up another batch of newly hatched chickens may be put to him and he will be as kind to them and take as much care of them as of the first, and so in succession.' Réaumur ('Art de Faire Eclore,' tom. ii, p. 8) bears testimony to similar facts and also to the propensity of Capons to sit. (See also 'Cottage Gardener,' 1860, p. 379."*)

In this connection I may also quote the following instance, which I find recorded among Mr. Darwin's MS notes:—

"April, 1862. We had a kitten which sucked its mother, and, when a month old, taken to ——— and sucked another cat; then to ——— and sucked two other cats, and then its instinct was confounded, and became mixed with reason or experience: for it tried repeatedly to suck three or four other kittens of its own age, which no one, as far as I am aware, ever saw any other kitten do. Thus born instinct may be modified by experience."

In his "Naturgeschichte der Säugethiere von Paraguay," p. 201, Dr. Reugger gives the following curious instance of interference with natural instincts brought about by changed conditions of individual life. Speaking of a kind of Cat, native in Paraguay, he says that there is no instance on record of the animal breeding when in captivity, and that on one occasion a female having been pregnant when captured and kept in confinement by Herr Nozeda, brought forth her young, but immediately afterwards devoured them. This, which took place in her own country, shows that even so well rooted an instinct as the maternal may be greatly altered in the individual by even a few months of change in the conditions of life. Similar facts in the case of the domestic Sow, pet Mice, and other animals exposed to the influence of domestication are, of course, very common.

It is needless, I think, to give further instances to prove the general principle that derangement of instinctive organization is apt to arise when an animal ceases to be in normal converse with its environment. But I may here adduce a curious instance of the derangement of the instinctive organization in an animal which was apparently in all respects in normal converse with its environment, and this to such an extent that it may properly be regarded as a case of insanity. But although perhaps pathological in nature, it is none the less available as showing the imperfection of instinct—the only difference between it and the cases previously

* *Ostronization, or the Caponizing of the Ostrich* (S. Breutnall, Port Elizabeth, 1883).

cited consisting in the changing causes being internal instead of external. The case was communicated to me by a lady, who, from its peculiar nature, desires me to withhold her name; but I quote the account in her own words:—

"A white fantail pigeon lived with his family in a pigeon-house in our stable-yard. He and his wife had been brought originally from Sussex, and had lived, respected and admired, to see their children of the third generation, when he suddenly became the victim of the infatuation I am about to describe. . . .

"No eccentricity whatever was remarked in his conduct until one day I chanced to pick up somewhere in the garden a ginger-beer bottle of the ordinary brown stone description. I flung it into the yard, where it fell immediately below the pigeon-house. That instant down flew paterfamilias, and to my no small astonishment commenced a series of genuflexions, evidently doing homage to the bottle. He strutted round and round it, bowing and scraping and cooing and performing the most ludicrous antics I ever beheld on the part of an enamoured pigeon. . . . Nor did he cease these performances until we removed the bottle; and, which proved that this singular aberration of instinct had become a fixed delusion, whenever the bottle was thrown or placed in the yard—no matter whether it lay horizontally or was placed upright—the same ridiculous scene was enacted; at that moment the pigeon came flying down with quite as great alacrity as when his peas were thrown out for his dinner, to continue his antics as long as the bottle remained there. Sometimes this would go on for hours, the other members of his family treating his movements with the most contemptuous indifference, and taking no notice whatever of the bottle. At last it became the regular amusement with which we entertained our visitors to see this erratic pigeon making love to the interesting object of his affections, and it was an entertainment which never failed, throughout that summer at least. Before next summer came round he was no more."

It is thus evident that the pigeon was affected with some strong and persistent monomania with regard to this particular object. Although it is well known that insanity is not an uncommon thing among animals, this is the only case I have met with of a conspicuous derangement of the instinctive as distinguished from the rational faculties—unless, indeed, we so regard the exhibitions of erotomania, infanticidal mania, &c., which occur in animals perhaps more frequently than they do in man.

But with reference to the imperfection of instinct, we have now some more important matters to consider than the mere enumeration of cases in which instinct may have been observed at fault. Let it first be observed that under the general heading "Imperfection of Instinct," we may include two very different classes of phenomena; for instincts may be imperfect because they have not yet been completely developed, or they may appear to be imperfect because not completely answering to some change in those circumstances of life with reference to which they have been fully developed. Now, if instincts have been developed at all, it is obvious that they must have passed through various stages of imperfection before they attained to perfection, and therefore we might expect to meet with some cases of instinct not yet perfected—cases, be it observed, which differ from those already mentioned, in that their faultiness arises, not from a novelty of experience with reference to which the instinct has not been developed, but from the fact of the instinct not being yet fully formed; and this ought more especially to be the case with instincts the perfection of which is not of vital importance to the species; for such instincts would not have been so rigorously trained or perfected by natural selection. A good illustration on this head seems to be afforded by the instinct of destroying the drones as exhibited by the hive-bee. Thus, to quote from "Animal Intelligence":—"Evidently the object of this massacre is that of getting rid of useless mouths; but there is the more difficult question as to why these useless mouths ever came into existence. It has been suggested that the enormous disproportion between the present number of males and the single fertile female, refers to a time before the social instincts became so complex or consolidated, and when, therefore, bees lived in lesser communities. Probably this is the explanation, although I think we might still have expected that before this period in their evolution had arrived bees might have developed a compensating instinct, either not to allow the queen to lay so many drone eggs, or else to massacre the drones while still in the larval state. We must remember, also, that among the wasps the males do work (chiefly domestic work, for which they are fed by their foraging sisters); so it is possible that in the hive-bee the drones were originally useful members of the community, and that they have lost their primitively useful instincts. But whatever the explanation, it is very curious that here, among the animals which are justly regarded as exhibiting the highest perfection of instinct, we meet with perhaps the most flagrant instance in the animal kingdom of instinct unperfected. It is the more remarkable that the drone-killing instinct should not have been better developed in the direction of killing the drones at the most profitable time—namely, in their larval or oval state—from the fact that in many respects it seems to have been developed to a high degree of discriminative refinement."

And, to take only one other illustration, Mr. Spalding writes:—

"Another suggestive class of phenomena that fell under my notice may be described as imperfect instincts. When a week old my turkey came on a bee right in its path—the first, I believe, it had ever seen. It gave the danger chirr, stood for a few seconds with outstretched neck and marked expression of fear, then turned off in another direction. On this hint I made a vast number of experiments with chickens and bees. In the great majority of instances the chickens gave evidence of instinctive fear of these sting-bearing insects; but the results were not uniform, and perhaps the most accurate general statement I can give is, that they were uncertain, shy, and suspicious. Of course to be stung once was enough to confirm their misgivings for ever. Pretty much in the same way did they avoid ants, especially when swarming in great numbers."

Similarly, and during the life-time of the individual, Mr. Spalding found an instinct in the course of development in the case already quoted of the turkeys catching flies. And precisely analogous facts may be noticed in the developing instincts of the child. Thus, for instance, the balancing of the head in an upright position may be said in man to be instinctive, for the power of doing so is first acquired about the tenth week, by constantly recurring efforts, and eventually becomes independent of intentional thought. Preyer describes the stages by which the latter, or completed, stage is reached through numberless gradations, the passage of which occupies about six weeks.* He says that the child first accidentally finds the comfort of the attitude, and so adopts it more and more constantly until through habit it becomes instinctive. He also gives exactly parallel facts in the case of learning to creep, sit, stand, walk, &c.†

Among animals in a state of nature we may, I think, regard all instincts which, so far as we can see, are trivial or useless, as instincts which are imperfect, in that they do not answer to any apparent needs in the animals' present conditions of life. Such instincts are not very numerous, and, as Mr. Darwin observes in the Appendix, they may be quoted as objections to his theory of the development of instinct by natural selection. I shall subsequently consider this difficulty, but here I have only to note the fact that instincts of this apparently purposeless kind occur, and that, *quâ* purposeless, they are imperfect. Such, for instance, is the instinct of a hen cackling when she has laid an egg, the cock-pheasant crowing when going to roost, cattle and elephants goring their sick or wounded companions, sundry instincts connected with excrements—such as burying them in earth, always depositing them in the same place, &c.—and other cases mentioned by Mr. Darwin in the Appendix.

But the most important class of considerations for us is one to which the foregoing may be said to lead up. We have seen that if instincts have been developed by evolution, we should expect to find cases in which they are in process of evolution, or not yet perfect; and we have also seen that this expectation is realized. Now in so far as instinct requires to be mixed with intelligence in order to be effective, it is as an instinct imperfect; it is as an instinct in course of formation, or at any rate not perfectly adapted to the possible circumstances of life. Therefore all cases of the education of instinct by intelligence—whether in the individual or the race—fall to be considered in the present connection. The consideration of this subject, however, lands us directly in a larger and deeper topic as to the origin and development of instinct in general. To this topic, therefore, we shall next address ourselves.

* *Die Seele des Kindes*, Leipzig, 1882, pp. 166–7.
† *Ibid.*, pp. 167–75.

CHAPTER XII.

INSTINCT *(continued)*.

ORIGIN AND DEVELOPMENT OF INSTINCTS.

INSTINCTS probably owe their origin and development to one or other of two principles.

I. The first mode of origin "consists in natural selection, or survival of the fittest, continuously preserving actions which, although never intelligent, yet happen to have been of benefit to the animals which first chanced to perform them. Thus, for instance, take the instinct of incubation. It is quite impossible that any animal can ever have kept its eggs warm with the intelligent purpose of hatching out their contents, so we can only suppose that the incubating instinct began by warm-blooded animals showing that kind of attention to their eggs which we find to be frequently shown by cold-blooded animals. Thus crabs and spiders carry about their eggs for the purpose of protecting them; and if, as animals gradually became warm-blooded, some species, for this or for any other purpose, adopted a similar habit, the imparting of heat would have become incidental to the carrying about of the eggs. Consequently, as the imparting of heat promoted the process of hatching, those individuals which most constantly cuddled or brooded over their eggs would, other things equal, have been most successful in rearing progeny; and so the incubating instinct would be developed without there ever having been any intelligence in the matter."*

II. The second mode of origin is as follows:—By the effects of habit in successive generations, actions which were originally intelligent become, as it were, stereotyped into per-

* Quoted from my own article on "Instinct," in the *Encyclopædia Britannica.*

manent instincts. Just as in the life-time of the individual adjustive actions which were originally intelligent may by frequent repetition become automatic, so in the life-time of the species actions originally intelligent may, by frequent repetition and heredity, so write their effects on the nervous system that the latter is prepared, even before individual experience, to perform adjustive actions mechanically which in previous generations were performed intelligently. This mode of origin of instincts has been appropriately called the "lapsing of intelligence."*

For the sake of subsequent reference, I shall allude to instincts which arise by way of natural selection, without the intervention of intelligence, as Primary Instincts, and to those which are formed by the lapsing of intelligence as Secondary Instincts.

Let us now consider the reasons which *à priori* lead us to assign the probable origin of instincts to these principles. Taking first the case of primary instincts, these reasons may be briefly rendered thus:—

(1.) Many instinctive actions are performed by animals too low in the scale to admit of our supposing that the adjustments which are now instinctive can ever have been intelligent. (2.) Among the higher animals instinctive actions are performed at an age before intelligence, or power of learning by individual experience, has begun to assert itself. (3.) Considering the great importance of instincts to species, we are prepared to expect that they must be in large part subject to the influence of natural selection. As Mr. Darwin observes, "it will be universally admitted that instincts are as important as corporeal structures for the welfare of each species under its present conditions of life. Under changed conditions of life it is at least possible that slight modifications of instinct might be profitable to a species; and if it can be shown that instincts do vary ever so little, then I can see no difficulty in natural selection preserving and continually accumulating variations of instinct to any extent that was profitable."

That instincts may arise by way of lapsed intelligence is rendered probable *à priori* by all the facts which show the resemblance between instincts and intelligent habits. To take only a few of these facts for the present purpose, I

* By Lewes, see *Problems of Life and Mind.*

cannot do better than confine myself to making a quotation from Mr. Darwin's MSS; for this will show how deep-seated and detailed is the resemblance between habit and instinct.

"In repeating anything by heart, or in playing a tune, every one feels that, if interrupted, it is easy to back a little, but very difficult suddenly to resume the thread of thought or action a few steps in advance. Now P. Huber has described a caterpillar which makes by a succession of processes a very complicated hammock for its metamorphosis; and he found that if he took a caterpillar which had completed its hammock up to, say the sixth stage of construction, and put it into a hammock completed up only to the third stage, the caterpillar did not seem puzzled, but repeated the fourth, fifth, and sixth stages of construction. If, however, a caterpillar was taken out of a hammock made up, for instance, to the third stage, and put into one finished to the ninth stage, so that much of its work was done for it, far from feeling the benefit of this, it was much embarrassed, and even forced to go over the already finished work, starting from the third stage which it had left off before it could complete its hammock. So, again, the hive-bee in the construction of its comb seems compelled to follow an invariable order of work. M. Fabre gives another curious instance how one instinctive action invariably follows another. A Sphex makes a burrow, flies away and seeks for prey, which it brings, paralyzed by having been stung, to the mouth of its burrow; but always enters to see that all is right within before dragging in its prey; whilst the Sphex was within its burrow, M. Fabre removed the prey to a short distance; when the Sphex came out it soon found the prey and brought it again to the mouth of the burrow; but then came the instinctive necessity of reconnoitering the just reconnoitered burrow; and as often as M. Fabre removed the prey, so often was all this gone over again, so that the unfortunate Sphex reconnoitered its burrow forty times successively! When M. Fabre altogether removed the prey, the Sphex, instead of searching for fresh prey and then making use of its completed burrow, felt itself under the necessity of following the rhythm of its instinct, and before making a new burrow, completely closed up the old one as if it were all right, although in fact utterly useless as containing no prey for its larva.*

* *Anns. des Sci. Nat.*, 4 ser., tome vi, p. 148. With respect to Bees, see

"In another way we perhaps see the relation of habit and instinct, namely in the latter acquiring great force if practised only once or twice for a short time; thus it is asserted that if a calf or infant has never sucked its mother, it is very much easier to bring it up by hand than if it has sucked only once.* So again Kirby† states that larva, after having 'fed for a time on one plant, will die rather than eat another, which would have been perfectly acceptable to them if accustomed to it from the first.'"

Such, then, are some of the *à priori* reasons for believing that instincts must have arisen from one or other of these two sources—natural selection or lapsing intelligence; it now remains to prove, *à posteriori*, that they have so arisen. I may first give a brief sketch of how this proof ought to proceed.

The proof, then, that instincts have had a primary mode of origin requires to show:—

I. That non-intelligent habits of a non-adaptive character occur in individuals.
II. That such habits may be inherited.
III. That such habits may vary.
IV. That when they vary the variations may be inherited.
V. That if such variations are inherited, we are justified in assuming, in view of all that we know concerning the analogous case of structures, that they may be fixed and intensified in beneficial lines by natural selection.

The proof that instincts have had a secondary mode of origin requires to show:—

VI. That intelligent adjustments when frequently performed by the individual become automatic, either to the extent of not requiring conscious thought at all, or, as consciously adjustive habits, not requiring the same degree of conscious effort as at first.
VII. That automatic actions and conscious habits may be inherited.

Primary Instincts.

Proceeding, then, to consider these sundry heads of proof, it is easy to establish Proposition I, inasmuch as the fact

Kirby and Spence, *Entomology*, vol. i, p. 497. For the hammock caterpillar, see *Mem. Soc. Phys. de Genève*, tome vii, p. 154.

* *Zoonomia*, p. 140. † *Intro. to Entomol.*, vol. i, p. 391.

which it states is a matter of daily observation. "Tricks of manner," indeed, are of such frequent occurrence in the nursery and schoolroom, that it usually entails no small labour on the part of elders to eradicate them, and when not eradicated in childhood they are apt to continue through life, unless afterwards conquered by the efforts of the individual himself. But in cases where the trick of manner is not obnoxious, or sufficiently unusual to call for checking, it is allowed to persist, and thus it is that almost every one presents certain slight peculiarities of movement which we recognize as characteristic.*

Such peculiarities of movement as we meet with them in ordinary life are slightly marked; but their significance in relation to instinct has been obtruded on my notice by observing them in the much more striking form in which they are presented by idiots. This is a class of persons which, as we shall find in my next work, is of peculiar interest in relation to mental evolution, because in them we have a human mind arrested in its development as well as deflected in its growth—therefore in many cases supplying to the comparative psychologist very suggestive material for study. Now one of the things which must most strike any one on first visiting an idiot asylum, is the extraordinary character and variety of the meaningless tricks of manner which are everywhere being displayed around him. These tricks, often ludicrous, sometimes painful, but usually meaningless, are always individual and wonderfully persistent. Generally speaking, the lower the idiot in the scale of idiotcy, the more pronounced is this peculiarity; so that if one sees a patient moving to and fro continually, or otherwise exhibiting "rhythmical movements," one may be pretty sure that the case is a bad one. But even among the higher idiots and "feeble-minded," strange and habitual movements of the hands, limbs, or features are exceedingly common.

Among animals similar facts are to be noticed. Scarcely any two sporting dogs "point" in exactly the same manner,

* Dr. Carpenter says (*Mental Physiology*, p. 373), "What particular 'trick' each individual may learn, depends very much upon accident. Thus, in the old times of dependent watch-chains and massive bunches of seals, these were the readiest playthings," &c. In view of the relation which such "tricks" bear to the formation of primary instincts, this remark has some importance; it shows that even aimless movements may be determined and rendered habitual by the conditions of the environment.

although every dog adheres to his particular attitude through life. Nearly all domestic animals exhibit slight but individually constant differences of movement when caressed, when they are threatened, when at play, &c. But perhaps a more striking view of this subject may be obtained by considering the sum of the neuro-muscular conditions, leading to individual peculiarities of movement, which we comprise under the term "disposition," or, if more prominent, "idiosyncracy." Thus many dogs develop the meaningless habit—which has all the strength of an incipient instinct, and in the case of the collie breed, as we shall subsequently see, inherited or innate—of barking round a carriage. Some cats take to "mousing" with avidity, while others can never be taught to care about the sport. All who keep pet birds—and indeed domestic animals of any kind—must have noticed the diversity of their dispositions in respect of play, boldness, amiability, &c.; and Mr. W. Kidd, who had a very large experience, is sure that the diversity of disposition in larks and canaries is displayed by nestlings reared from the nest.*

Almost innumerable instances might be given of individual variations in the instincts of nest-building.† Even as

* See *Gardener's Chronicle*, 1851, p. 181, which is referred to in this connection in Mr. Darwin's MSS.

† For example, the Nut-hatch usually builds in the hollow branch of a tree, plastering up the opening with clay; but Mr. Hewetson found a pair which for many years occupied a hole in a wall (*Yarrel's Birds*), and Mr. Bond describes another nest placed in the side of a hay-stack, built up with a mass of clay weighing no less than eleven pounds, and the nest measuring thirteen inches in height (*Zoologist*, 2nd ser., p. 2850). The golden-crested Wren, also, frequently exhibits variations in the structure and situation of its nest (*Hist. Brit. Birds*, 4th ed., vol. i, p. 450). The Golden Eagle builds in precipitous crags of rock; but Mr. D. E. Knox (*Antumns on the Spey*, 1872, pp. 141-3), describes a nest which he himself examined on a fir-tree, not above twenty feet from the ground. Couch says that "more than one pair of birds will sometimes unite in occupying one nest, and either rear their broods in common, or one of them will surrender the future care of them to the other (*Illustrations of Instinct*, p. 233). Mr. S. Stone, writing of the Missel-thrush says, "From what has been written, it appears plain that some individuals use clay or plaster in the construction of the nest, while others contrive to do without it, which agrees with my own observation, for although I have found nests which did not contain plaster, the greater part of those which have fallen in my way—and they have been not a few—certainly have had a plastering of some kind between the twigs and lichens outside and the fine grasses which invariably constitute the lining; this has been more especially the case when the bird has selected as a site the horizontal branches of a tree (*Field*, Jan. 8, 1861. This is a clipping which I find among Mr. Darwin's MS notes). As

low down in the psychological scale as the insects, we are not without evidence of individual variations of instinct. Thus, for instance, Forel observed great diversities of building among the *F. truncicola*—the nests being sometimes domed, sometimes made under stones, and sometimes excavated in the wood of old trees. Likewise, Büchner observes, "one ant will let herself be killed rather than let go the pupa which she holds, while another will let them fall and run away like a coward," and similar statements are made by Moggridge.

But as showing strongly marked individual differences of disposition in animals, and also that such differences may lead to useless or capricious actions having all the strength of incipient instincts, I think a good class of cases to select are those in which one animal conceives a strong though senseless attachment to another animal of a different species. Thus, for instance, I once found a wounded widgeon on the shore, and took it home to my poultry yard. After a time its wounds healed, and I then cut its wings to keep it as a pet. The bird soon became perfectly tame, and then conceived a strong, persistent, and unremitting attachment to a peacock which also belonged to the establishment. Wherever the peacock went the widgeon followed like a shadow, so that during the day time the one bird was never seen without the other being in close attendance. If a separation were forcibly effected, the distress of the widgeon was very great, and she would whistle incessantly till restored to her old place waddling behind the tail of the peacock. This devoted attachment was the more remarkable from the fact that it was not in the smallest degree reciprocated by the peacock. He never paid the slightest heed to his constant companion, nor, indeed, did he seem to notice that she was always just behind him. At night he used to roost upon the gable of a cottage. The poor widgeon could not fly to accompany him, and even if she could would probably not have been able to sit upon the gable; but she always kept as near him as circumstances would permit, for as soon as he flew up to his gable she would squat herself down upon the ground just

observed in the text, such instances might be multiplied indefinitely; but as a considerable number of additional and well selected cases are given in Mr. Darwin's essay at the end of this book, it is needless for me to adduce any further illustrations.

below it—a devotion which eventually cost her her life, as she thus fell a prey to a prowling cat. Now here we have a curious case of a bird that had been wild, taking a violent fancy for the wholly useless companionship of another and very dissimilar bird; for it should be added that she chose the peacock as the object of her persistent regard out of a large number of other kinds of domestic birds which lived about the place.

Similarly, cats often like to associate with horses, and in some cases with dogs, birds, rats, and other unlikely creatures. Dogs not unfrequently make friendships with a variety of animals, and in a case recorded by F. Cuvier a terrier found so much delight in the companionship of a caged lion, that when the lion died the dog pined away and died also. Thompson gives cases in which horses have become "extremely attached to dogs and to cats, and seemed pleased to have them placed on their backs in their stalls."* Rengger mentions a monkey which was so fond of a dog that it cried with grief during the absence of its friend, caressed it on its return, and assisted it in all its quarrels with other dogs. "A peccari in the menagerie at Paris formed a strong attachment with one of the keeper's dogs, and a seal in the same place allowed a little water-dog to play with it and to take fish from its mouth, which it always resented if this were attempted by the other seals in the same tank. Dogs have lived on terms of friendship with gulls and ravens . . . . and a rat has been known to accompany his master in his walks," &c., &c.†

Colonel Montagu, in the Supplement to his "Ornithological Dictionary," p. 165, relates the following singular instance of an attachment which took place between a Chinese goose and a pointer. "The dog had killed the male bird, and had been most severely punished for the misdemeanour, and finally the dead body of his victim was tied to his neck. The solitary goose became extremely distressed for the loss of her partner and only companion; and probably having been attracted to the dog's kennel by the sight of her dead mate, she seemed determined to persecute the dog by her constant attendance and continual vociferations; but after a little time a strict friendship took place between these incongruous animals. They fed out of the same trough, lived

* Thompson, *Passions of Animals*, pp. 360–1. † *Ibid.*

under the same roof, and in the same straw bed kept each other warm; and when the dog was taken to the field, the lamentations of the goose were incessant."

The same author gives cases of attachment between a pigeon and a fowl, a terrier and a hedgehog, a horse and a pig, a horse and a hen, a cat and a mouse, a fox and harriers, an alligator and a cat, &c., all as having fallen under his own observation. (*Ibid.*, p. 162.)

It is not impossible that the so-called "domestic pets" which are kept by many species of ants* may really be useless adjuncts to the hive, capricious love of association having perhaps in these ants become by inherited habit truly instinctive. This, at any rate, must be the explanation of the fact that birds of different species will, even in a state of nature, occasionally associate, as is the case with Guinea-fowls and partridges, and, according to Yarrell, with partridges and landrails. Such unusual cases among birds in a state of nature are of special interest, because they may then properly be regarded as the beginnings of such a firmly set and truly instinctive association as that which obtains between rooks and starlings, &c.†

Enough has now been said in support of Proposition I, viz., *that non-intelligent habits of a non-adaptive character occur in individuals.* We shall next proceed to Proposition II, viz., *that such habits may be inherited.*

That this is the case with tricks of manner in man is a matter to be observed in almost every family, and was long ago pointed out by John Hunter. Mr. Darwin in his MSS gives a case which he himself observed, "and can vouch for its perfect accuracy." "A child who as early as between her fourth and fifth year, when her imagination was pleasantly excited, and at no other time, had a most peculiar trick of rapidly moving her fingers laterally with her hands placed on the side of her face; and her father had precisely the

* See *Animal Intelligence*, pp. 83–4.

† Prof. Newton, F.R.S., informs me that "bands of the Golden-crested Wren may frequently be observed in winter consorting with bands of the Coal-Titmouse, and in a less degree with those of the Long-tailed Titmouse; while parties of Redpoles and Siskins will for a time join their company, or *vice versâ*. The flocking together of Rooks and Daws is, of course, an everyday occurrence, as is also for some months the association of Starlings with them, and in many cases the combination of all with Lapwings.

same trick under the same frame of mind, and which was not quite conquered even in old age: in this instance there could not possibly have been any imitation."*

That the more frequent and more pronounced tricks of manner which are manifested by idiots are likewise inherited is highly probable; but I have no evidence on this point, as idiots in civilized countries are not allowed to propagate.

In the case of animals, however, the evidence is abundant. Thus, again to quote from Mr. Darwin's MSS, "the Rev. W. Darwin Fox tells me that he had a Skye terrier bitch which when begging rapidly moved her paws in a way very different from that of any other dog which he had ever seen; her puppy, which never could have seen her mother beg, now when full grown performs the same peculiar movement exactly in the same way."†

As regards the inheritance of disposition, we have only to look to the sundry breeds of dogs to see how marked differences of this kind may become signally distinctive of different breeds. It will be remembered that at present we are only concerned with the inheritance of useless, unintelligent, or non-adaptive habits, and therefore have here nothing to do with the useful and intelligent habits which are bred into our various races of dogs by means of artificial selection combined with training. But even in the case of purely meaningless traits of character, which are of no use either to the animals themselves or to man, we find the influences of heredity at work. Thus, for instance, the useless and even annoying habit of barking round a carriage, which occurs among sundry breeds of dogs, is particularly pronounced in the collie, and is truly innate or not dependent on imitation. This is shown by the fact that collies which from puppyhood have never seen other dogs bark at horses, will nevertheless spontaneously begin to do so.‡ Several other useless traits of character or disposition peculiar to different breeds might be mentioned; but I shall pass on to the most remarkable instance

* This case is stated in different words in *Variation of Animals and Plants*, &c., vol. i, pp. 450–1.

† Here, however, I may remark that I have noticed several Syke terriers perform these movements while begging, so that the action seems to be due to some race-distinction of a psychological kind, and not merely to an individual peculiarity. It therefore leads on to the class of cases next considered in the text.

‡ See *Nature*, vol. xix, p. 234.

I have met with in dogs of the inheritance of a thoroughly senseless psychological peculiarity. I refer to the instance which was communicated some years ago to Mr. Darwin by Dr. Huggins, F.R.S., and which I shall quote in his own words.

"I wish to communicate to you a curious case of an inherited mental peculiarity. I possess an English mastiff, by name Kepler, a son of the celebrated Turk out of Venus. I brought the dog, when six weeks old, from the stable in which he was born. The first time I took him out he started back in alarm at the first butcher's shop he had ever seen. I soon found he had a violent antipathy to butchers and butchers' shops. When six months old a servant took him with her on an errand. At a short distance before coming to the house she had to pass a butcher's shop; the dog threw himself down (being led with a string), neither coaxing or threats would make him pass the shop. The dog was too heavy to be carried, and as a crowd collected, the servant had to return with the dog more than a mile, and then go without him. This occurred about two years ago. The antipathy still continues, but the dog will pass nearer to a shop than he formerly would. About two months ago, in a little book on dogs, published by Dean, I discovered that the same strange antipathy is shown by the father, Turk. I then wrote to Mr. Nicholls, the former owner of Turk, to ask him for any information he might have on the point. He replied, 'I can say that the same antipathy exists in King, the sire of Turk, in Turk, in Punch (son of Turk out of Meg), and in Paris (son of Turk out of Juno). Paris has the greatest antipathy, as he would hardly go into a street where a butcher's shop is, and would run away after passing it. When a cart with a butcher's man came into the place where the dogs were kept, although they could not see him, they all were ready to break their chains. A master-butcher, dressed privately, called one evening on Paris' master to see the dog. He had hardly entered the house before the dog (though shut in) was so much excited that he had to be put into a shed, and the butcher was forced to leave without seeing the dog. The same dog at Hastings made a spring at a gentleman who came into the hotel. The owner caught the dog and apologised, and said he never knew him to do so before, except when a butcher came to his house. The gentleman at once said that was his business.'"

We see, then, that non-intelligent habits of non-adaptive or useless character may be strongly inherited by domestic animals. As showing that the same is true of breeds or strains in wholly wild animals, I may quote Humboldt, who says,* that the Indians who catch monkeys to sell them "knew very well that they can easily succeed in taming those which inhabit certain islands; while monkeys of the same species, caught in the neighbouring continent, die of terror or rage when they find themselves in the power of man:" and in his MSS I find that Mr. Darwin has a note saying, "divers dispositions seem to run in families of crocodiles." But one of the most curious instances that I have met with of the commencement of a racial and useless deviation from a strong ancestral instinct, is one which is communicated to Mr. Darwin in a letter from Mr. Thwaits, who writes from Ceylon under the date 1860, and whose letter I find among Mr. Darwin's MSS. Mr. Thwaits here says that his domestic ducks quite lost their natural instincts with regard to water, which they never enter unless driven. The young birds, when forcibly placed in a tub of water are "quite alarmed," and have to be quickly taken out again "or they would drown in their struggling." Mr. Thwaits adds that this peculiarity does not extend to all the ducks in the island, but only occurs in one particular breed or strain.

In Mr. Darwin's MSS I also find the following remarks: "So many independent authors have stated that horses in different parts of the world inherit artificial paces, that I think the fact cannot be doubted. Dureau de la Malle asserts that these different paces have been acquired since the time of the Roman classics, and that from his own observation they are inherited.† . . . . Tumbler pigeons offer an excellent instance of an instinctive action, acquired under domestication, which could not have been taught, but must have appeared naturally, though probably afterwards vastly improved by the continued selection of those birds which showed the strongest propensity—more especially in

* *Personal Narrative*, vol. iii, p. 383.

† After giving numerous references on this point in a footnote, Mr. Darwin concludes the latter thus:—"I may add that I was formerly struck by no horse on the grassy plains of La Plata having the natural high action of some English horses." For a number of other instances of hereditary transmission of qualities in the case of the Horse, see *Variation of Animals and Plants*, &c., vol. i, pp. 454–6.

ancient times in the East, when flying pigeons was much esteemed. Tumblers have the habit of flying in a close flock to a great height, and as they rise tumbling head over tail. I have bred and flown young birds, which could not possibly have ever seen a tumbler; after a few attempts even they tumbled in the air. Imitation, however, aids the instinct, for all fanciers are agreed that it is highly desirable to fly young birds with first-rate old ones. Still more remarkable are the habits of the Indian sub-breed of tumblers, on which I have given details in a former chapter, showing that during at least the last 250 years these birds have been known to tumble on the ground, after being slightly shaken, and to continue tumbling until taken up and blown upon. As this breed has gone on so long, the habit can hardly be called a disease. I need scarcely remark that it would be as impossible to *teach* one kind of pigeon to tumble as to *teach* another kind to inflate its crop to the enormous size which the pouter pigeon habitually does."*

This case of the tumblers and pouters is singularly interesting and very apposite to the proposition before us, for not only are the actions utterly useless to the animals themselves, but they have now become so ingrained into their psychology as to have become severally distinctive of different breeds, and so not distinguishable from true instincts. This extension of an hereditary and useless habit into a distinction of race or type is most important in the present connection. If these cases stood alone they would be enough to show that useless habits may become hereditary, and this to an extent which renders them indistinguishable from true instincts.†

In the Appendix several instructive cases of the same kind will be found, such as that of the Abyssinian pigeon, which, when fired at, "plunges down so as almost to touch the sportsman, and then mounts to an immoderate height;"‡ the biscacha, which "almost invariably collects all sorts of

* For further particulars on the instinct of tumbling, see *Variation of Animals and Plants*, vol. i, p. 219, and 230.

† Some years ago the Ratels which were confined in one cage at the Zoological Gardens acquired the apparently useless habit of perpetually tumbling head over heels. If their progeny were to be exposed for a number of generations to similar conditions of life, they would probably develope a true instinct of turning somersaults analogous to that of the tumbler-pigeon.

‡ I have frequently noticed a similar propensity in the Lapwing.

rubbish, bones, stones, dry dung, &c., near its burrow;" the guanacoes which "have the habit of returning (like flies) to the same spot to drop their excrement;" horses, dogs, and the hyrax, showing a somewhat similar and equally useless propensity; hens cackling over their eggs, &c., &c. So that I think the evidence is abundant in support of the proposition that senseless or useless habits may be inherited, and thus become racial characteristics, or purposeless instincts.

Passing on, then, to Propositions III and IV,—viz., *that such habits may vary, and that when they vary the variations may be inherited*—the truth of these facts has already been made apparent. The paces of the horse in different parts of the world are so many race-characteristics of the animals; the ground-tumblers display an inherited variation as compared with the air-tumblers, and if tumblers are not allowed to exercise their art, it undergoes the variation of becoming obliterated—just as we shall presently see is the case with many true instincts. The different dispositions of the same species of monkeys on different islands, prove that the ancestral disposition must have varied in the progeny, and have then continued to be inherited in its varied states along the several lines of descendants.

From the exclusive nature of the requirement, it is not easy to find many examples of inherited varieties of useless habits, nor is it important that I should give a number of illustrations on this head. There is abundant evidence that non-intelligent and purposeless habits are inherited, and this is the main point; for that such habits, when inherited, should vary, is a matter of certainty, seeing, as we presently shall, that such is the case even with intelligent and useful habits. If the latter are liable to vary in their course of inheritance, *à fortiori* the former must be similarly liable, inasmuch as they arise in a manner analogous to fortuitous "sports" of structure (which are always eminently variable), and afterwards have no check imposed on their variability either by intelligence or by selection.

Similarly Proposition V requires very little to be said in the way of proof. If among a number of meaningless habits, all more or less hereditary and more or less variable, any one should happen from the first to be, or afterwards to vary so as to become accidentally beneficial to the animal, then we are bound to believe that natural selection would fix this habit, or its beneficial variations. And the proof that such a process has taken place is given by the fact of their being many instincts — such as the incubating instinct before alluded to—which cannot conceivably have been developed in any other way. Whether or not this instinct began in habits adapted to the protection of the eggs, it is certain that it cannot have begun with any intelligent reference to hatching them; and it is no less certain that before the instinct attained its present degree of perfection, it must have passed through many stages of variation, few if any of which can have been due to intelligent purpose on the part of the birds. And further proof is rendered, as I have also previously observed, by the fact that many instincts are displayed by animals too low in the zoological scale to admit of our supposing that they can ever have been due to intelligence. To give only one illustration, the larva of the caddice fly lives in water and constructs for itself a tubular case made of various particles glued together. If during its construction this case is found to be getting too heavy—*i.e.*, its specific gravity greater than that of the water—a piece of leaf or straw is selected from the bottom of the stream to be added to the structure; and conversely, if the latter is found to be getting too light, so as to show a tendency to float, a small stone is morticed in to serve as ballast.* In such a case as this it seems impossible that an animal so low in the zoological scale can ever have consciously reasoned—even in the most concrete way—that some particles have a higher specific gravity than others, and that by adding a particle of this or that substance, the specific gravity of the whole structure may be adjusted to that of the water. Yet the actions involved are no less clearly something more than reflex; they are instinctive, and can only have been evolved by natural selection. Similarly, Professor Duncan suggests, in a lecture before the British Association, 1872, that the instinct of the Odynerus—which forms a tubular ante-chamber and provision-chamber filled with stung grubs for the future use of offspring which it never saw—probably arose in this way. M. Fabre has observed that *Bembex indica* lays an egg in a chamber,

* *A Monographic Revision and Synopsis of the Trichoptera of the European Fauna*, 1881, by Robert M'Lachlan, F.R.S.

and that the egg hatches very shortly. The insect then visits its living offspring every day, bringing it small larvæ stung to keep them quiet. Now this instinct may have been altered in Odynerus by a delay arising in the time of hatching, and a series of victims having been therefore placed in the provision-chamber in obedience to the primitive instinct, which has thus become modified into a new one.

Numerous other instincts will be found mentioned in the Appendix, the origin of which can only be attributed to the uncompounded influence of natural selection. I feel, therefore, that it is needless for me to adduce further illustrations, and so shall here conclude my observations on instincts of the primary class.

Secondary Instincts.

Coming now to the second series of propositions, we shall find that their proof casts a good deal of reflected light upon those which we have just considered—light which tends still further to demonstrate the latter.

First, then, we have to show that "*intelligent adjustments, when frequently performed by the individual, become automatic, either to the extent of not requiring conscious thought at all, or, as consciously adjustive habits, not requiring the same degree of conscious effort as at first.*

The latter part of this proposition has already been proved in an earlier chapter of this book. That "practice makes perfect" is a matter, as I have previously said, of daily observation. Whether we regard a juggler, a pianist, or a billiard player, a child learning his lesson, or an actor his part by frequently repeating it, or any one of a thousand other illustrations of the same process, we see at once that there is truth in the cynical definition of a man as "a bundle of habits." And the same, of course, is true of animals. "Training" an animal is essentially the same process as educating a child, and, as we shall presently have occasion to show, animals in a state of nature develop special habits in relation to local needs.

The extent to which habit or repetition may thus serve to supersede conscious effort is a favourite theme among psychologists; and one or two instances have already been given in the chapter on the Physical Basis of Mind. To this point, therefore, I need not recur.

It remains to mention another class of acquired mental habits, and one which is still more suggestive in relation to instinct, inasmuch as the habits are purely mental, and not associated with mechanically distinctive movements. Thus, as Professor Alison remarks,* the sense of modesty in man is not a true instinct, because it is neither innate nor is it exhibited by all the members of the species—being, in fact, only displayed by the civilized races. Yet, although merely a taught habit of mind, among morally cultured persons it is in strength and precision indistinguishable from a true instinct. Similarly, though in a lesser degree, the influences of refinement and good taste, operating upon the individual from childhood, produce such a powerful and unremitting influence, that the extreme nicety, spontaneity, and readiness of adjustment to highly complex conditions which result are recognized even in ordinary conversation as akin to the promptings of instinct; for we commonly say that a man has "the instincts of a gentleman," or that so and so is "underbred." This latter term, however, introduces us to the division of our subject which we have to consider under the next heading—namely, the extent to which habits of mind, intentionally or intelligently acquired by the individual, may be transmitted to progeny. To this branch of our discussion, therefore, we shall now pass.†

Accepting, then, Proposition VI as beyond dispute, we have here to substantiate Proposition VII, viz., *That automatic actions and conscious habits may be inherited.*

Now we have already seen that this is certainly the case

* Article "Instinct," *Todd's Cyclo. of Anat.*, vol. iii, 1839.

† Mr. Darwin's MS points out that persons of weak intellect are very apt to fall into habitual or automatic actions, and these, from not being performed under the mandates of the will, are more nearly allied to reflex actions than are properly voluntary or deliberate movements. This correlation is also to be observed in animals, and the MS gives a case which Mr. Darwin observed of an idiotic dog, whose instinct of turning round before lying down (a remnant, probably, of the instinct of forming a bed in long grass) was so strongly developed, or so little checked by intelligence, "that he has been counted to turn round twenty times before lying down."

This action of turning round may certainly be regarded as the survival of a secondary instinct. Now secondary instincts are formed by a descent from intelligent action, through habitual action, towards reflex action; therefore it is interesting that when, as in such a case as this, they are fully formed as instincts, they are found to resemble automatic habits in showing most unrestricted play when intelligence is enfeebled or idiotic.

with automatic actions which have arisen accidentally, or without intelligent purpose; and it would be anomalous were the fact otherwise with automatic actions which have been acquired consciously. The evidence that the fact is not otherwise is considerable.

First we may take the case of man. "On what a curious combination of corporeal structure, mental character, and training," says Mr. Darwin, "must hand-writing depend! Yet every one must have noted the occasional close similarity of the hand-writing in father and son, although the father had not taught the son . . . . Hofacker, in Germany, remarks on the inheritance of hand-writing; and it has been even asserted that English boys, when taught to write in France, naturally cling to their English manner of writing." Dr. Carpenter says he is "assured by Miss Cobbe that in her family a very characteristic type of hand-writing is traceable through *five* generations;" and in his own family there occurred a curious case of a gentleman who inherited a "constitutional" character of hand-writing, and lost his right arm by an accident; "in the course of a few months he learnt to write with his left hand, and before long the hand-writing of the letters thus written came to be indistinguishable from that of his former letters." This case reminds me of a fact which I have frequently observed—and which has doubtless been observed by others—viz., that if I write in any unusual direction (as, for instance, on the perpendicular face of a recording cylinder), the hand-writing is unaltered in character, although both the hand and the eye are working in a most unusual manner; so strong is the *mental* element in hand-writing. Similarly, as observed in a previous chapter, if one takes a pencil in each hand and writes the same word with both hands simultaneously—the left hand writing from right to left—on holding the backward written word before a mirror, the hand-writing may at once be recognized.

Many other instances might be given of the force of inheritance in the mental acquisitions of man.* But turning

* See Carpenter, *Mental Physiology*, pp. 393-4, where he discusses and gives cases of hereditary aptitude for music and painting. Also Galton's *Hereditary Genius*, for high mental qualities running in families, either in the same or in analogous lines of activity; and Spencer (*Psychology*, i, p. 422) for race-characteristics of psychology in man. The effects of "good breeding" or "blood" in bequeathing hereditary disposition and refinement have already

now to the more important case of animals, I shall give only a few examples among almost any number that I could quote. Thus, in Norway, the ponies are used without bridles, and are trained to obey the voice; as a consequence a race-peculiarity has been established, for Andrew Knight says "the horse breakers complain, and certainly with very good reason, that it is impossible to give them what is called a mouth; they are nevertheless exceedingly docile, and more than ordinarily obedient, when they understand the commands of their masters."* Again, Mr. Lawson Tait tells me that he had a cat which was taught to beg for food like a terrier, so that she developed the habit of assuming this posture—so very unusual in a cat—whenever she desired to be fed. All her kittens adopted the same habit under circumstances which precluded the possibility of imitation; for they were given away to friends very early in life, and greatly surprised their new owners when, several weeks afterwards, they began spontaneously to beg.†

In order to show that the same principles apply to animals in a state of nature, it will be enough to adduce the one instance of hereditary wildness and tameness, for this instance affords evidence of the most conclusive kind. Wildness or tameness simply means a certain group of ideas or disposition, having the character of an instinct, so that we may properly speak of a wild animal as "instinctively afraid" of man or other enemy, and of a tame one as instinctively the reverse. Indeed, one of the most typical and remarkable illustrations of instinct that could be given is that of the inborn dread of enemies, as exhibited, for instance, by chickens at the sight of a hawk, by horses at the smell of a wolf, by monkeys at the appearance of a snake, &c. Now, fortunately, there is material for amply proving both that these instincts may be lost by disuse, and, conversely, that they may be acquired as instincts by the hereditary transmission of ancestral experience.

been alluded to, and I think observation will show that the same applies to the sense of modesty.

* *Phil. Trans.*, 1839, p. 369.

† Inasmuch as the action of "begging" is so unusual in the Cat, the above case of its hereditary transmission is more remarkable than the similar cases which occur in the Dog; see Lewes, *Problems of Life and Mind*, vol. i, p. 229, and Fiske, *Cosmic Philosophy*, vol. ii, p. 150. and more especially a case recorded by Mr. L. Hurt, in *Nature* (Aug. 1, 1872) of a Skye terrier

The proof that instinctive wildness natural to a species may be lost by disuse is strikingly rendered by the case of rabbits. As Mr. Darwin remarks, "hardly any animal is more difficult to tame than the young of the wild rabbit; scarcely any animal is tamer than the young of the tame rabbit; but I can hardly suppose that domestic rabbits have often been selected for tameness alone; so we must attribute at least the greater part of the inherited change from extreme wildness to extreme tameness, to habit and long-continued close confinement;* and in his MSS he adds, "Captain Sulivan, R.N., took some young rabbits from the Falkland Islands, where this animal has been wild (*i.e.*, feral) for several generations, and he is convinced that they are more easily tamed than really wild rabbits in England. The facility of breaking in the feral horses in La Plata can, I think, be accounted for on the same principle of some little of the effects of domestication being long inherent in the breed." Similarly Mr. Darwin points out in his MSS that there is a great contrast between the natural tameness of the tame duck and the natural wildness of the wild.† The still more remarkable contrasts which are presented between our domestic dogs, cats, and cattle I shall consider later on; for in them it is probable that the principle of selection has

belonging to him which had great difficulty in acquiring by tuition the accomplishment of begging, but afterwards habitually practised it as a general expression of desire. Mr. Hurt then adds, "One of his daughters, who has never seen her father, is in the constant habit of sitting up, although she has never been taught to do so, and has not seen others sit up."

* *Origin of Species*, p. 211.

† With reference to these points I may here appropriately quote the following note, which occurs among Mr. Darwin's MSS.

"'The wild rabbit,' says Sir J. Sebright (*On Instincts*, 1836, p. 10) 'is by far the most untameable animal that I know, and I have had most of the British Mammalia in my possession. I have taken the young ones from the nest, and endeavoured to tame them, but could never succeed. The domestic rabbit, on the contrary, is perhaps more easily tamed than any other animal, excepting the dog.' We have an exactly parallel case in the young of the wild and tame Duck."

I may also quote the following interesting corroboration of the above statement with reference to ducks, from a letter recently published in *Nature*, by Dr. Rae, F.R.S. (July 19, 1883):—"If the eggs of a wild duck are placed with those of a tame one under a hen to be hatched, the ducklings from the former, on the very day they leave the egg, will immediately endeavour to hide themselves, or take to the water, if there is any water, should any person approach, whilst the young from the tame duck's eggs will show little or no alarm, indicating in both cases a clear instance of instinct or 'inherited memory.'"

played an important part, and at present we are confining our attention to the evidence concerning the formation of *secondary* instincts, or the mere lapsing of intelligence into instinct without the aid of selection.

We see, then, that the instinct of wildness may be eradicated by mere disuse, without any assistance from the principle of selection, and further, that this effect persists, or becomes but gradually obliterated, through successive generations of the animals when feral, or restored to their aboriginal conditions of life. Conversely, it has now to be shown that instincts of wildness may be acquired by the hereditary transmission of novel experiences, also without the aid of selection. This is shown conclusively by the original tameness of animals in islands unfrequented by man, gradually passing into an hereditary instinct of wildness as the special experiences of man's proclivities accumulate; for although selection may here play a subordinate part, it must be a very subordinate one. Pages might be filled with facts on this head from the writings of travellers, but to economize space I cannot do better than refer to Mr. Darwin's remarks, with their appended references in his chapter at the end of this volume. To these remarks, however, I may add that the development of fire-arms, together with the growth of sporting interests, has given game of all kinds an instinctive knowledge of what constitutes "safe distance," as every sportsman can testify; and that such instinctive adaptation to newly developing conditions may take place without much aid from selection is shown by the short time, or the small number of generations, which is sufficient to allow for the change—witness, for instance, the following, which I quote from the paper on "Hereditary Instinct" by the careful observer, Andrew Knight:—"I have witnessed, within the period above mentioned, of nearly sixty years, a very great change in the habits of the Woodcock. In the first part of that time, when it had recently arrived in the autumn, it was very tame; it usually chuckled when disturbed, and took only a very short flight. It is now, and has been during many years, comparatively a very wild bird, which generally rises in silence, and takes a comparatively long flight, excited, I conceive, by increased hereditary fear of man."*

* *Phil. Trans.*, 1837, p. 369.

But the force or influence of heredity in the domain of instinct (whether of the primary or secondary class) is perhaps most strongly manifested in the effects of crossing. It is not, indeed, easy to obtain this class of evidence in the case of wild species, because hybrid forms in a state of nature are rare. But when a wild species is crossed with a tame one, it usually happens that the hybrid or mongrel progeny present a blended psychology. And still more cogent is the evidence of such blending when two different breeds of domesticated animals are crossed, having diverse hereditary habits, or as Mr. Darwin calls them, "domestic instincts." Thus a cross-breed between a setter and a pointer will blend the movements and habits of working peculiar to these two breeds; Lord Alford's celebrated strain of greyhounds acquired much courage from a single cross with a bull-dog;* and a cross with a beagle "generations back will give to a spaniel a tendency to hunt hares."†

Again, Knight says:—"In one instance I saw a very young dog, a mixture of the Springing Spaniel and Setter, which dropped upon crossing the track of a Partridge, as its male parent would have done, and sprang the bird in silence; but the same dog, having a couple of hours afterwards found a Woodcock, gave tongue very freely, and just as its female parent would have done. Such cross-bred animals are, however, usually worthless, and the experiments and observations I have made upon them have not been very numerous or interesting."

On this point Mr. Darwin writes:—"These domestic instincts, when thus tested by crossing, resemble natural instincts, which in like manner become curiously blended together, and for a long time exhibit traces of the instincts of either parent; for example, Le Roy describes a dog, whose great-grandfather was a wolf, and this dog showed a trace of its wild parentage only in one way, by not coming in a straight line to his master when called."‡ Some further remarks on this subject will be found in Mr. Darwin's appended essay on instinct; and here I may fitly conclude the present chapter by quoting the following paragraph which occurs in another part of his MSS.

* *Youatt on Dog*, p. 311.
† Blaine, *Rural Sports*, p. 863, quoted by Darwin.
‡ *Origin of Species*, p. 210.

"In Chapter VII I have given some facts showing that when races or species are crossed there is a tendency in the crossed offspring, from quite unknown causes, to revert to ancestral characters. A suspicion has crossed me that a slight tendency to primeval wildness sometimes thus appears in crossed animals. Mr. Garnett in a letter to me states that his hybrids from the musk and common duck 'evinced a singular tendency to wildness.' Waterton ('Essays on Natural History,' p. 197) says that in his duck, a cross between the wild and the tame, 'their wariness was quite remarkable.' Mr. Hewitt, who has bred more hybrids between pheasants and fowls than any other man, in letters to me speaks in the strongest terms of their wild, bad, and troublesome dispositions; and this was the case with some which I have seen. Captain Hutton made nearly the same remark to me in regard to the crossed offspring from a tame goat and a wild species from the western Himalaya. Lord Powis' agent, without my having asked him the question, remarked to me that the crossed animals from the domestic Indian Bull and common cow 'were more wild than the thorough-bred breed.' I do not suppose that this increased wildness is invariable; it does not seem to be the case, according to Mr. Eyton, with the crossed offspring from the common and Chinese geese; nor, according to Mr. Brent, with crossed breeds from the Canary."

CHAPTER XIII.

INSTINCT *(continued)*.

BLENDED ORIGIN, OR PLASTICITY OF INSTINCT.

FROM the foregoing discussion it may, I think, be taken as established:—

1st. That propensities or habitual actions may originate and be inherited without education from parents or otherwise, as in the case of "tricks of manner," peculiar dispositions, tumbling of tumbler pigeons, &c.; in such cases there need be no intelligence concerned in the propensity or action, but if such propensities or actions occur in nature (and, as we have seen, there can be no doubt that they do), those which happen to be of benefit to the animals performing them, will be fixed and improved by natural selection; when thus fixed and improved they constitute what I have called instincts of the primary class.

2nd. That adjustments originally intelligent may by frequent repetition become automatic, both in the individual and in the race; as instances of such "lapsed intelligence" in the individual I have given the highly co-ordinated and laboriously acquired actions of walking, speaking, and others; as instances of the same thing in the race I have dwelt on the hereditary character of handwriting, artistic talent, &c., and in the case of animals, on peculiar habits—such as grinning in dogs, begging in cats—being transmitted to progeny, as well as the more instructive facts with regard to the loss of wildness by certain domesticated animals, and the gradual acquisition of this instinct by animals inhabiting islands previously unfrequented by man. All these and other such cases have been chosen as illustrations, because in none of them can the principle of selection have operated in any considerable degree.

Although for the sake of clearness I have so far kept separate these two factors in the formation of instinct, it has now to be shown that instincts are not necessarily confined to one or other of these two modes of origin exclusively; but, on the contrary, that instincts may have, as it were, a double root—the principle of selection combining with that of lapsing intelligence to the formation of a joint result. Thus, hereditary proclivities or habitual actions, which were never intelligent but, being useful, were originally fixed by natural selection, may come to furnish material for further improvement, or be put to improved uses, by intelligence; and, conversely, adjustments originally due to lapsed intelligence may come to be greatly improved, or put to improved uses, by natural selection.

As an example of the first of these complementary cases —or that of a primary instinct modified and improved by intelligence—let us regard the case of the caterpillar which, before changing into a crysalis, crosses a small space with a web of silk (to which the crysalis can be firmly suspended), but which when placed in a box covered with a muslin lid perceives that this preparatory web is unnecessary, and therefore attaches its crysalis to the already woven surface supplied by the muslin;* or let us regard the case of the bird described by Knight, which observed that, having placed her nest upon a forcing house, she did not require to visit it during the day when the heat of the house was sufficient to incubate the eggs, but always returned to sit upon the eggs at night when the temperature of the house fell.† In both these cases of primary instincts modified by intelligent adaptation to particular circumstances—and hundreds of others might be added—it is evident that if the particular circumstances were to become general, the adaptation to them, becoming likewise general, would in time become instinctive by lapsed intelligence: if muslin and forcing houses were to become normal additions to the environment of the caterpillar or the bird, the former would now cease to build its web, and the latter cease to incubate her eggs by

* See Kirby and Spence, *Entomology*, vol. ii, p. 476. It is evident that the weaving of a web by a caterpillar adapted to the needs of its future condition as a crysalis, must be due to instinct of the primary kind, inasmuch as no individual caterpillar prior to the formation of such a structure can have known by experience what it is to be a crysalis.

† *Loc. cit.*

day; in each case a secondary instinct would become blended with a previously existing primary one, so producing a new instinct with a double root or origin.

Conversely, as an example of a primary instinct becoming similarly blended with a previously existing secondary, let us take the following:—

The grouse of North America display the curious instinct of burrowing a tunnel just below the surface of the snow. In the end of this tunnel they sleep securely; for, when any four-footed enemy approaches the mouth of the tunnel, the bird, in order to escape, has only to fly up through the thin covering of snow. Now in this case the grouse probably began to burrow for the sake of protection, or concealment, or both; and, if so, thus far the burrowing was probably an act of intelligence. But the longer the tunnel the better would it have served the purposes of escape, and therefore natural selection would almost certainly have tended to preserve the birds which made the longest tunnels, until the utmost benefit that length of tunnel could give had been attained.*

Thus then we see that in the formation of instincts there are two great principles in action, which may operate either singly or in combination; these two principles being the lapsing of intelligence and the agency of natural selection. In the previous chapter we were engaged in considering instincts which are due to either one or other of these principles alone; in the present chapter we shall consider instincts which are due to the joint operation of both principles.

Now it is clear at a glance that if even in fully formed instincts we often find, as in the above examples, a "little dose of judgment," it becomes difficult to estimate the importance, either of this little dose of judgment becoming habitual by repetition, and so improving the previous instinct, or of its becoming mixed with the influence of natural selection. For, taking the latter case alone, if, as we have seen, intelligent actions may by repetition become automatic (secondary instincts), and if they may then vary and have their variations fixed in beneficial lines by natural selection, how much more scope may be given to natural selection in

* The facts of this case have been told me by Dr. Rae, F.R.S.

this further development of an instinct, if the variations of the instinct are not wholly fortuitous, but arise as intelligent adaptations of ancestral experience to the perceived requirements of individual experience.

Trusting then it is sufficiently clear that the two principles which may operate either singly or together in forming instincts, may operate together whichever of the two may happen to have, in any particular case, the historical priority, I may in future neglect to entertain the question of such priority; without considering whether in this and that case selection was prior to lapsing of intelligence, or lapsing of intelligence was prior to selection, it will be enough to prove that the two principles are conjoined.

To prove this we have to show, much more copiously than has been done in the above two or three illustrations, not only, as was proved in the previous chapter, that fully formed instincts may vary, but further that their variation may be determined by intelligence.

Plasticity of Instinct.

In former publications I have used this term to express the modifiability of instinct *under the influence of intelligence.* I shall now give some chosen instances of such modifiability, and then proceed to indicate the causes which most frequently lead to intelligence thus acting upon instinct. It is of importance that I should begin by rendering the fact of the plasticity of instinct beyond question, not only because it is still too much the prevalent notion that instincts are unalterably fixed, or rigidly opposed to intelligent alteration under changed conditions of life; but also because it is this principle of plasticity that largely supplies to natural selection those variations of instinct in beneficial lines, which are necessary to the formation of new instincts of a primo-secondary kind.

Huber observes: "How ductile is the instinct of bees, and how readily it adapts itself to the place, the circumstances, and the needs of the community."

If this may be said of the animals in which instinct has attained its highest perfection and complexity, even without evidence we might be prepared to expect that instinct is everywhere ductile. Moreover the bees constitute a good class to choose for our present purpose, because, as I have shown in "Animal Intelligence," their wonderful instinct of making hexagonal cells can only be regarded as an instinct of the primary kind; yet, as we shall see, though so well fixed an instinct of the primary kind, it may be greatly modified by an intelligent appreciation of novel circumstances.

Kirby and Spence, detailing the observations of Huber, write as follows:—

"A comb, not having been originally well fastened to the top of his glass hive, fell down during the winter amongst the other combs, preserving, however, its parallelism with them. The bees could not fill up the space between its upper edge and the top of the hive, because they never construct combs of old wax, and they had not then an opportunity of procuring new; at a more favourable season they would not have hesitated to build a new comb upon the old one; but it being inexpedient at that period to expend their provision of honey in the elaboration of wax, they provided for the stability of the fallen comb by another process. They furnished themselves with wax from the other combs by gnawing away the rims of the cells more elongated than the rest, and then betook themselves in crowds, some upon the edges of the fallen comb, others between its sides and those of the adjoining combs, and there securely fixed it by constructing several *ties* of different shapes between it and the glass of the hive; some were pillars, some buttresses, and others beams artfully disposed and adapted to the localities of the surfaces joined. Nor did they content themselves with repairing the accidents which their masonry had experienced; they provided against those which might happen, and appeared to profit by the warning given by the fall of one of the combs to consolidate the others and prevent a second accident of the same nature.

"These last had not been displaced, and appeared solidly attached by their base: whence Huber was not a little surprised to see the bees strengthen their principal points of connexion by making them much thicker than before with old wax, and forming numerous ties and braces to unite them more closely to each other and to the walls of their habitation. What was still more extraordinary, all this happened in the middle of January, at a period when the bees ordinarily cluster at the top of the hive, and do not engage in labours of this kind. . . . .

"Having placed in front of a comb which the bees were constructing a slip of glass, they seemed immediately aware that it would be very difficult to attach it to so slippery a surface, and, instead of continuing the comb in a straight line, they bent it *at a right angle*, so as to extend beyond the slip of glass, and ultimately fixed it to an adjoining part of the woodwork of the hive which the glass did not cover. This deviation, if the comb had been a mere simple and uniform mass of wax, would have evinced no small ingenuity; but you will bear in mind that a comb consists on each side or face of cells having between them bottoms in common; and if you take a comb, and, having softened the wax by heat, endeavour to bend it in any part at a right angle, you will then comprehend the difficulties which our little architects had to encounter. The resources of their instinct, however, were adequate to the emergency. They made the cells on the *convex* side of the bent part of the comb much *larger*, and those on the *concave* much *smaller* than usual; the former having three or four times the diameter of the latter. But this was not all. As the bottom of the small and large cells were as usual common to both, the cells were not regular prisms, but the smaller ones considerably wider at the bottom than at the top, and conversely in the larger ones! What conception can we form of so wonderful a flexibility of instinct? How, as Huber asks, can we comprehend the mode in which such a crowd of labourers, occupied at the same time on the edge of a comb, could agree to give it the same curvature from one extremity to the other; or how they could arrange together to construct on one face cells so small, while on the other they imparted to them such enlarged dimensions? And how can we feel adequate astonishment that they should have the art of making cells of such different sizes correspond?" *

Other observations of Huber show that even under ordinary circumstances bees are frequently in the habit of altering the construction of their cells. Thus, for instance, the cells which are destined to receive drones requiring to be considerably larger than those which are destined to receive neuters, and the rows of all the cells being continuous, where

* Kirby and Spence, *loc. cit.*, pp. 485–495.

a transition takes place from one class of cell to the other, a complex geometrical problem arises how to unite hexagonal cells of a small with others of a large diameter, without leaving any void spaces or interfering with the regularity of the comb. Without occupying space with what would necessarily be a rather lengthy exposition of the manner in which the bees solve the problem, it is enough to say that in passing from one form of cell to the other, they require to construct a great many rows of intermediate cells which differ in form, not only from the ordinary cells, but from each other. When the bees arrive at any stage in this process of transition, they might stop at that stage and continue to build the whole of their comb upon this pattern. But they invariably proceed from one stage to another until the transition from small hexagons to large hexagons, or *vice versâ*, is effected. On this subject Kirby and Spence remark: "Réaumer, Bonnet, and other naturalists cite these irregularities as so many examples of imperfections. What would have been their astonishment if they had been aware that part of these anomalies had been *calculated* (? adaptive); that there exists as it were a moveable harmony in the mechanism by which the cells are composed! . . . It is far more astonishing that they know how to quit their ordinary routine when circumstances require that they should build male cells: that they should be instructed to vary the dimensions and the shape of each piece so as to return to a regular order; and that, after having constructed thirty or forty ranges of male cells, they again leave the regular order in which they were formed, and arrive by successive diminutions at the point from which they set out . . . . Here again, as observed in a former instance, the wonder would be less if *every* comb contained a *certain* number of transition and of male cells, constantly situated in *one* and the *same* part of it; but this is far from being the case. The event which alone, at whatever period it may happen, seems to determine the bees to construct male cells, is the oviposition of the queen. So long as she continues to lay the eggs of workers, not a male cell is provided; but as soon as she is about to lay male eggs, the workers seem aware of it, and you then see them form their cells irregularly."

Here, then, we have concerted variation in the mode of constructing the cells of a normal and definite kind, and we find that in this case the variation is determined by an event (the oviposition of male eggs) which we may suppose all the bees simultaneously to perceive. But in the present connection the important thing to note is that during even the ordinary work of bees occasion frequently arises to modify the construction of their cells, so that the instincts of the animal are not, as it were, rigidly set to the undeviating formation of the ordinary cell; there is a "moving harmony" in the operation of the instinct which secures plasticity in the formation of the comb, so that when occasion arises the "moving harmony" as it were, changes its key; and it does so in obedience to an intelligent perception of the exigencies of the occasion.

The same thing is shown in a higher degree by some other experiments of Huber, which consisted in making the bees deviate from their normal mode of building their combs from above downwards, to building them from below upwards, and also horizontally. Without describing these experiments in detail, it is enough to say that his contrivances were such that the bees had either to build in these abnormal directions or not to build at all; and the fact that under such circumstances they built in directions which none of their ancestors or none of themselves had ever built before, is good evidence of a primary instinct being greatly modified by intelligence—better evidence, be it observed, of modification than that which is furnished by the previously cited cases, inasmuch as bees often require in a state of nature to change the shape of their cells, but cannot ever have required to reverse the direction of building them.

The same remarks apply to the following observations, which are also due to Huber. A very irregular piece of comb, when placed on a smooth table, tottered so much that the humble bees could not work on so unsteady a basis. To prevent the tottering, two or three bees held the comb by fixing their front feet on the table, and their hind feet on the comb. This they continued to do, relieving guard, for three days, until they had built supporting pillars of wax. "Now," as Mr. Darwin observes in his MSS, "such an accident as this could hardly have occurred in nature."

Some other humble bees when shut up, and so prevented from getting moss wherewith to cover their nests, tore threads from a piece of cloth, and "carded them with their feet into a fretted mass," which they used as moss. Again, Andrew Knight observed that his bees availed themselves of a kind of cement made of wax and turpentine, with which he had covered decorticated trees—using this material instead of their own propolis, the manufacture of which they discontinued;* and more recently it has been observed that bees, "instead of searching for pollen, will gladly avail themselves of a very different substance, namely, oatmeal." †

Again, *Osmia aurulenta* and *O. bicolor* are species of bees which construct tunnels in hard banks of earth or clay, in which they afterwards deposit their eggs—one in each partitioned cell. But when they find tunnels ready-made (as in the straws of a thatched roof) they save themselves the trouble of employing their instincts in the way of tunnel-making—merely building transverse partitions in the tube to form a series of separate cells. It is specially remarkable that when they thus utilize the whorl of an empty snail-shell, the number of cells which they partition off is regulated by the size of the shell, or the length of the whorl. Moreover, if the whorl proves too wide near the orifice of the shell for its walls to constitute the boundaries of a single cell, the bee will build a partition at right angles to the plane of the others, so forming a double cell, or two cells side by side.‡

Now, in all these cases it is evident that if, from any change of environment, such accidental conditions were to occur ordinarily in a state of nature, the bees would be ready

* *Phil. Trans.*, *loc. cit.*

† *Origin of Species*, p. 228. It is interesting in connection with these facts to note how singularly well they happen to meet a criticism of Kirby and Spence, which was advanced before they had been observed, with the object of discrediting the view of instinct being modified by intelligence. These authors ask (*loc. cit.*, vol. ii, p. 497), why, if such were the case, should not bees sometimes be found to use mud or mortar instead of precious wax or propolis: "Show us," they say, "but one instance of their having substituted mud for propolis . . . . and there could be no doubt of their having been guided by reason." It is curious that this demand should so soon have been met by so apposite an observation. Doubtless mud is not so good a material for the purposes required as propolis, but as soon as the bees are furnished with a substance that is as good, they are ready enough to prove their "reason," even to the satisfaction of what was supposed, *à priori*, a crucial test. This case should serve as a warning against the use of the question-begging argument, which where any degree of evidence is presented of intelligence compounded with instinct, forthwith raises the standard and says—Show us an animal doing this or that, which would be still more remarkable, and then we shall be satisfied.

‡ See F. Smith, *Catol. Brit. Hymenoptera*, pp. 159-60.

to meet them by intelligent adjustment, which, if continued sufficiently long and aided by selection, would pass into true instincts of building combs in new directions, of supporting combs during their construction, of carding threads of cloth, of substituting cement for propolis, or oatmeal for pollen.

Were it necessary, other instances of the plasticity of instinct could be drawn from bees and likewise from ants,* but quitting now the Hymenoptera, I shall pass to other animals.

Dr. Leech gives,† on the authority of Sir J. Banks, a case of a web-spinning spider which had lost five of its legs, and, as a consequence, could only spin very imperfectly. It was observed to adopt the habits of the hunting spider, which does not build a web, but catches its prey by stalking. This change of habit, however, was only temporary, as the spider recovered its legs after moulting. But it seems evident from this case that, so far as the plasticity of instinct is concerned, the web-spinning spider would be ready at any time to adopt the habit of hunting, if for any reason it should not be able to build a web—and this even by way of sudden transition in the life-time of an individual.

Coming now to vertebrated animals, we may easily find that the same principles obtain in them. And here, for the sake of brevity, I shall confine myself to instances drawn from the oldest, most constant, and, therefore, presumably the most fixed of the instincts which vertebrated animals display, viz., the maternal.

With regard to Birds, I showed in the preceding chapter that individual variations of nest-building are not uncommon. We have now to remark that such variations, or deviations from the ancestral modes, are not always the result of mere caprice, but sometimes of intelligent purpose. In order to

* See *Animal Intelligence*, from which I may specially quote the following, in order to show briefly that ants quite as much as, or more than bees, present a "moving harmony" in the construction of their architecture:—"The characteristic *trait* of the uilding of ants," says Forel, "is the almost complete absence of an unchangeable model peculiar to each species, such as is found in wasps, bees, and others. The ants know how to suit their indeed little perfect work to circumstances, and to take advantage of each situation. Besides, each works for itself on a given plan, and is only occasionally aided by others when they understand its plan" (p. 129).

† *Transactions Linn. Soc.*, vol. xi, p. 393. This case is briefly alluded to by Mr. Darwin in the Appendix.

show this, it will be sufficient to state the following instances.

Thread and worsted are now habitually used by sundry species of birds in building their nests, instead of wool and horsehair, which in turn were no doubt originally substitutes for vegetable fibres and grasses; this is specially noticeable in the case of the tailor-bird and Baltimore oriole, and Wilson believes that the latter improves in nest-building by practice—the older birds making the better nests. The common house-sparrow furnishes another instance of intelligent adaptation of nest-building to circumstances; for in trees it builds a domed nest (presumably, therefore, the ancestral type), but in towns avails itself by preference of sheltered holes in buildings, where it can afford to save time and trouble by constructing a loosely formed nest. A similar case is furnished by the gold-crested warbler, which builds an open cup-shaped nest where foliage is thick, but makes a more elaborate domed nest with a side entrance, where the site chosen is more exposed. Moreover, the chimney and house-swallows have taken to building in chimneys and under the roofs of houses by way of an intelligent or plastic change of instinct, and in America this change has taken place within the last three centuries or less. Indeed, according to Captain Elliott Coues, all the species of swallow on the American continent (with one possible exception) have modified the structure of their nests in accordance with the novel facilities afforded by the settlement of the country; for he writes:—

"Various species, indeed, now regularly accept the artificial nesting-places man provides, whether by design or otherwise. Such is notably the case with several kinds of Wrens, with at least one kind of Owl, with one Bluebird, the Pewit Flycatcher, and especially the House-sparrow. Various other birds occasionally avail themselves of like privileges, still retaining in the main their original habits. But in no other case than that of the Swallows is the modification of habit so profound, or so nearly without exception throughout the entire family. . . . All of our Swallows have been modified by human agency, excepting the Bank Swallow. . . . . Some of them, like the Purple Martin and the Violet-green Swallow, are still surviving their apprenticeship under the new *régime*, which the settlement of the country has brought about. . . . Those whose acquired habits have become thoroughly ingrained are now pretty constant in their adherence to a single plan of architecture; but the Violet-green Swallow, for instance, at present nests in a very loose fashion, according to circumstances." *

The statement made in 1870 by the distinguished naturalist Pouchet to the effect that within the same interval of half a century the house-swallow had materially altered its mode of nest-building at Rouen,† was subsequently shown by M. Noulet to be erroneous;‡ but this passage which I have quoted from Captain Elliott Coues is sufficient to show that facts analogous to those stated by M. Pouchet have occurred among many species of the swallow tribe.

In "Animal Intelligence" I gave some cases of the remarkable intelligence which is displayed by certain birds when they remove their eggs or their young from places where they have been disturbed (pp. 288–9), and I added the remark that it is easy to see that if any particular bird is intelligent enough, as in the cases quoted, to perform this adjustive action of conveying young—whether to feeding-grounds, as in the case of the hen, or from sources of danger, as in the case of partridges, blackbirds, and goat-suckers—inheritance and natural selection might develop the originally intelligent adjustment into an instinct common to the species. And it so happens that this has actually occurred in at least two species of birds—viz., the woodcock and wild duck, both of which have been repeatedly observed to fly with their young to and from their feeding-ground.

Since writing the above, I have found among Mr. Darwin's MSS a letter from Mr. Haust, dated New Zealand, December 9th, 1862, and stating that the "Paradise Duck," which naturally or usually builds its nest along the rivers on the ground, has been observed by him on the east of the island, when disturbed in their nests upon the ground, to build 'new ones on the tops of high trees, afterwards bringing their young ones down on their backs to the water," and exactly the same thing has been observed of the wild ducks of Guiana.§ Now, if intelligent adjustment to peculiar circumstances is

* *Birds of Colorado Valley*, pp. 292–4. † *Comptes Rendus*, lxx, p. 492.

‡ *Ibid.*, lxxi, p. 78. In the first edition of *Animal Intelligence* I quoted this statement of Pouchet without knowing that it had been questioned.

§ See *Geol. Journ.*, vol. iv, p. 325.

thus adequate, not only to make a bird transport her young upon her back, or, as in the case of the woodcock, between her legs, but even to make a web-footed water-fowl build her nest on a high tree, I think we can have no doubt that, if the need of such adjustment were of sufficiently long continuance, the intelligence which leads to it would eventually produce a remarkable modification in the ancestral instinct of nest-building.

Lastly, " a curious example of a recent change of habits has occurred in Jamaica. Previous to 1854, the palm swift (*Tachornis phœnicobea*) inhabited exclusively the palm trees in a few districts of the island. A colony then established themselves in two cocoa-nut palms in Spanish Town, and remained there till 1857, when one tree was blown down, and the other stripped of its foliage. Instead of now seeking out other palm trees, the swifts drove out the swallows who built in the piazza of the House of Assembly, and took possession of it, building their nests on the tops of the end walls and at the angles formed by the beams and joists, a place which they continue to occupy in considerable numbers. It is remarked that here they form their nests with much less elaboration than when built in the palms, probably from being less exposed."*

Turning now from the instinct of nidification to that of incubation, I shall give the results of some observations and experiments which I made several years ago and published in " Nature," from which I quote the account. In these cases the plasticity of the maternal instinct was shown by the fact that the instinct was directed in all its force to the young of other animals, although there is ample evidence to prove that the foster-mothers perceived the unnatural character of their brood. Indeed, it is just because of this evidence that I quote these cases in the present connection, for otherwise they might rather be taken to exemplify non-intelligent variations of instinct with which we were concerned in the last chapter. But inasmuch as the intelligence of the animals was displayed by the manner in which they adapted their ancestral instincts to the requirements of their adopted progeny, the cases become available rather as proof of the intelligent variation of instinct.†

* Wallace, *Natural Selection*, Chapter VI, where see for some of the preceding and also for other instances.

† The yearning for progeny which arises from the parental instinct being

"Spanish hens, as is notorious, scarcely ever sit at all; but I have one purely bred one just now that sat on dummies for three days, after which time her patience became exhausted. However, she seemed to think that the self-sacrifice she had undergone during those three days merited some reward, for on leaving the nest, she turned foster-mother to all the Spanish chickens in the yard. They were sixteen in number, of all ages, from that at which their own mothers had just left them up to full-grown chickens. It is remarkable, too, that although there were Brahma and Hamburg chickens in the yard, the Spanish hen only adopted those of her own breed. It is now four weeks since this adoption took place, but the mother as yet shows no signs of wishing to cast off her heterogeneous brood, notwithstanding that some of her adopted chickens have grown nearly as large as herself.

"The following, however, is a better example of what may be called plasticity of instinct. Three years ago I gave a pea-fowl's egg to a Brahma hen to hatch. The hen was an old one, and had previously reared many broods of ordinary chickens with unusual success even for one of her breed. In order to hatch the pea-chick she had to sit one week longer than is requisite to hatch an ordinary chick, but in this there is nothing very unusual, for, as Mr. Spalding observes, the same thing happens with every hen that hatches out a brood of ducklings.* The object with which I made this experiment, however, was that of ascertaining whether the period of maternal care subsequent to incubation admits, under peculiar conditions, of

unsatisfied, induces even such an intelligent animal as man to adopt progeny; and the proverbial passion of old maids for keeping cats, dogs, and other domestic animals, is probably analogous to the cases given in the text of female animals adopting the young of other species.

In this connection I may quote the following account which I have received from a friend, whom I know to be an accurate and conscientious observer; for it shows that even among birds in a state of nature the yearning for progeny may induce them to adopt the young of other species, just as in the cases of birds in a state of domestication which are about to be given in the text:—

"In July, 1878, I found a wren's nest with young birds, which were being fed by a wren and a sparrow. I made sure that the young birds were wrens, and I noticed that the sparrow continued to feed them after they had left the nest. The behaviour of the two birds was very dissimilar, the wren being bold and its visits to the nest incessant, whereas the sparrow was very shy and its visits much less frequent."

* The greatest prolongation of the incubatory period I have ever known was in the case of a pea-hen, which sat very steadily on addled eggs for a period of four months, and had then to be forced off in order to save her life.

being prolonged; for a pea-chick requires such care for a very much longer time than does an ordinary chick. As the separation between a hen and her chickens always appears to be due to the former driving away the latter when they are old enough to shift for themselves, I scarcely expected the hen in this case to prolong her period of maternal care, and indeed only tried the experiment because I thought that if she did so, the fact would be the best one imaginable to show in what a high degree hereditary instinct may be modified by peculiar individual experiences. The result was very surprising. For the enormous period of eighteen months this old Brahma hen remained with her ever-growing chicken, and throughout the whole of that time she continued to pay it unremitting attention. She never laid any eggs during this lengthened period of maternal supervision, and if at any time she became accidentally separated from her charge, the distress of both mother and chicken was very great. Eventually the separation seemed to take place on the side of the peacock; but it is remarkable that although the mother and chicken eventually separated, they never afterwards forgot each other, as usually appears to be the case with hens and their chickens. So long as they remained together, the abnormal degree of pride which the mother showed in her wonderful chicken was most ludicrous; but I have no space to enter into details. It may be stated, however, that both before and after the separation the mother was in the habit of frequently combing out the top-knot of her son—she standing on a seat or other eminence of suitable height, and he bending his head forward with evident satisfaction. This fact is peculiarly noteworthy, because the practice of combing out the top-knot of their chickens is customary among peahens. In conclusion, I may observe that the peacock reared by this Brahma hen turned out a finer bird in every way than did any of his brothers of the same brood which were reared by their own mother, but that on repeating the experiment next year with another Brahma hen and several pea-chickens, the result was different, for the hen deserted her family at the time when it is natural for ordinary hens to do so, and in consequence all the pea-chickens miserably perished."*

I allude to the following instructive case from Jesse's "Gleanings,"† because it has been independently and uncon-

* *Nature*, Oct. 28, 1875. † Vol. i, p. 98.

sciously corroborated in every detail by a correspondent, Mrs. L. MacFarlane, of Glasgow. Indeed, the similarity is so precise, that I think the two descriptions must refer to the same incident; but as to this I cannot be sure, because upon my writing to Mrs. MacFarlane to enquire, she answers that she is not able to inform me. However, this point is immaterial, for my correspondent had the story at first hand from the lady to whom the birds belonged (and with whom she was intimately acquainted), so that if the case is not the same as the one narrated by Jesse, its repetition is so exact that the same description applies to both the cases.

"A hen, who had reared three broods of ducks in three successive years, became habituated to their taking to the water, and would fly to a large stone in the middle of the pond, and quietly and contentedly watch her brood as they swam about it. The fourth year she hatched her own eggs, and finding that her chickens did not take to the water as the ducklings had done, she flew to the stone in the pond, and called them to her with the utmost eagerness. This recollection of the habits of her former charge is not a little curious."

My correspondent, Mrs. MacFarlane, also gives me another closely similar but even more remarkable case, which was observed by her sister, Miss Mackillar, of Tarbert, Cantyre. In this case a hen had also reared three successive broods of ducklings in successive years, and then hatched out a brood of nine chickens. The season being late, she was confined for some weeks till the chickens became strong enough to face the cold weather. Then, in the words of my correspondent, "the first day she was let out she disappeared, and after a long search my sister found her beside a little stream which her successive broods of ducklings had been in the habit of frequenting. She had got four of her chickens into the stream, which was fortunately very shallow at the time. The other five were standing on its margin, and she was endeavouring by all sorts of coaxing hen-language, and by pushing each chicken in turn with her bill, to get them into the water also."

From these cases it is evident that in a portion of the lifetime of an individual hen there may be laid, by intelligent observation and memory, the basis of a new instinct, adapted to an immense and sudden change in the habits of progeny: and that in all the foregoing cases the foster-mother was not blind to the unnatural character of her brood, is proved by the fact of her having adapted her actions to their peculiar requirements. But to test the degree to which such adaptation might go, I tried the experiment of selecting the two most diverse kinds of animals I could think of, and giving the young of the one to be reared as foster-children by the other. The animals which I selected for this purpose were a ferret and a hen. The following was the result of the experiment as published at the time in "Nature."*

"A bitch ferret strangled herself by trying to squeeze through too narrow an opening. She left a very young family of three orphans. These I gave, in the middle of the day, to a Brahma hen, which had been sitting on dummies for about a month. She took to them almost immediately, and remained with them for rather more than a fortnight, at the end of which time I had to cause a separation, in consequence of the hen having suffocated one of the ferrets by standing on its neck. *During the whole of the time that the ferrets were left with the hen, the latter had to sit upon the nest;* for the young ferrets, of course, were not able to follow the hen about as young chickens would have done, in accordance with the strong instinct of following with which Mr. Spalding has shown young chickens to be endowed. The hen, as might be expected, was very much puzzled at the lethargy of her offspring. Two or three times a day she used to fly off the nest, calling upon her brood to follow; but, on hearing their cries of distress from cold, she always returned immediately and sat with patience for six or seven hours more. It only took the hen one day to learn the meaning of these cries of distress; for after the first day she would always run in an agitated manner to any place where I concealed the ferrets, provided that this place was not too far away from the nest to prevent her from hearing the cries of distress. Yet I do not think it would be possible to conceive of a greater contrast than that between the shrill piping note of a young chicken and the hoarse growling noise of a young ferret. On the other hand, I cannot say that the young ferrets ever seemed to learn the meaning of the hen's clucking. During the whole of the time that the hen was allowed to sit upon the ferrets she used to comb out their hair with her bill, in the same way as hens in general comb out the

* Vol. xi, p. 553.

feathers of their chickens. While engaged in this process, however, she used frequently to stop and look with one eye at the wriggling nest-full with an enquiring gaze expressive of astonishment. At other times, also, her family gave her good reason to be surprised; for she used often to fly off the nest suddenly with a loud scream, an action which was doubtless due to the unaccustomed sensation of being nipped by the young ferrets in their search for the teats. It is further worth while to remark that the hen showed so much uneasiness of mind when the ferrets were taken from her to be fed, that at one time I thought she was going to desert them altogether. After this, therefore, the ferrets were always fed in the nest, and with this arrangement the hen was perfectly satisfied—apparently because she thought that she had some share in the feeding process. At any rate she used to cluck when she saw the milk coming, and surveyed the feeding with evident satisfaction.

"Altogether I consider this a very remarkable case of the plasticity of instinct. The hen, it should be said, was a young one, and had never reared a brood of chickens. A few months before she reared the young ferrets, she had been attacked and nearly killed by an old ferret which had escaped from its hutch. The young ferrets were taken from her several days before their eyes were open.

"In conclusion, I may add that a few weeks before trying this experiment with the hen, I tried a similar one with a rabbit which had littered six days before . . . . Unlike the hen, however, the doe perceived the imposture at once, and attacked the young ferret so savagely that she broke two of its legs before I could remove it. To have made the experiment parallel with the other, however, the two mammalian mothers should have littered on the same day."

Lastly, turning to the Mammalia, a friend of the Rev. Mr. White, of Selborne, gave him an account of a leveret which he saw reared by a cat.* Prichard gives an account of a cat that reared a puppy,† and from among many analogous instances that might be rendered, I shall only quote the following, as it is remarkable on account of displaying voluntary adoption by a cat of the young of animals which her other instincts and constant practice had taught her to regard as prey.

* Bingley, *Animal Biography*, i, 269. † *Nat. Hist. of Mankind*, i, 102.

"Some years ago the late Hon. Marmaduke Maxwell of Terregles took me to his stable to show me a cat which was at the time bringing up a family of young rats. The cat some weeks previously had had a litter of five kittens; three were taken away and destroyed shortly after their birth; next day it was found that the cat had replaced her lost kittens by three young rats, which she nursed with the two remaining kittens. A few days afterwards the two kittens were taken away, and the cat very shortly replaced them by two more young rats, and at the time I saw them the young rats—which were confined in an empty stall—were running about quite briskly, and about one-third grown. The cat happened to be out when we went into the stable, but came in before we left; she immediately jumped over the board into the stall and lay down: her strange foster-family at once ran under her, and commenced sucking. What renders the circumstance more extraordinary is, that the cat was kept in the stable as a particularly good ratter."*

* Mr. P. Dudgeon, *Nature*, vol. xx, p. 77.

CHAPTER XIV.

INSTINCT *(continued)*.

MODES IN WHICH INTELLIGENCE DETERMINES THE VARIATION OF INSTINCT IN DEFINITE LINES.

WE have now seen that instincts may have what I term a blended origin—or, in other words, that intelligent adjustment by going hand in hand with natural selection, must greatly assist the latter principle in the work of forming instincts, inasmuch as it supplies to natural selection variations which are not merely fortuitous, but from the first adaptive. I shall next show what I conceive to be the chief modes in which intelligence thus operates, or co-operates with selection, in the formation of instincts.

Of course in general terms it is easy to see that the mode in which intelligence thus co-operates is by enabling an animal to perceive that, owing probably to some change in its environment, it may best adapt itself to the existing conditions of its life by deviating in some degree from its ancestral instincts (as when the tailor-bird seeks for threads of cotton instead of fibres of grass wherewith to sew its nest), or by intelligent observation giving rise to adjustive actions, which by repetition lead to an instinct *de novo* (as in the case of the honey-guide, which has acquired the remarkable instinct of attracting the attention of man, and leading him to the nests of bees).* But with animals, as with men, original ideas are not always forthcoming at the time they are wanted, and therefore it is often easier to imitate than to invent. Thus, the first mode which I shall consider whereby intelligence may change or deflect an instinct, is that of imitation. For although it is true that the initial stage of such deflection occurs in the "original ideas," nothing

* See *Animal Intelligence*, p. 315.

further remains to be said of these. If they occur similarly and simultaneously in a large number of individuals, as may be the case where the new adjustment is simple and obvious, there may be no need of imitation to assist in changing the instinct. But in other cases I am inclined to think that imitation may play an important part in this matter. I must confess, however, that in searching for evidence of one *species* of animal imitating the beneficial habits of another, I have been surprised at the rarity of its occurrence, although, as I shall presently show, there is abundant evidence of one *individual* imitating the habits of another individual—whether of its own or of other species, and whether the action imitated is beneficial or useless. This difference, I think, is probably to be explained by the reflection that in all cases where such imitation between species and species may have obtained in the past, we should now only see an instinct common to the two species, and therefore should have no evidence that it was not always common. Consequently, it is only in cases where the imitation by one species of the habits of another is in its earlier phases that we can find evidence of the fact. The following are the only cases of such imitation that I have been able to meet with; but to them I add a number of cases of individual imitation, because this must evidently form the groundwork of imitation among species.

I quote the following from Mr. Darwin's MSS:—

"From some experiments which I was making, I had occasion very closely to watch some rows of the tall kidney-bean, and I daily saw innumerable hive-bees alighting as usual on the left wing-petel, and sucking at the mouth of the flower. One morning, for the first time, I saw several humble-bees (which had been extraordinarily rare all summer) visiting these flowers, and I saw them in the act of cutting with their mandibles holes through the under side of the calyx, and thus sucking the nectar: all the flowers in the course of the day became perforated, and the humble-bees in their repeated visits to the flowers were thus saved much trouble in sucking. The very next day I found all the hive-bees, without exception, sucking through the holes which had been made by the humble-bees. How did the hive-bees find out that all the flowers were bored, and how did they so suddenly acquire the habit of using the holes? I never saw, though I have long attended to the subject, or heard of hive-bees themselves boring holes. The minute holes made by the humble-bees were not visible from the mouth of the flower, where the hive-bees had hitherto invariably alighted: nor do I believe, from some experiments which I have made, that they were guided by the scent of the nectar escaping through these orifices more readily than through the mouth of the flower. The kidney-bean is also an exotic. I must think that the hive-bees either saw the humble-bees cutting the holes, and understood what they were doing, and immediately profited by their labour; or that they merely imitated the humble-bees after they had cut the holes, and when sucking at them. Yet I feel sure that if anyone who had not known this previous history had seen every single hive-bee, without a moment's hesitation, flying with the utmost celerity and precision from the under side of one flower to another, and then rapidly sucking the nectar, he would have declared that it was a beautiful case of instinct."

Mr. Darwin in his MSS has also the following observations concerning the subject of imitation:—"It is difficult to determine how much dogs learn by experience and imitation. I apprehend there can be little doubt that the manner of attack of the English Bull-dog is instinctive (Rollin, 'Mem., &c.,' tom. iv, p. 339). I believe that certain dogs in South America without education rush at the belly of the stag which they hunt, and that certain other dogs when first taken out run round the heads of Peccaris. We are led to believe that these actions are imitative when we hear from Sir J. Mitchell ('Australia,' vol. i, p. 292), that his dogs did not learn how safely to seize the Emu by the neck, until the close of his second expedition. On the other hand Mr. Couch ('Illustrations of Instinct,' p. 191) gives the case of a dog who learned, after a single battle with a Badger, the spot where it would inflict a fatal bite, and it never forgot the lesson. In the Falkland Islands it seems that the dogs learned from each other the best way of attacking the wild cattle (Sir J. Ross, 'Voyage,' vol. ii, p. 246)."

Again, Mr. Darwin points out that many species of wild animals certainly learn to understand and to profit by the danger cries and signals employed by other species, and this is a kind of imitation.* He also adduces a good deal of

* Thus, for instance, he says that "the inhabitants of the United States like to have martins build on their houses, as their cry when a hawk

evidence to show that birds of different species, whether in a state of nature or domestication, frequently imitate one another's song; and singing is certainly instinctive, for Couch says that he knew a gold-finch, which had never heard the song of its own species, nevertheless singing this song, though tentatively and imperfectly.*

Yarrell tells of a hawfinch that learnt the song of a blackbird, though afterwards it quite forgot this song, which could not have happened with its natural music,† a fact which shows that although imitation is able largely to modify instinct, its effects are not so deeply engrained as those which are stamped by heredity. Even the sparrow, which naturally can scarcely be said to have a song, will learn the song of a linnet,‡ and Dureau de la Malle gives the case of wild blackbirds in his garden learning a tune from a caged bird;§ similarly, he taught a starling the Marseillaise, and from this bird all the other starlings in a canton to which he took it learned the air. In this way, too, many birds acquire the song of their foster-parents of other species.‖ Lastly, a number of observations on wild birds in America imitating each other's music have lately been published by Mr. E. E. Fish.¶

It is certain, however, that some birds have a much greater aptitude than others, both for learning and retaining the songs of different species. Thus a blackbird [starling?] has been known so well to imitate the crowing of a cock as to deceive the cocks themselves,** while Yarrell says the same thing of a starling's power of imitating the cackling of a hen.†† Of course such facts are notorious as regards the Mocking-bird (*Turdus polyglottus*), and also, at least when in a state

appears serves to alarm the chickens, though the latter are not aborigines of the country." And many similar instances might be given.

* *Illustrations of Instinct*, p. 113. See also Bechstein, *Stübenvögel*, 4th ed., p. 7.

† *Brit. Birds*, vol. i, p. 486.

‡ *Descent of Man*, p. 370.

§ *Anns. des Sc. Nat.*, 3rd series. 2 vol. Tome x, p. 118.

‖ Barrington, *Phil. Trans.*, 1773, p. 264.

¶ *Bulletin of the Buffalo Society of Nat. Sc.* 1881, pp. 23-6.

** *Loundoun's Mag. Nat. Hist.*, vol. iv, p. 433.

†† *Loc. cit.*, vol. i, p. 204; also in 4th ed., vol. ii, pp. 229-30, where it is said on the authority of sundry observers, that starlings in a state of nature also imitate the kestrel, wryneck, partridge, moorhen, coot, oyster-catcher, golden plover, redshank, curlew, whimbrel, herring-gull, quail, and corn-crake, while Professor Newton tells me that at Cambridge he has heard the starlings very perfectly imitating the quacking of ducks.

of confinement, of parrots, jays, jackdaws, and starlings; and these facts are rendered more remarkable from the additional fact that none of these birds have any proper song of their own, and might therefore be supposed not to have a developed ear for bird-music. Still more remarkable, however, is the fact that these birds are able correctly to imitate songs having a proper musical notation, and that they both learn such songs more readily and retain them better than even those singing-birds which are most apt at learning tunes. For Bechstein says that even the Bullfinch requires nine months of regular and continued instruction to become firm in its performance, and that very frequently all instruction is forgotten in moulting. Couch, indeed, says that with all such birds "it is with them as with the human race; those which are quick at attaining are also rapid in losing their acquirements," and conversely; but clearly this statement is no more true of birds than it is of "the human race." For of any of the songless birds above named it would be a sign of unusual dulness to require nine months of continuous instruction in a single tune, and, on the other hand, they do not so readily forget what they learn. But the most remarkable extension of the power of vocal imitation which these birds display is unquestionably that of uttering articulate words. This subject will require to be considered more fully in my next work. Meanwhile it is enough merely to mention it with reference to the wonderful power and precision of the imitation which is betokened in thus modifying the instinct of uttering a caw or scream, into the singing of a definite tune or the speaking of articulate words.

The habit displayed by cats, and even young kittens, of washing their faces might well be deemed instinctive, and so, most probably, it is; but that it may also be acquired by imitation is proved by the fact that puppies when brought up by a cat perform the same movements. This was first observed by Audouin,* and has since been independently corroborated by several observers, of whom I may mention the following:—

Dureau de la Malle gives the case of a terrier which belonged to himself, and which from the time of its birth was brought up with a kitten six months its senior. For two years the terrier had no association with other dogs. Soon

* *Anns. des Sc. Nat.*, tom. xxii, p. 397.

the terrier began to bound like a cat, and to roll a mouse or a ball with his fore-paws; he also licked his paws and rubbed them over his ears. Yet if a strange cat came into the garden he chased it away.* Prichard gives another case of a dog reared by a cat learning to lick its paws and wash its face,† and a precisely similar case is communicated to me by Mrs. M. A. Baines. Another precisely similar case I find recorded in Mr. Darwin's MSS as communicated to him by Professor Hoffmann of Giessen. Again, the late Dr. Routh, President of Magdalen College, Oxford, observed that his King Charles terrier (which had been suckled and reared by a cat from the age of three days) was as afraid of rain as was the foster-mother; that he would never, if he could possibly avoid it, set his paw in a wet place; that he licked his feet two or three times a day for the purpose of washing his face, which process he performed "in the true cattish position, sitting upon his tail;" that "he used to watch a mouse-hole for hours together;" and had "in short all the ways, manners, and dispositions of his wet-nurse."‡ Lastly, another case is recorded in "Nature"§ of a dog belonging to Mr. C. H. Jeens, which, having been reared by a cat from the age of one month, used to catch mice, and when it caught one to treat it "after the well-known manner of cats, allowing it to run a distance, then pouncing upon it, and so on for many minutes." Conversely Dr. E. Darwin records the case of a cat learning from a dog the medicinal use of the herb *Agrostis canina.* I think it is probable that the following facts, which I quote from Mr. Darwin's MSS, are also, in part at least, to be attributed to imitation, though here the imitation is within the limits of the same species.

"It has been stated that lambs turned out without their mothers are very liable to eat poisonous herbs; and it seems to be certain that cattle, when first introduced into a country, are killed by eating poisonous herbs which the cattle already naturalized there have learnt to avoid."‖

It seems needless to give further instances of imitation

* *Anns. des Sc. Nat.*, tom. xxii, p. 388.

† *Nat. Hist. of Mankind*, 3rd ed., vol. i, p. 102.

‡ Miss Mitford's *Life and Letters*, vol. ii, p. 277.

§ *Nature*, vol. viii, p. 79.

‖ See *Annls. and Mag. of Nat. Hist.* 2nd ser., vol. ii, p. 364; and *Stillingfleet's Tracts*, p. 350. In regard to Lambs, see *Youatt on Sheep*, p. 404.

among animals, but it may be said in general that, as the faculty of imitation depends on observation, it is found in greatest force, as we should expect, among the higher or more intelligent animals—reaching its maximum in the monkeys, where, as is well known, it passes into ludicrous extremes. And in this connection it is interesting to observe that a child begins to imitate very early in life, and that the faculty goes on developing during the first year or eighteen months, after which it remains stationary for a time, and is then of much service in developing language.* With growing intelligence, this faculty subsequently declines, and in after life may be said to stand in an inverse relation to originality or the higher powers of the mind. Therefore among idiots below a certain grade (though of course not too low), it is usually very strong and retains its supremacy through life, while even among idiots of a higher grade, or the "feeble-minded," a tendency to undue imitation is a very constant peculiarity. The same thing is conspicuously observable in the case of many savages; so that in view of all these facts we must conclude that the faculty of imitation is one very characteristic of a certain area of mental evolution, and therefore that within the limits of this area it must conduce in no small degree to the formation of instinct.†

* See Preyer, *loc. cit.*, pp. 176–182, where a number of detailed observations on this head are given. He says that the first imitative movement begins as early as the fifteenth week in protruding the lips when anyone performs this action before the child. [This action seems to come naturally to young children, and may I think probably have some hereditary connection with the same movement as so strongly pronounced in the orang outang. For a picture of such protrusion in this animal, see Darwin. *Expression of Emotions*, p. 141.] Towards the end of the first year imitative movements become more numerous and more quickly learnt, and the child takes active pleasure in their performance. At twelve months Preyer observed his child repeating in its dreams imitative movements which had made a strong impression on it while awake, —*e.g.*, blowing with the mouth. Later still, complicated imitative movements are performed for mere amusement, as is apparently the case with monkeys.

† With reference to imitation in connection with instinct, I think it is desirable here again to express my opinion already given in *Animal Intelligence*, on the theory published by Mr. Wallace, in his *Natural Selection*, that the nidification of birds is due to the young birds consciously imitating the structure of the nests in which they have themselves been reared—the characteristic nidification of each species of bird being thus maintained. I have advanced in *Animal Intelligence* sundry general considerations, which I thought sufficient to negative this theory on *à priori* grounds; but since then I have found among Mr. Darwin's MSS a letter which describes the results of the test experiment which Mr. Wallace himself suggests. This experiment is to rear young birds from the egg in an artificial nest or incubator unlike

But the influence of this faculty in the formation of instinct proceeds further than we have yet noted. For among the more intelligent animals it is played upon for this very purpose by the animals themselves; the parents of each successive generation intentionally educate their young in the performance of quasi-instinctive actions. Thus, for instance, old hawks purposely educate the instinctive faculties of their young, so as more quickly to bring these instincts into a state of perfection. For the manner in which hawks swoop upon their prey must certainly be regarded as instinctive; yet La Malle observed,* and the observation was afterwards corroborated by Brehm,† that the old birds perfected the natural instincts of their young ones in teaching them "dexterity, as well as judgment of distances, by first dropping through the air dead mice and sparrows, which the young generally failed to catch, and then bringing them live birds and letting them loose."‡

And analogous facts are to be observed in the case of old birds teaching the young ones to fly. We have already seen that Mr. Spalding proved such teaching to be unnecessary in the sense of not being required to develop the power of flight. This is instinctive, so that the young bird, whether or not instructed by its parents, would fly. Yet the instruction must be of some use, as in some species, at any rate, it is

the natural nest, and then observe whether when adult these birds will instinctively build the nest characteristic of their species. Now I find among Mr. Darwin's MSS a letter to him from Mr. Weir, which seems to set any such question at rest. Writing under the date May, 1868, Mr. Weir says as the result of a large experience of birds kept by him in aviaries:—"The more I reflect on Mr. Wallace's theory that birds learn to make their nests because they have been themselves reared in one, the less inclined do I feel to agree with him." He gives the following fact, which seems to be conclusive against this theory:—"It is usual with many Canary fanciers to take out the nest constructed by the parent birds, and to place a felt nest in its place, and when the young are hatched and old enough to be handled, to place a second clean nest, also of felt, in the box, removing the other, and this is done to avoid acari. But I never knew that canaries so reared failed to make a nest when the breeding time arrived. I have on the other hand marvelled to see how like a wild bird their nests are constructed. It is customary to supply them with a small set of materials, such as moss and hair; they use the moss for the foundation, and line with the finer materials, just as a wild goldfinch would do, although, making it in a box, the hair alone would be sufficient for the purpose. I feel convinced nest building is a true instinct."

* *Anns. de Sc. Nat.*, tom. xxii, p. 406.

† *Mag. Nat. Hist.*, vol. ii, p. 402.

‡ *Descent of Man*, p. 73.

laboriously given;* and the only use it can be is that of developing the powers of flight more rapidly than they would develop if not thus assisted.

Similarly, the singing of birds is certainly instinctive; yet it is improved by imitation and practice—the young birds listening to the old and profiting by their instruction, as is proved by the cases previously cited of birds which had never heard the songs of their own species yet singing their songs, but doing so "tentatively and imperfectly."

Again, although terriers take to hunting rabbits instinctively, it is usual, as I have myself observed, for their parents to teach them, or lead on their natural instincts by imitation, whereby the hereditary aptitude develops more quickly than it would if left to itself.

The Duke of Argyll† give a curious case, which he "knows to be authentic," of a golden eagle in the possession of Mr. W. Pike, Glendarry, Co. Mayo, which in the spring of 1877 laid three eggs. These Mr. Pike took away, and substituted for them two goose eggs. The eagle hatched out the two eggs. One of the goslings died, and was torn up by the eagle to feed the survivor, "who, to the great tribulation of its foster-parent, refused to touch it. . . . The eagle, however, in the course of time, taught the goose to eat flesh, and (the goose having free exit and ingress to the eagle's cage) always called it by a sharp bark whenever flesh is given to it, when the goose hastens to the cage, and greedily swallows all the flesh, &c., which the eagle gives it."

Again, there is evidence to show that the knowledge which animals display of poisonous herbs is of the nature of a mixed instinct, due to intelligent observation, imitation, natural selection, and transmission; for, as Mr. Darwin points out in the Appendix, "lambs turned out without their mothers are very liable to eat poisonous herbs; and it seems to be certain that cattle, when first introduced into a country, are killed by eating poisonous herbs, which the cattle already introduced have learnt to avoid."‡

In this case there is indeed no evidence of the young

* Sir H. Davy gives an account of such laborious instruction as witnessed by himself in the case of the golden eagle. See *Animal Intelligence*, p. 290.

† *Nature*, vol. xix, p. 554.

‡ *Youatt on Sheep*, p. 404; and *Anns. and Mag. Nat. Hist.*, 2nd ser., vol. ii, p. 364, &c.

P 2

being intentionally instructed by the old, but they are instructed by themselves, *i.e.*, by their individual experience. And this is, after all, the most important point, or the point to which the intentional education by parents is subsidiary. I shall therefore give a few more instances to show that many instincts (usually those of obviously secondary origin) are first manifested by young animals in an imperfect, or not fully evolved condition, and afterwards become perfected in the school of individual experience. Such cases stand in marked antithesis to those of the congenitally perfect instincts already alluded to, which have been so well investigated by Mr. Spalding.

It is unquestionably a true instinct that leads a ferret to thrust its long canines through the medulla oblongata of its victim; but Professor Buchanan states* that young ferrets, "instead of having for their single object to put themselves into a position to inflict the death wound, engage in conflict with rats;" yet they had the proper instinct, though not in complete working order, for they attacked properly the medulla oblongata of dead rats. Similarly I myself observed with the ferrets which I reared under a hen, that when half-grown and put to a rabbit for the first time, they clearly knew that their attack should be directed against one end of the rabbit, but were not quite certain which; for after some time of indecision they in the first instance attacked the rump, and only after finding this of no use tried the proper place. But of more interest still in this connection was the behaviour of these ferrets when half-grown towards a fowl. They had been taken away from their foster-mother, the hen, some weeks previously, but still no doubt retained a recollection of her. Therefore, when presented with another hen, their hereditary instincts prompted attack, while their individual associations inhibited the prompting. There was therefore a manifest conflict of feelings, which had its expression in a prolonged period of indecision. And although eventually the hereditary instincts prevailed over the associations formed by individual experience, the prolonged hesitation proved that the latter exerted a strong modifying force.

Mr. Darwin says in his MSS that in 1840 he saw some chickens which had been hatched without a mother, and

* *Anns. and Mag. Nat. Hist.*, vol. xviii, p. 378.

"when exactly four hours old they ran, jumped, chirped, scratched the ground, and cuddled together as if under the hen; all actions beautifully instinctive." After giving this as an instance of what I have called pure instinct, he proceeds by way of comparison to say, "It might have been thought that the manner in which fowls drink, by filling their beaks, lifting up their heads, and allowing the water to run down by its gravity, would have been specially taught by instinct; but this is not so, for I was most positively assured that the chickens of a brood reared by themselves generally required their beaks to be pressed into a trough, but if there were older chickens present, who had learnt to drink, the younger ones imitated their movements, and thus acquired the art."

Upon the whole, then, with reference to the modes whereby intelligence operates in modifying instinct, we may say that in all cases when it does so, there must first be intelligent perception of the desirability of the modification on the part of certain individuals, who modify their actions accordingly. In some cases the principle of imitation probably assists in changing the instinct by inducing other individuals of the same species, and living in the same area, to follow the example of their more intelligent companions; or the principle of imitation may come in at an earlier stage, the habits of one species *suggesting* to the members of another species the modification of an instinct. Lastly, intelligence may operate by the intentional tuition of young by their parents.

But perhaps the best evidence of the extreme modification which instinct may be made to undergo by the effects of individual experience, or of changed conditions of life, is that which is afforded by the enormous mass of facts to which we are naturally led on by some of the cases just given; I mean the facts connected with the domestication of animals. For the effects of domestication in modifying instincts are quite as strongly shown as are its effects in modifying structures, as was long ago observed by Dr. E. Darwin. So important and extensive a class of facts, however, require to be considered by themselves. I shall therefore now proceed to do this without any further special reference to the effects of imitation or of education operating upon instinct during the lifetime of the individual.

CHAPTER XV.

INSTINCT *(continued)*.

DOMESTICATION.

FROM the nature of the case it is not to be expected that we should obtain a great variety of instances among wild animals of new instincts acquired under human observation, seeing that the conditions of their life as a rule remain pretty uniform for any periods over which human observation can extend. But fortunately, from a time anterior to the beginning of history, mankind, in the practice of domesticating animals, has been engaged on making what we may consider a gigantic experiment on this subject. Seeing that the animals chosen for this purpose have been bred and reared under human care for a series of innumerable generations, and that in some cases the members of certain "breeds" are persistently selected and trained to perform certain kinds of work, we should expect, if instincts arise by secondary means in conjunction with primary, to find evidence, not only of the dwindling of natural instincts, but also of the formation of new and special instincts. For it is evident that artificial education and artificial selection by man are influences the same in kind, though not in degree, as those of natural education and natural selection, to the combined operation of which our theory ascribes the formation of instincts. We might therefore, as I have said, expect to find among our domestic animals some evidence of the formation of what we may call artificial, or in Mr. Darwin's phraseology, domestic instincts. And such evidence we do find.

Taking first the case of the impairment or loss of natural instincts, I have already alluded to the striking example supplied by the hereditary tameness of domesticated animals. More, however, now remains to be said on this point, for it will be remembered that previously our attention was confined to cases in which this loss is to be attributed to changed experience alone, without the aid of selection, or to primary means unassisted by secondary. In this connection I adduced the cases of the Rabbit and the Duck; I shall now adduce the cases in which artificial selection has probably assisted mere disuse in obliterating natural wildness.

The most remarkable of these instances is perhaps that supplied by the Cat, inasmuch as the nearest congener of this animal—the wild cat—is the most obstinately untameable of all animals. The case of the Dog, however, is in this connection scarcely less remarkable, seeing that fierceness and distrust are such constant features in the psychology of all the wild races. Probably, too, if there were such an animal now in existence as the truly wild Horse, we should find its disposition to resemble that of the Zebra, Quagga, or Wild Ass, the latter of which, though not so untractable as either of the former, is nevertheless a very different animal in this respect from our proverbially patient donkey. Similarly, as Handcock observes, "In the wild state kine possess acuteness of sight and smell, and a spirit of fierceness in defending their young, which disappear when, by domestication, we have reduced them to a condition in which the former of these qualities would be of no value, and the latter dangerous to themselves and others." This consideration led Handcock to add the shrewd remark, "Upon the whole it seems to be established as a principle that, where there is no room for the exercise of pure instinct, either by man's interposition or otherwise, it will languish, like all the natural senses."*

So much, then, to prove that instinctive wildness is eradicated from all species which have been sufficiently long exposed to the influences of domestication. I shall now give a few facts to show that the power of domestication thus to reduce or destroy the innate tendencies of wild animals extends to still more special lines of psychological formation.

Mr. Darwin says† "All wolves, foxes, jackals, and species of the cat genus, when kept tame, are most eager to attack poultry, sheep, and pigs; and this tendency has been found incurable in dogs which have been brought home as puppies

* *Zoological Journal*, p. 320.

† *Origin of Species*, p. 211.

from such countries as Tierra del Fuego and Australia, where the savages do not keep these domestic animals.* How rarely, on the other hand, do our civilized dogs, even when quite young, require to be taught not to attack poultry, sheep, and pigs. No doubt they occasionally do make an attack, and are then beaten; and, if not cured, are destroyed; so that habit and some degree of selection have probably concurred in civilizing by inheritance our dogs. On the other hand, young chickens have lost, wholly by habit, that fear of the dog and cat which no doubt was originally instinctive in them; for I am informed by Captain Hutton that the young chickens of the parent stock, the *Gallus bankiva*, when reared in India under a hen, are at first excessively wild. So it is with young pheasants reared in England under a hen. It is not that chickens have lost all fear, but fear only of dogs and cats; for if the hen gives the danger-chuckle, they will run (more especially young turkeys) from under her, and conceal themselves in the surrounding thickets." The MS adds, "Pigeons are not as constantly kept as poultry, and every fancier knows how difficult it is to keep his favourites safe from their incorrigible enemy—the cat."

As additional evidence that instincts may be lost, or as Handcock says, "languish" under domestication, it is enough to point to the instinct of incubation having become aborted in the Spanish hen; and to the maternal instincts having similarly dwindled in cattle in certain parts of Germany, where for hundreds of generations it has been the custom to remove the calves from the mothers immediately after birth.† The same authority says that sheep will allow strange lambs to suck them in countries where it has long been the custom to change lambs, which is not the case with other sheep.

* In the MSS detailed evidence on this point is given, from which I quote the following:—

"This was the case with a native dog from Australia, whelped on board ship, which Sir J. Sebright tried for a year to tame, but which 'if led near sheep or poultry became quite furious.' So again Captain FitzRoy says that not one of the many dogs procured from the natives of Tierra del Fuego and Patagonia which were brought to England could easily be prevented from indulgence in the most indiscriminate attack on poultry, young pigs, &c.' (Colonel H. Smith, on *Dogs*, 1840, p. 214; and Sir J. Sebright, on *Instinct*, p. 12. Also Waterton's *Essay on Nat. Hist.*, p. 197, for extreme wildness of young pheasants at sight of a dog.)" And the MSS also contain a letter from Sir James Wilson, giving Mr. Darwin an account of a tamed Dingo, which obstinately persisted in killing poultry and ducks whenever he got loose.

† Stuorn, *Ueber Racen*, &c., s. 82.

Lastly, according to Mr. J. Shaw, "where the dog is valued solely for food, as in the Polynesian Islands and China, it is described as an extremely stupid animal,"* and White says, in his "Natural History of Selborne,"† that these dogs have lost some of what we must regard as their strongest instincts, for "though they are so strictly carnivorous animals, from having been for so many generations fed on vegetable food they have lost their instinctive taste for flesh."

Thus much, then, for what we may call the negative influence of domestication, or its power of destroying natural instincts. We shall now turn to the still more striking and suggestive side of the subject, viz., the positive influence of domestication in developing new instincts not natural to the species, but artificially produced by accumulative instruction through successive generations, combined with selection. And here I shall confine myself to the species of domestic animal in which these effects have been most conspicuous, viz., the Dog. Doubtless the reason why these effects are most conspicuous in the case of this animal is because his utility to man has always depended mainly upon his intelligence, so that man has here persistently directed the influences of domestication towards an artificial shaping of that intelligence. For it is in this connection of interest to observe that the only features in the primitive psychology of the dog which have certainly remained unaffected by contact with man, are those features which, being neither useful nor harmful to man, have never been either cultivated or repressed. Such is the case, for example, with the instincts of covering excrement, rolling in filth, turning round and round to make a bed, hiding food, &c.‡

As evidence of the positive influences of domestication on the psychology of the dog, I may first draw attention to what occurs to me as a very suggestive case. One of the most dis-

* This sentence occurs as a quotation in a letter by Mr. Shaw to Mr. Darwin, but the reference is not supplied.

† Letter 57.

‡ La Malle says that it is not until dogs are ten or twelve months old that they begin to bury superfluous food. This, if true, would point to the conclusion that the instinct was one lately acquired in the history of the wild species, and therefore presumably is not so firmly fixed as the instincts of wildness, fierceness, attacking poultry, and so on, which have been so completely eradicated by human agency.

tinctive peculiarities of the psychology of the dog is the high degree in which there are developed the ideas of ownership and property—ideas which have of course been bred into canine intelligence by man. Most carnivorous animals in their wild state have an idea of property as belonging to captors, and the manner in which certain predacious Carnivora take possession of more or less definite areas as their hunting-grounds implies an incipient notion of the same thing. From the germ thus supplied by nature the art of man has operated in the case of the dog, till now the idea of defending his master's property has become in this animal truly instinctive. Without any training, and even sometimes against training, many dogs will bark and fly at strangers passing the gates or doors which bound their master's premises. Instances without number might be multiplied to show the careful vigilance of dogs over property entrusted to their charge; but, as the fact is so well known, space need not here be occupied with its proof. I shall, however, give one or two observations which I myself made in this connection on a terrier which I reared from puppyhood, because I am perfectly certain that in this case the idea of protecting property was innate or instinctive, and not due to individual instruction. I have seen this dog escort a donkey which had baskets on its back filled with apples. Although the dog did not know that he was being observed, he accompanied the donkey all the way up a long hill for the express purpose of guarding the apples. For every time that the donkey turned back his head to take an apple out of the baskets, the terrier sprang up and snapped at his nose; and such was the vigilance of the dog that, although his companion was keenly desirous of tasting some of the fruit, he never allowed him to get a single apple during the half hour that they were left together. I have also seen this terrier protecting meat from other terriers, which lived in the same house with him, and with which he was on the best of terms. More curious still, I have seen him seize my wrist-bands while they were being worn by a friend to whom I had temporarily lent them—no doubt recognizing them as mine by his sense of smell, which was exceedingly good.

Akin to this inborn idea of protecting the property of his master, is the idea which the dog has of himself as constituting a part of that property—*i.e.*, the idea of ownership as extended to himself. That this idea is likewise inborn I have observed in the case of a very young Newfoundland puppy which was given to me when scarcely able to toddle, but which nevertheless at once followed me through tolerably crowded streets. Yet this puppy can scarcely have known me from any of the other persons he met, and therefore he can only have followed me from his instinctive idea of ownership, and his consequent fear of getting lost. This abstract idea of ownership is well developed in many, if not in most dogs; so that, for instance, it is not at all an unusual thing to find that if a master consigns his dog to the care of a friend previously unknown to the animal, the latter will feel quite safe under the charge of one whom he has seen to be his master's friend. For the time being the allegiance of the animal is transferred, and he feels to his master's friend, not as to a stranger, but as to a deputed owner. It is not, I think, improbable that what appears to be the acquired instinct of barking is, as it were, an offshoot from this acquired instinct of property, and of protecting self as property, by drawing the attention of a master to the approach of strangers or enemies.

Mr. Darwin has made a strong point of other and still more special "domestic instincts" of the dog, which are perhaps even more interesting than those above mentioned, from the fact of their having been intentionally bred into the animals by continued training with selection; I allude to the instincts of the sheep-dog, retriever, and pointer. He briefly alludes to these cases in the "Origin of Species" (p. 209), but dwells more fully upon them in his uncondensed MSS, from which therefore I shall quote.

"Look at the several breeds of Dogs, and see what different tendencies are inherited, many of which cannot, from being utterly useless to the animal, have been inherited from their one or several wild prototypes. I have talked with several intelligent Scotch shepherds, and they were unanimous in saying that occasionally a young sheep dog without any instruction will naturally take to run round the flock, and that all thorough-bred dogs can be easily taught to do this; and although they intensely enjoy the exercise of their innate pugnacity, yet they do not worry the sheep, as any wild canine animal of the same size would do. Look again at the Retriever, which so naturally takes to bringing back any object to his master. The Rev. W. D. Fox informs me that he taught in a single morning a Retriever six months old to fetch and carry well, and in a second morning to return on the path to search for an object left purposely behind and not seen by the dog. Yet I know from experience how difficult it is to teach the habit at least to terriers.

"Let us consider one other case, though so often quoted, that of the Pointer. I have myself gone out with a young dog for the first time, and his innate tendency was shown in a ludicrous manner, for he pointed fixedly not only at the scent of game, but at sheep and large white stones; and when he found a lark's nest, we were actually compelled to carry him along; he backed the other dogs. . . . The silence of Pointers, also, is the more remarkable, as all who have studied these dogs agree in classing them as a sub-breed of Hound, which gives tongue so freely. But the tendency in the young Pointer to back other dogs, or to point without perceiving any scent of game when they see other dogs point, is perhaps the most singular part of his inborn propensities.*

"Now if we were to see one kind of wolf, in a state of nature, running round a herd of deer, and skilfully driving them whither he liked, and another species of wolf, instead of chasing its prey, standing silent and motionless on the scent for more than half an hour with the other wolves of the pack all assuming the same statue-like attitude and cautiously approaching, we should surely call these actions instinctive. The chief characteristics of instinct seem to be fulfilled in the pointer. A young dog cannot be supposed to know why he points, any more than a butterfly why it lays its eggs on a cabbage. . . . . It seems to me to make no essential difference that pointing is of no use to the dog, only to man; for the habit has been acquired through artificial selection and training for the good of man, whereas ordinary instincts are acquired through natural selection and training exclusively for the animal's own good. The young pointer often points without any instruction, imitation, or experience; though, no doubt, as we have also seen sometimes to be the

* "With respect to the inherited tendency to back, see St. John's *Wild Sport of the Highlands*, 1846, p. 116; Colonel Hutchinson on *Dog Breaking*, 1850, p. 144; and Blaine, *Ency. of Rural Sports*, p. 791.—Besides the tendency to point, pointers inherit a peculiar manner of quartering their ground."

case with true instincts, he often profits by these aids. Moreover, each breed of dogs delights in following his inborn propensity.

"The most important distinction between pointing, &c., and a true instinct, is that the former is less strictly inherited, and varies greatly in the degree of its inborn perfection: this, however, is just what one might have expected; for both mental and corporeal characters are less true in domestic animals than in those in a state of nature, inasmuch as their conditions of life are less constant and man's selection and training far less uniform, and have been continued for an incomparably shorter period, than is the case in nature's productions."

Although the familiar fact of young pointers pointing instinctively does not need further corroboration, I shall quote a brief passage from the paper of Mr. Andrew Knight on "Hereditary Instincts,"* because it shows, as in the case of "backing," to what extreme nicety of detail the hereditary knowledge may in some cases extend.

"It is well known that very young pointers, of slow and indolent breeds, will point partridges without any previous instruction or practice. I took one of them to a spot where I had just seen a covey of small partridges alight in August, and amongst them I threw a piece of bread to induce the dog to move from my heels, which it had very little disposition to do at any time, except in search of something to eat. On getting among the partridges, and perceiving the scent of them, its eyes became suddenly fixed, and its muscles rigid, and it stood trembling with anxiety for several minutes. I then caused the birds to take wing, at sight of which it exhibited strong symptoms of fear and none of pleasure. A young Springing Spaniel, under the same circumstances, would have displayed much joy and exultation, and I do not doubt but that the young pointer would have done so too, if none of its ancestry had ever been beaten for springing partridges improperly."

From this same paper I must quote the following and more or less analogous cases:—

"A young Terrier whose parents had been much employed in destroying Polecats, and a young Springing Spaniel whose

* *Phil. Trans.*, 1837, p. 367.

ancestry through many generations had been employed in finding Woodcocks were reared together as companions, the Terrier not having been permitted to see a Polecat or any other animal of similar character, and the Spaniel having been prevented seeing a Woodcock or other kind of game. The Terrier evinced, as soon as it perceived the *scent* of the Polecat, very violent anger; and as soon as it *saw* the Polecat attacked it with the same degree of fury as its parents would have done. The young Spaniel, on the contrary, looked on with indifference, but it pursued the first Woodcock it ever saw with joy and exultation, of which its companion, the Terrier, did not in any degree partake. . . In several instances young and wholly inexperienced dogs appeared very nearly as expert in finding Woodcocks as their experienced parents.

"Woodcocks are driven in frosty weather, as is well known, to seek their food in springs and rills of unfrozen water, and I found that my old dogs knew about as well as I did the degree of frost which would drive the woodcocks to such places; and this knowledge proved very troublesome to me, for I could not sufficiently restrain them. I therefore left the old experienced dogs at home, and took only the wholly inexperienced young dogs; but to my astonishment some of them, in several instances, confined themselves as closely to the unfrozen grounds as their parents would have done. When I first observed this I suspected that woodcocks might have been upon the unfrozen ground during the preceding night, but I could not discover (as I think I should have done had this been the case) any traces of their having been there; and as I could not do so, I was led to conclude that the young dogs were guided by feelings and propensities similar to those of their parents."

Elsewhere in his essay this author remarks, "It may, I think, be reasonably doubted whether any dog having the habits and propensities of the Springing Spaniel would ever have been known, if the art of shooting birds on the wing had not been acquired."

Lastly, with reference to those artificial instincts of the dog, which are of this highly specialized nature—amounting, in fact, to hereditary memory of a most minute kind—I may allude to a remark made by Professor Hermann, that sporting dogs appear, when first taken out to hunt, and therefore previous to any individual experience, to anticipate the effects of a gun in bringing down a bird.*

Suggestive, however, as is the formation by man of such special canine instincts as we have now considered, we have in them only, as it were, small details of the modification which human agency has produced in the psychology of the dog. It is, indeed, not more true that man has in a sense created the remarkable structure of the greyhound or the bulldog, than that he has implanted the no less remarkable instincts of the pointer or the retriever; but we should gain a very inadequate conception of the profound influence which he has exercised in moulding the mind of this animal were we to confine our attention to such special cases as these.

If we contrast the psychology of "the friend of man" with that of any of the wild breeds, we see at once, not only that the animal has had many of its natural instincts suppressed and many artificial instincts imposed, but also that it has acquired, as Sir J. Sebright has observed, "an instinctive love of man." But the general affection, faithfulness, and docility of the dog, are too proverbial to need special exposition. We have merely to observe that these qualities, so unlike anything with which we meet in wolves, foxes, jackals, and wild dogs generally, can only be attributed to prolonged contact with, and selection by, his human masters; so that as the domestic dog is at present constituted these artificially imposed qualities usually lead the animal to entertain higher affection and faithfulness towards man than towards its own kind. It may not be superfluous in this connection again to point out that among wild animals we do not unfrequently find a disposition to associate with members of other species, even when no actual benefit arises from the association; and in this accidental or useless proclivity we may distinguish the germ which in the case of the dog has been cultured into what we see—amply justifying the remark of the old writer quoted by Darwin, "A dog is the only thing on this earth that luvs you more than he luvs himself."

Not only affection, faithfulness, and docility, but likewise all other emotional qualities of the dog which are useful to man have been developed by man to the extraordinary degree which we observe. It would be superfluous to cite,

* *Handbuch der Physiologie*, Bd. II, Theil II, pp. 282-3.

or even to give references to cases illustrating the exalted level to which sympathy has attained. This, together with the intelligent affection from which it springs, gives rise to a love of approbation and dread of blame, which as far as they go are in no way distinguishable from the same feelings as they occur in man himself. To this subject I shall have to return when in my next work we come to treat of the genesis of Conscience.

Again, as Mr. Grant Allen has pointed out, the sense of *dependence* which a dog shows is very instructive. "The original dog, who was a wolf or something very like it, could not have had any such artificial feeling. He was an independent, self-reliant animal. . . . But at least as early as the days of the Danish shell-mounds, perhaps thousands of years earlier, man had learned to tame the dog." Therefore, as a result of continuous education, selection, and breeding, although "among a few dogs, like those of Constantinople, the instinct may have died out by disuse, ... when a dog is brought up from puppyhood under a master, the instinct is fully and freely developed, and the masterless condition is thenceforth for him a thwarting and disappointing of all his natural feelings and affections."*

Indeed, so strong are the combined effects of long-continued breeding and individual education, that they may overcome the strongest of natural instincts and desires—witness a dog which will starve rather than steal, and also the recorded cases in which even the maternal instinct has been overborne by the desire of serving a master. To give only one example of this surprising fact, I shall quote from the "Shepherd's Calendar" of the poet Hogg:—

A collie belonged to a man named Steele, who was in the habit of consigning sheep to her charge without supervision. On one occasion, says Hogg, "whether Steele remained behind or took another road, I know not, but, on arriving home late in the evening, he was astonished to hear that his faithful animal had never made her appearance with the drove. He and his son, or servant, instantly prepared to set out by different paths in search of her, but on their going out into the streets, there was she coming with the drove, not one missing, and, marvellous to relate, she was carrying a young pup in her mouth. She had been taken in travail on

* *Evolutionist Abroad*, p. 182, *et seq.*

the hills, and how the poor beast had contrived to manage her drove in her state of suffering is beyond human calculation, for her road lay through sheep the whole way. Her master's heart smote him when he saw what she had suffered and effected; but she was nothing daunted, and, having deposited her young one in a place of safety, she again set out full speed to the hills and brought another and another, till she brought her whole litter, one by one; but the last one was dead."

There is still one respect—and this a most suggestive one—in which artificial instincts resemble natural instincts, over and above that of obliteration by disuse or acquirement by training and selection. In order to show this it will be sufficient to quote the following passage from Mr. Darwin's MSS, part of which has already been published in the "Variation of Animals and Plants under Domestication" (vol. i, p. 43):—

"It is well known that when two distinct species are crossed, the instincts are curiously blended, and vary in the successive generations, just like corporeal structures. To give an example: a dog kept by Jenner (Hunter's "Animal Economy," p. 325), which was grandchild, or had a quarter-blood of the jackal in it, was easily startled, was inattentive to the whistle, and would steal into fields and catch mice in a peculiar manner. Now I could give numerous examples of crosses between breeds of dogs, both having artificial instincts, in which these instincts have been most curiously blended, as between the Scotch and English sheep-dog, pointer and setter: the effect, moreover, of such crosses can sometimes be traced for very many generations, as in the courage acquired by Lord Orford's famous greyhounds from a single cross with the bull-dog ("Youatt on the Dog," p. 31). On the other hand, a dash of the greyhound will give a family of sheep-dogs a tendency to hunt hares, as I was assured by an intelligent shepherd."

Our *à posteriori* proof of Proposition VII is now concluded, and with its proof our considerations on the origin and development of instinct are drawing to a close. For we have now seen that instincts may arise under the influence of natural selection alone, under that of lapsing intelligence alone, or under both these influences combined. And in proving that habits intelligently acquired may, like habits acquired without intelligence, be inherited, we have also proved, as in the analogous case of primary instincts, that these habits in the course of generations may vary, that their variations may be inherited, and that the favourable variations may be fixed and further intensified by natural or artificial selection. For it is only by granting all these statements that we can possibly explain many of the foregoing facts. Clearly man could never have produced the artificial instincts of the dog, unless he had practically recognized the facts of variability and inheritance—a recognition which is forcibly expressed in the immense difference between the market value of a pointer or setter of important pedigree, and a pointer or setter whose parentage is unknown. As Thompson well says:—"It would be necessary to recommence the business of training with each successive generation, if the bodily and mental changes which the animals have undergone in the continued process of domestication had not become so engrafted as to be propagated with them. These acquired characteristics have gathered fresh strength in each succeeding generation, till at length they have assumed a permanent stamp." And if artificial selection is of such high importance in the formation of domestic instincts, much more must natural selection be of importance in the formation of natural instincts.

CHAPTER XVI.

INSTINCT *(continued)*.

LOCAL AND SPECIFIC VARIETIES OF INSTINCT.

I HAVE now shown that instincts may arise through the influence of natural selection, or of lapsing intelligence, or of both these principles combined; and that even fully formed instincts are liable to change when changing circumstances require. The most striking evidence on this head, or that of the mutability of fully formed instincts, is perhaps the evidence given in the last chapter, showing the influence of domestication both in obliterating the strongest of natural and in creating the most fantastic of artificial instincts. But inasmuch as we have previously seen that any considerable change in the circumstances to which an instinct is appropriate, is apt to throw the machinery of that instinct out of gear, the evidence of the mutability of instinct drawn from the effects of domestication may be open to the criticism that the changes produced are of an unnatural character, or due to an impairment of the normal apparatus of instinct. I do not myself think that if this criticism were raised it would be one of any force, seeing that domestication not only has the negative effect of impairing or destroying natural instincts, but also, as I have said, the positive effect of creating artificial instincts. Still it is desirable to supplement the evidence drawn from the facts of domestication with further evidence drawn from the field of nature; for here, at least, no criticism of the kind which I have suggested can be advanced. I propose, therefore, in this chapter to consider all the facts which I have been able to collect, tending to show that among animals in a state of nature instincts undergo transformations which are precisely analogous to those that they undergo among animals in a state of domestication. The kind of evidence on which I rely to show this is two-fold—1st the occurrence among wild animals of local varieties of instinct, and 2nd the similar occurrence of specific varieties.

Local Varieties of Instinct.

Under the first of these two divisions I shall seek to show that the mutability of instinct finds a most marked and suggestive expression in certain cases where wild animals of the same species living in different parts of the world (and therefore exposed to different environments), present differences in their instinctive endowments of a marked and constant kind. One class of such cases has already been given with reference to the acquisition of an instinctive fear of man by those animals in a state of nature which inhabit localities frequented by man: but as the subject appears to me an important one—seeing that a definite local variety is on its way to becoming a new instinct—I shall now give all the best instances which I have been able to collect.

Beginning with insects, Kirby and Spence state on the authority of Sturm that the dung-beetle, which rolls up pellets or little balls of dung, saves itself the trouble of making the pellets when it happens to live on sheep-pastures; for it then "avails itself of the pellet-shaped balls ready made to its hands which the excrement of the sheep supplies." Here we have intelligent adaptation to peculiar conditions, and so the case might have been quoted as one of the plasticity of instinct; but as sheep-pastures are definite local areas, I have quoted it as a case of the local variation of instinct. All cases of such local variation must have some determining cause, and doubtless most frequently this cause is intelligent adaptation to peculiar local conditions. Therefore I have chosen this case to lead off with just because it might equally well be quoted in this or in the previous chapter.

Again it is stated by Lonbiere, in his history of Siam, "that in one part of that kingdom, which lies open to great inundations, all the ants made their settlements upon trees; no ants' nests are to be seen anywhere else." And Forel states a closely similar fact with reference to a species of European ant, *Lasius acerborum*, which on the plains is never found to build under stones, while in the Alps it frequently builds under the same stones as the *Myrmica.*

With regard to Bees, it appears that both in Australia and California, the hive-bees when introduced "retain their industrious habits only for the first two or three years. After that time they gradually cease to collect honey till they become wholly idle."* Again, Mr. Packard, jun., records some observations† which were made by the Rev. L. Thompson, whom he designates "a careful observer," of bees (*Apis mellifica*) eating moths which were entrapped in certain flowers. On the fact being communicated to Mr. Darwin, he wrote, that he "had never heard of bees being in any way carnivorous, and the fact is to me incredible. Is it possible that the bees opened the bodies of the *Plusia* to suck the nectar contained in their bodies? Such a degree of reason would require confirmation, and would be very wonderful." But whatever the object of the bees may have been, their actions, which are described as "suddenly darting" and "furious," certainly display some marked variation of instinct under the guidance of intelligence. Moreover, the explanation entertained by Messrs. Thompson and Packard—viz., that the bees were partly carnivorous, is perhaps not so "incredible" as it appeared to Mr. Darwin, if we remember that wasps are unquestionably apt to develop carnivorous tastes.‡

Turning now to local variations of instinct in Birds, I may first allude to the following instances in the Appendix, which although not adduced in this connection by Mr. Darwin, are no less apposite to it.

"It is notorious that the same species of bird has slightly different vocal powers in different districts; and an excellent observer remarks, 'an Irish covey of partridges springs without uttering a call, whilst on the opposite coast the Scotch covey shrieks with all its might when sprung.'§ Bechstein says that from many years' experience he is certain that in

* *Animal Intelligence*, p. 188, where see for references to Dr. E. Darwin, Kirby and Spence, and later writers on this matter.

† *American Naturalist*, Jan. 1880.

‡ See, *e.g.*, *Nature*, vol. xxi, pp. 417, 494, 538, and 563, detailing observations of the fact by Sir D. Wedderburn, Messrs. Newall, F.R.S., Lewis Bod, and W. G. Smith.

§ W. Thompson, in *Nat. Hist. Ireland*, vol. ii, p. 65, says he has observed this, and that it is well known to sportsmen.

the Nightingale a tendency to sing in the middle of the night or in the day runs in families and is strictly inherited."*

Professor Newton informs me that the Ring-plover on the extensive sand-dunes of Norfolk and Suffolk habitually displays a very curious and instructive case. These birds naturally build on the sea-shore, depositing their eggs in a hollow which they scoop out in the shingle. The sea has retreated for miles from the extensive sand-dunes in question, which have become covered with grass. Apparently the Ring-plovers have gone on breeding for numberless generations on the site which was at one time the sea-coast, the distance between them and the sea having therefore gradually increased more and more.† Hence the birds are now living on wide grassy surfaces instead of on shingle, but their instinct of laying their eggs on stones remains; so that after having scooped out a hollow in the ground, they collect small stones from all quarters and deposit them in the hollow. This has the effect of rendering their nests very conspicuous, and the fact shows in a striking way how a fixed ancestral instinct may, while in the main persisting under changed conditions of life, nevertheless so vary in reference to these changed conditions as to constitute the beginning of a new instinct.

For further instances of local variation in the instincts of nest-building, I may in this connection again refer to the highly instructive cases previously mentioned in illustration of the plasticity of instinct under the moulding influence of intelligence.‡ I allude to the fact that on the American Continent various species of birds—notably a kind of Owl, a Blue-bird, the Pewit Flycatcher, several species of Wren, and nearly all the species of Swallow—have adapted the structure of their nests to the artificial nesting-places provided by man, in just the same way (though more gradually and on a much larger scale) as did the colony of Palm-swifts in Jamaica. But with still more special refer-

* *Stuben-Vögel*, 1840, s. 323; see on different powers of singing in different places, s. 205 and 265.

† That this is the explanation is not merely probable *à priori*, but receives additional corroboration from the fact that these same sand-dunes are now the habitat of a species of lepidopterous insect which elsewhere is found upon the coast.

‡ See above, p. 210. Compare also many of the cases given in the Appendix.

ence to the local variation of instinct, I may here quote a further statement from Captain Coues' work previously cited; for it shows that even on different parts of the American Continent the same species of birds exhibit these differences in their mode of nest-building. He says:—"There is no question of the fact that some of the Swallows which in the East now invariably avail themselves of the accommodation man furnishes, in the West live still in holes in trees, rocks, or the ground;" and he proceeds to give several special instances.* Lastly, the fact has already been noted that House-sparrows exhibit a similar local variation of instinct wherever they come into contact with the dwellings of man.†

Passing on now to other animals, we find several instructive cases of the local variation of instinct among the Mammalia. Thus the curious habit has been observed among cattle inhabiting certain districts of sucking bones. Archbishop Whately made this the subject of a communication to the Dublin Natural History Society many years ago. Recently it has been observed by Mr. Donovan of cattle in Natal,‡ and by Mr. Le Conte, of cattle in the United States.§ Probably this habit is induced by the absence of some constituent of food in the grass which is supplied by the bones, and therefore if the habit happened to prove beneficial to the cattle (instead of deleterious as Whately asserts), it is easy to see that cattle in a state of nature might become transmuted from herbivorous to omnivorous, or even purely carnivorous. Probably the ancestors of the Pig have passed through the former of these stages. On the other hand, the Bear seems to be in process of becoming omnivorous from the contrary direction—being carnivorous in its affinities, but not infrequently adopting the habit of eating grass and herbs.

And in this connection I may refer to an interesting case of transition from herbivorous to carnivorous habits which was published at the Academy of Natural Science of Phila-

* *Op. cit.*, p. 394. This fact, I think, tends to confirm the statement of Mr. Edward (*Zool.*, p. 6842) that on the coast of Banffshire the house-swallow presents a local instinct of building in caves and on projecting rocks.

† When house-sparrows build in trees—which they occasionally do and which must be regarded as reversion to primitive instinct—"the structure is very large, more than a yard in circumference, and covered with a dome." (*Yarrel's British Birds*, 4th Ed., Pt. X, p. 90.)

‡ *Nature*, vol, xx, p. 457. § *Ibid.*

delphia on February 18th, 1873, by Mr. W. K. G. Gentry. A rodent popularly known as the Chickaree (*Scinus hudsonius*), which like most of its kind is normally herbivorous, has adopted in the neighbourhood of Mount Airy a habit common among the *Mustelidæ*, of climbing trees for the purpose of catching birds and sucking their blood. Mr. Gentry suggests that this transition from herbivorous to carnivorous habits may have arisen from the propensity shown by some squirrels of sucking the eggs of birds—the passage from this habit to that of sucking the blood of birds being but small. Lastly, in this connection I may adduce a precisely analogous case of a marked local variation of instinct taking place in a species of bird.

Mr. I. H. Potts, writing from Ohinitahi to "Nature" (February 1st, 1872), says that the mountain parrot (*Nestor notabilis*) was then exhibiting a "progressive development of change in habits from the simple tastes of a honey-eater to the savageness of a tearer of flesh." For "the birds come in flocks, single out a sheep at random, and each alighting on its back in turn, tears out the wool, and makes the sheep bleed, till the animal runs away from the rest of the sheep. The birds then pursue it, and force it to run about till it becomes stupid and exhausted. If in that state it throws itself down, and lies as much as possible on its back to keep the birds from picking the part attacked, they then pick a fresh hole in its side, and the sheep, when so set upon, in some instances dies. . . . Here we have an indigenous species making use of a recently imported aid for subsistence, at the cost of a vast change in its natural habits." Since this account was written the change of habits in question has grown to become a very serious matter to the sheep-farmers. It appears that the birds prefer the fat parts of their victims, and have learnt to bore into the abdominal cavity straight down upon the fat of the kidneys, thus of course killing the sheep.

Another case of local variation of instinct is furnished by the statement of Adamson, that in the island of Sor rabbits do not burrow. This statement, however, although accepted by Dr. E. Darwin, has not, so far as I know, been either confirmed or refuted. But with reference to variations in the instinct of burrowing, I may allude with more confidence to the case given by Mr. Darwin in the Appendix on the authority of Dr. Andrew Smith, viz., "that in the uninhabited parts of South Africa the hyænas do not live in burrows, whilst in the inhabited and disturbed parts they do. Several mammals and birds usually inhabit burrows made by other species, but when such do not exist they excavate their own habitations."

In "Animal Intelligence" I stated, under the authority of Dr. Newbury's Report on the Zoology of Oregon and California, that the beavers in those districts exhibit the peculiarity of never constructing dams, and seeing that the building of these structures may be regarded as one of the strongest instincts manifested by the species, I supposed the failure of the Oregon and Californian beavers in manifesting this instinct to constitute a remarkable case of the local variation of instinct. Professor Moseley, however, who has travelled in Oregon, now writes me that this absence of beaver dams is in his opinion due simply to the severity with which the animals are trapped. "What few beavers that remain are too constantly liable to interruption to be able to construct dams, or for this to be worth their while. They thus live a more or less vagrant life about the streams." It will be observed, however, that Professor Moseley speaks of "the few beavers that remain," whereas Dr. Newbury says of the same districts:—"We found the beavers in numbers of which, when applied to beavers, I had no conception." Therefore I infer that since the time when Dr. Newbury's Report was published, the number of the beavers must have been greatly reduced by trapping. But if so, at the time when the Report was published, Professor Moseley's explanation of the absence of dams can scarcely have applied to the facts of the case. Hence, I am still disposed to think that we have in this case an instance of the local variation of instinct—seeing that the variation of habit was remarkable even before the introduction of the disturbing elements to which Professor Moseley now alludes. Be this as it may, however, it is certain that the solitary beavers of Europe present a striking local variation of instinct, not only in having lost their social habits, but also in having ceased to build either lodges or dams.

The last instance of the local variation of instinct which I have to adduce is one which has already attracted a good deal of attention; I refer to the barking of dogs.* The habit

* A somewhat analogous instance seems to be supplied by the "cat-a-

of barking, although perhaps acquired as a result of domestication, is so innate and general among most of the breeds, that it deserves to be regarded as an instinct. Yet Ulloa noticed that in Juan Fernandez the dogs did not attempt to bark till taught to do so by the importation of some dogs from Europe—their first attempts being strange and unnatural. Linnæus records that the dogs of South America did not bark at strangers. Hancock says that European dogs when conveyed to Guinea "in three or four generations cease to bark, and only howl like the dogs natives of that coast." Lastly, it is now well known that the dogs of Labrador are silent as to barking. So that the habit of barking, which is so general among domestic dogs as to be of the nature of an instinct, is nevertheless seen to vary with geographical position.

Specific Variations of Instinct.

To the above instances of the local variations of instinct, I shall now add a few cases of what we may call specific variations of instinct—that is to say, instincts which occur in a species of a character strikingly different from the instincts which occur in the rest of the genus. After what has been said on the local variations of instinct, the attesting value of the cases which we are about to consider must be evident. For we should expect that if the conditions which determine a local variation of instinct are constant over a sufficient length of time, the variation should become fixed by heredity, and so give rise to a change of instinct in the species affected—which change ought to become observable in the contrast exhibited by the instincts of this species and those of the rest of its allies. This head of evidence becomes of special value when we remember that it is the nearest approach we can hope to obtain of anything resembling a palæontology of instincts. Instincts, unlike structures, do not occur in a fossil state, and therefore in the course of their modification they do not leave behind them any permanent record, or tangible evidence, of their transformations. But we obtain evidence of transformation almost as conclusive in the cases to which I now allude; for if a living species

wallings" of cats; for, according to Roulin (quoted by Dr. Carpenter in *Contemp. Rev.*, vol. xxi, p. 311), the domestic cats in South America do not make these sounds.

inhabiting a certain restricted area exhibits a marked departure from the instincts elsewhere characteristic of its genus, we can scarcely question that the departure is indeed a *departure*—*i.e.*, that originally the instincts were the same as those occurring in the rest of the genus, but that owing to peculiar local conditions, local variations of instinct arose and were continued till they became hereditary, and so led to a *parting away* of the instincts of this species from those of its allies.*

For the sake of brevity I shall here confine my instances to those which may be drawn from Birds.

The following concise statement of facts relating to the strong instinct of parasitism in the only two genera of birds where it is known to occur, is quoted from an Editorial note in "Land and Water" (Sep. 7, 1867), and displays very remarkable and instructive cross-relations as regards the existence and absence of this instinct in the sundry species composing these two genera.

"The only non-cuculine genus of birds known up to the present time, which has the habit of entrusting its egg to the charge of strangers, is that of the cow-buntings (*Molothrus*), and the parasitic habit of *M. pecoris* of North America has been amply described by the ornithologists cited by our correspondent. There are several other species of this genus, and the same parasitic habit was observed in another of them by Mr. Darwin. The *Molothri* are birds belonging to the great American family of *Cassicidæ*, which corresponds to that of *Sturnidæ* in the Old World; and they are nearly akin to the troopials (*Agelaius*). It is remarkable that not any of the various American *Cuculidæ* are parasitic; whereas several genera of this family inhabiting the major continent and its islands, with Australia, are now well known to be so.

* From the above remarks it will appear that I do not agree with Mr. Darwin in his view, expressed in the Appendix, that cases of specific variation of instinct are difficulties in the way of his theory of the gradual development or evolution of instincts. On the contrary, for the reasons given above, I regard such cases as corroborations of this theory. The source of this difference of opinion is, that while Mr. Darwin is above all things anxious to find evidence of connecting links in the formation of an instinct, I feel that to expect such evidence in every case of instinct would be unreasonable, if not inconsistent with the theory that innumerable instincts owe their present existence to the destruction through natural selection of the animals which presented them in a lesser degree of perfection. I shall recur to this point in a future chapter.

First, there are the very numerous species of true *Cuculus*, with its immediate sub-divisions, inhabitants chiefly of Southern Asia, Africa, and Australia. Secondly, the crested cuckoos (*Coccystes*), exemplified by *C. glandarius*, which is common enough in Spain, and has been known to stray into this country. This bird deposits its eggs in the nests of magpies and crows. Another species (*C. melanoleucus*), which is very common in India, selects for this purpose the nests of a particularly noisy and familiar group of birds in that part of the world, often called 'dirt-birds' (*Malacocercus*); and as the latter lay a spotless blue egg, similar in colour to that of the hedge-chanter (*Accentor modularis*) of Europe, the egg of the particular cuckoo which seeks their nests is of a nearly similar spotless greenish-blue colour. Another very common Indian bird of this family is the koël (*Eudynamis orientalis*), the male of which is coal-black, with a ruby eye, and the female beautifully speckled. A pair, in fine condition, may now be seen in one of the aviaries in the Zoological Gardens. The Indian koël invariably deposits its egg in a crow's nest, and the egg is not unlike that of a crow in its colouring and markings. Several other species of koël inhabit the Asiatic islands, and there is one in Australia; and as the koëls are not migratory birds, it follows that the parasitic habit is independent of any migratory necessity. That extraordinary cuculine bird, the Australian channel-bill (*Scythrops novæ-hollandiæ*), is known to be parasitic, for the young have been repeatedly seen tended and fed by birds of other species; and therefore it is a *lapsus pennæ* on the part of Mr. Gould, in his 'Handbook of the Birds of Australia,' describing a specimen of it as having been an 'incubating female!" But the coucals (*Centropus*), very common and conspicuous birds in Southern Asia, Africa, and Australia, are not parasitic; neither, we have reason to believe, are the extensive malkoha series (*Phœnicophaus* and kindred genera), which inhabit the same geographical area. Among the American *Cuculidæ*, the species of *Coccyzus* are nearly akin to the crested cuckoos (*Coccystes*) of the major continent; and these, like the parasitic *Cuculidæ*, produce their eggs at considerable intervals, so that eggs and young of different ages are found in the same nest; while more advanced young, that had quitted the nest, are still fed by their parents while keeping to the immediate vicinity of the nest; as may likewise be observed of the screech-owls (*Strix*, as now limited). In the ani (*Crotophaga*), which have much in common with the coucals of the major continent, while in other respects their habits are very peculiar for birds of this family, 'an immense nest of basket-work' is formed by the united labours of a flock of them, usually on a high tree, where 'many parents bring forth and educate a common family.' Mr. Richard Hill, whose statements in Jamaican ornithology are worthy of unlimited confidence, writes Mr. Gosse, observes: 'Some half-dozen of them together build but one nest, which is large and capacious enough for them to resort to in common, and to rear their young ones together.' All of these diversified facts must be borne in mind by naturalists who would try to assign a reason for the parasitic habits of various *Cuculidæ*, as also those of the 'cow-buntings,' which have no other trait in common with the parasitic genera of *Cuculidæ*."

The Upland Goose of South America furnishes an admirable case of a fixed specific variation of instinct. These birds are true geese with well webbed feet; yet they never enter the water except perhaps for a short time after hatching their eggs, when they do so for the protection of their young. Similarly, Mr. Darwin's MS says of the Upland Geese of Australia, which also have well webbed feet, that "they are long-legged, run like gallinaceous birds, and seldom or never enter the water: Mr. Gould informs me that he believes they are perfectly terrestrial, and I am told that at the Zoological Gardens these birds and the Sandwich Islands Goose seem quite awkward in the water." The MS also points out that "the long-legged Flamingo likewise has webbed feet, yet lives on marshes, and is said seldom even to wade except in very shallow water. The Frigate bird with its extremely short legs never alights on the water, but picks up its prey from the surface with wondrous skill; yet its four toes are all united by a web; the web, however, is considerably hollowed out between the toes, and so tends to be rudimentary.

"On the other hand, there does not exist a more thoroughly aquatic bird than the Grebe, but its toes are only widely bordered by membrane. The water-hen may be constantly seen swimming about and diving with perfect ease; yet its long toes are bordered by the merest fringe of membrane. Other closely allied birds belonging to the genera *Crex*, *Passa*, &c., can swim well, and yet have scarcely any traces of web;

moreover their extremely long toes seem admirably adapted to walk over the softest swamps and floating plants; yet the common corncrake belongs to one of these very genera, and having the same structure of feet, haunts meadows, and is scarcely more aquatic than a quail or partridge."

The MS goes on to detail other and analogous cases, such as that of the Ground-woodpecker, Ground-parrots, and Tree-frogs, which have abandoned their arboreal habits; in all which cases the generic structures specially adapted to arboreal habits remain. Similarly the swallow-tailed Hawk is mentioned as catching flies on the wing like a swallow; a Petrel—" those more aërial of birds "—which has assumed the habits of an Auk; the Water-ouzel, a member of the Thrush family, which runs along the bottom of streams using its wings for diving and its feet for grasping stones under the water, "and yet the keenest observer could never have foretold this singular manner of life from the most careful examination of its structure."

All the above cases are given by Mr. Darwin, not in relation to Instinct, but to enforce his argument on adaptive structures being developed by natural selection instead of designed in special creation. But I have used them in relation to the development of Instinct, because, if we already believe in the natural evolution of organic structures, such cases as these afford the best possible evidence of the variation of instinct. As evolutionists we could have no stronger testimony to the previous though now obsolete instincts of a species, than that which is supplied by the presence of peculiar though useless structures which in allied species are correlated with particular instincts. For we must always remember, as previously observed, that instincts are never, like structures, fossilized, and therefore that we can never obtain direct historical evidence of their transmutation. But the best substitute for this evidence is, I think, such testimony as I have adduced of persisting structures pointing to obsolete instincts. Similar evidence in kind, though not quite so strong in degree, is furnished by cases in which one species of a genus, or one genus in a family, exhibits an instinct peculiar to that species or genus—*i.e.*, cases in which the instinct does not occur in allied species and genera; for this shows, if we already accept the doctrine of the transmutation of species, that the peculiar instinct must have arisen in the particular species or genus in question, after that species or genus had branched off from the more ancestral type. Now such cases of specific instinct are by no means rare—cases, I mean, like that of the Californian Woodpecker (*Melanerpes formicivorus*), which displays the curiously distinctive instinct of storing acorns in the crevices of the bark of the yellow pine (*Pinus ponderosa*) for future food, while no other species of woodpecker shows any tendency to such a habit.* But such cases of instinct peculiar to one species or genus are so common that I feel it would be needless to enumerate them, in view of the more conclusive cases just given—cases more conclusive because the obsolete instincts happen to have been of a kind requiring special corporeal structures for their operation, which now survive their ancestral uses.† Lastly, we must not forget the important fact that we are far from being wholly without evidence of the transmutation of instinct taking place under actual observation—as in the case of the ducks in Ceylon having quite lost their natural instincts with regard to water (in this resembling the upland geese), sparrows and swallows building on houses instead of on trees, insects, birds and mammals which normally feed on vegetable substances suddenly becoming carnivorous, &c., &c.; for all these cases of *local* varieties of instinct are really so many cases of *racial* varieties, and the step between this and *specific* varieties is clearly not a large one.

* According to Mr. C. J. Jackson (*Proc. Boston Nat. Hist. Soc.*, vol. x, p. 227) the acorns selected for storing are only those which are infested with maggots, which serve as food for the young in the following spring; and the acorns are driven into holes specially prepared for them, and which fit so well that the maggots when they come to maturity are unable to escape—being therefore imprisoned in a larder until they are required for the use of the young birds. See also J. K. Lord's *Naturalist in Vancouver's Island*, vol. i, pp. 289–92, and *The Ibis*, 1868.

† The most suggestive of this class of cases are those in which the species which exhibits the instinct peculiar to itself happens to have become dispersed over wide geographical areas since the instinct arose, and being therefore now found in different parts of the world, living under different conditions of life, and yet retaining the same peculiar instinct. Thus, for instance, "in all quarters of the globe species of trap-door spiders are found occurring in more or less localized areas," and the harvesting ants of Europe and America belong to the same genus. The South American Thrush lines its nest with mud in the same way as does our own Thrush, the Hornbills of Africa and of India in their nidification show the same peculiar instinct of imprisoning their hens in holes of trees with plaster, &c., &c.

CHAPTER XVII.

INSTINCT *(continued)*.

EXAMINATION OF THE THEORIES OF OTHER WRITERS ON THE EVOLUTION OF INSTINCT, WITH A GENERAL SUMMARY OF THE THEORY HERE SET FORTH.

MILL, from ignoring the broad facts of heredity in the region of psychology, may be said to deserve no hearing on the subject of instinct; and the same, though in a lesser degree, is to be remarked of Bain. Herbert Spencer, and his expositor Fiske, express with strong insistence the view that natural selection has been of very subordinate importance as an evolving source of instinct. Lewes virtually ignores natural selection altogether, but nevertheless is not in agreement with Spencer, inasmuch as Spencer regards instinct as "compound reflex action," and the precursor of intelligence, while, as we have already seen, Lewes regards it as "lapsed intelligence," and therefore necessarily the successor of intelligence. Thus, while Lewes maintains that all instincts must originally have been intelligent, Spencer maintains that no instinct need ever have been intelligent.* The deliverance of Darwin upon this subject I shall render bye-and-by.

The position of Mr. Spencer is severely logical, and this renders easy the definition of the points wherein I here disagree with him. His argument is that instinctive actions grow out of reflex, and in turn pass into intelligent actions, so that in his terminology an instinctive action need never have been intelligent, and an intelligent action need never become instinctive. He is express in saying that although "in its higher forms, Instinct is probably accompanied by a

* *I.e.*, no true instinctive action occurring in all individuals of a species; he recognizes the principle of lapsing intelligence in individuals.

rudimentary consciousness," nevertheless this consciousness is not essential to the formation of the instinct; but, on the contrary, is an effect of the growing complexity of the instinct—"the quick succession of changes in a ganglion, implying as it does perpetual experiences of differences and likenesses, constitutes the raw material of consciousness; the implication is that as fast as instinct is developed, some kind of consciousness becomes nascent."

Now, although we have seen in a previous chapter that this view contains much truth—and truth that is of special value in relation to the development of Consciousness—it appears to me impossible to obtain by it a complete explanation of the phenomena of instinct. Multitudes of facts of the kind which I have given may be rendered to prove that many of the higher instincts can only have arisen by way of "lapsed intelligence;" so that if I were called upon to adopt either the extreme view of Spencer, which abolishes intelligence and even consciousness as a factor in the formation of instinct, or the extreme antithetical view of Lewes, which ignores reflex action with natural selection as other factors in the process; I should feel less difficulty in choosing the latter than the former. Not only do many of the higher instincts bear internal evidence of having been at some period of their history determined by intelligence, and not only do many of these higher instincts now show themselves to be plastic under an admixture with "a little dose of judgment," but the examples of instinct which are chosen by Mr. Spencer are not, strictly speaking, examples of instinct at all. They are chosen as illustrations because they are the simplest cases of what is ordinarily called instinct, and so lie nearest to reflex action; if, however, we pause to examine any of them, we find that they are not true instincts, but cases of more or less elaborate neuro-muscular adjustment, or, in his own words, of "compound reflex action." And the fact that he defines or "describes" instinct as compound reflex action does not carry any proof that his doctrine is correct. To call a spade a club, and then argue that because it is a club it cannot be a spade, is futile; the question consists in the validity of the definition. Now it is just because we cannot draw a line between simple reflex action and "compound reflex action," so as to say that the one is mechanical and the other instinctive, that I have drawn the line at consciousness, and denominate all actions which occur below this line (howsoever compound) reflex, while reserving the term instinctive for habitual actions (howsoever simple) into which there enters this element of consciousness. And in doing this I feel certain that I am not merely imparting clearness to our classification, but also following the dimly intended meaning of the term instinct as ordinarily used. No one thinks of sneezing, or of the convulsions produced by tickling, as examples of instinctive actions; yet they are "compound reflex actions" to a degree of compounding not easily paralleled, and certainly much more so than any of the non-psychical adjustments which are given by Mr. Spencer as illustrations of instinct.

These illustrations have reference to polyps and creatures with rudimentary eyes, wherein the reactions to stimuli described appear to me, as I have said, in no way to deserve to be called instinctive. For instance, he shows how it is possible that without survival of the fittest and without intelligent adjustment, "psychical states being habitually connected, must, by repetition in countless generations, become so coherent that the special visual impression will directly call forth the muscular actions by which prey is seized. Eventually, the sight of a small object in front will cause the various motions requisite for the capture of prey." But even in this, the most extreme case supposed, if there is not and never has been any consciousness concerned, the complex adjustment is in no way distinguishable from a reflex action. When I observed jelly-fish crowding into the path of a sunbeam shining through a darkened tank, and saw that they did so in order to follow the crustaceans on which they feed and which always seek the light, I described the case as one of reflex action, the development of which had no doubt been largely assisted by natural selection; and I should still regard it as a misnomer to call it a case of instinct. For, on the one hand, such cases are not nearly so complex in the neuro-muscular machinery which they betoken as are many or most of the reflex actions exhibited by the higher animals, and, on the other hand, if we were to call them instincts, so also should we require to call every other case of reflex action. It is, indeed, impossible, as I said at the commencement of these chapters on Instinct, always in particular cases to draw the line between instinct and reflex action; but, as I likewise said, "this is altogether a separate matter" and has nothing to do with defining what instinct is. And certainly, as I there showed, instinct is something more than reflex action; "there is in it the element of mind."

Moreover, if we were to classify these and all other cases of still more compound reflex action under the designation of instinct, there would be no category left in which to place all cases of true instinct, *i.e.*, cases where consciousness is necessary to the performance of an action which but for the occurrence of consciousness would be properly classified as a reflex action. Of course if we choose we may altogether ignore the distinction which the occurrence of consciousness in an action imposes, and so classify all reflex actions and all instinctive actions under one denomination; but this is not what Mr. Spencer professes to do. He draws a distinction between reflex action and instinct; but he does not draw it at consciousness; and the result is that while no real distinction is drawn between the two (for compound reflex action is still nothing more than a mechanical advance upon simple reflex action), the great distinction which actually exists is ignored. Let us take an illustration. The giving of suck to young by mammals must be regarded as a truly instinctive act. Why? I answer, for one reason, because the animal which performs the action is conscious of performing it. If, on the other hand, the young animal which is taking the suck is too young (as in the case of the Kangaroo) to be reasonably supposed conscious of performing its part in the process, I should say that the action of the young animal is to be regarded as reflex. But Mr. Spencer would classify both these actions under the common designation of instinctive. Suppose, then, that this is done, and what should we say to this case from among the polyps? McCready describes a species of Medusa which carries its larvæ on the inner side of its bell-like body. The mouth and stomach of the Medusa hang down like the tongue of a bell, and contain the nutrient fluids. McCready observed this depending organ to be moved first to one side and then to the other side of the bell, in order to give suck to the larvæ on the sides of the bell—the larvæ dipping their long noses into the nutrient fluids which that organ of the parent's body contained. Now if this case occurred in any of the higher animals, where we might suppose intelligent consciousness of its occurrence to be present, it would properly be regarded as a case of instinct. But as it occurs in an animal so low in the scale as a jelly-fish, we are not warranted in assuming the presence of an intelligent perception of the process, and therefore in my view we must classify the case, not as one of instinct, but as one of reflex action, which, like all other cases of complex reflex action, has probably been developed by natural selection. But it would follow from Mr. Spencer's view that the case must be classified as one of instinct, and therefore as presenting no point of psychological distinction from that of giving suck in the case of a mammal. Surely it is a more philosophical mode of constructing a psychological classification, to acknowledge the great distinction which the presence of a psychical element makes between two such cases as these; and, if so, the distinction stated in its simplest terms is the one which I have already stated—viz., that while the stimulus to a reflex action is at most a sensation, the stimulus to an instinctive action can only be a perception.

In my opinion, then, Mr. Spencer's theory of the formation of instincts is seriously at fault in that it fails to distinguish the most essential feature of instinct; moreover it does not recognize the important principle of the lapsing of intelligence, and thus fails to account for the very existence of that whole class of instincts which I have called secondary. Next I have to show that this theory is further defective in that it fails to recognize sufficiently the other and no less important principle of natural selection, and so in large measure fails to account for the existence of that whole class of instincts which I have called primary. Thus, he says expressly with reference to instinct, "while holding survival of the fittest to be always a co-operating cause, I believe that in cases like these it is not the chief cause."* Now it so happens that the "cases" of which he is speaking are those of the artificial instincts of pointers, retrievers, and other domestic animals; hence by "survival of the fittest," we must understand artificial selection (which is here the analogue of natural selection among wild animals), and therefore the remark happens to be particularly unfortunate in the connection in which it occurs, seeing it is perfectly certain that but for the most careful and continued selection by man, our pointers and retrievers would never have come into existence. But

* *Principles of Psychology*, i, p. 423.

even as regards the instincts of wild animals the judgment in question appears to me no less objectionable. How, for instance, are we to account by any process of "direct equilibration" for the incubating instinct, cell-making instinct, the instinct of cocoon-spinning, not to mention all the other primary instincts which I have considered, nor again to repeat all the proof of the variability and heredity of acquired habits?

Still, having thus shown as clearly as I can that in my opinion Mr. Spencer certainly attributes much too little to the influence of natural selection in the formation of instincts, and also that I think he has committed a still graver oversight by altogether ignoring the influence of lapsing intelligence, I shall next show that his argument is of use in discovering another consideration which, for the sake of avoiding confusion, I have hitherto suppressed. His argument briefly stated is that instincts may arise independently both of natural selection and of lapsing intelligence, by "direct equilibration" alone; he supposes them to arise immediately out of reflex action. Now, although we have seen that if such is the case they ought not to be called instincts, *unless* they present a mental constituent, still they must be called instincts if, as he further supposes, the growing complexity of the reflex process culminates in evolving such an element. We have already seen, while treating of the dawn of consciousness, that this most probably is the way in which the mind-element arose, and, if so, Mr. Spencer's argument does present a possible third mode in which many of the simpler instincts—or instincts of the lowest animals—may have taken origin. This third mode, it will be observed, is the converse or opposite of that which we have called the lapsing of intelligence; it is a mode which leads up to or culminates in consciousness (when for the first time the action ceases to be reflex and becomes instinctive), instead of descending or becoming degraded into unconsciousness. Now, that such a process may take place, is, I think, on *à priori* grounds very probable, although from the nature of the case it is not possible to find proof of its occurrence; for if it does occur, it can only do so among the lowest animals, where we are not able to obtain evidence of consciousness even if incipiently present. Therefore, as the process can only refer to the genesis of actions which occupy the doubtful border-land between the reflex and the instinctive, this possible third mode in which rudimentary instincts may arise need not claim consideration with reference to the origin of instincts in general, although the subject is, as we have seen, of much importance in relation to the origin of consciousness.

It only remains to point out that if instincts ever do arise by way of this third mode, the implication would appear to be, as Mr. Spencer admits, that "survival of the fittest must always be a co-operating cause." I should, however, even here be inclined to go further, and to say that survival of the fittest must in this co-operation be of more than the subordinate importance which Mr. Spencer attributes to it. For instance, taking again the case of the Medusæ seeking the light, and supposing the action to have become dimly conscious and so incipiently instinctive; when the tendency to seek the light first began to manifest itself, and the individuals which sought the light were thereby enabled to procure more food than those which did not, natural selection would at once begin to develop the reflex association between luminous stimulation and movement towards light. Here, in fact, the intervention of any other cause of a directly equilibrating kind seems out of the question, inasmuch as, apart from high intelligence, which *ex hypothesi* is absent, there could be no bond of union between the stimulus supplied by light and the obtaining of food in the light. Only by natural selection could such a bond have here been established; and the same considerations apply to many or most of the quasi-instinctive actions exhibited by low animals.

So much then for the view which would regard all instincts as outgrowths of reflex action. But scarcely less objectionable is the other extreme view which would regard all instincts as outgrowths of intelligence. This, as I have said, is the view expressed by Lewes, and also, I may add, by the Duke of Argyll, who seems never to have read Mr. Darwin's doctrine of the development of instincts by natural selection.* But be individual opinion what it may, surely it is sufficiently evident, as pointed out at the commencement of our discussion, that to assign all instincts to an intelligent

* See *Contemporary Review*, November, 1880, where the Duke argues that the origin of many instincts is hopelessly obscure, because they cannot be explained by the unaided principle of lapsing intelligence—without once alluding to the immense field of possibilities which is opened up by the introduction of the principle of natural selection.

origin is a hopeless attempt at making a valid explanation of one thing a satisfactory explanation of another.

Recognizing, then, in the light of all the foregoing facts, both the principles which are concerned in the development of instincts, I shall now pass on to state the opinion of Mr. Darwin.

In the "Origin of Species" he writes (pp. 206–7), "If we suppose any habitual action to become inherited—and it can be shown that this does sometimes happen—then the resemblance between what originally was a habit and an instinct becomes so close as not to be distinguished. If Mozart, instead of playing the pianoforte at three years old with wonderfully little practice, had played a tune with no practice at all, he might truly be said to have done so instinctively.* But it would be a serious error to suppose that the greater number of instincts have been acquired by habit in one generation, and then transmitted by inheritance to succeeding generations. It can be clearly shown that the most wonderful instincts with which we are acquainted, namely, those of the hive-bee and of many ants, could not possibly have been acquired by habit.

"It will be universally admitted that instincts are as important as corporeal structures for the welfare of each species, under its present conditions of life. Under changed conditions of life, it is at least possible that slight modifications of instinct might be profitable to a species; and if it can be shown that instincts do vary ever so little, then I can see no difficulty in natural selection preserving and continually accumulating variations of instinct to any extent that was profitable. It is thus, I believe, that all the more complex and wonderful instincts have originated. As modifications of corporeal structures arise from, and are increased by, use or habit, and are diminished or lost by disuse, so I do not doubt it has been with instincts. But I believe that the effects of habit are in many cases of subordinate importance to the effects of natural selection of what may be called spontaneous variations of instincts;—that is of variations

* From this it will be observed that by the phrases "inherited habit," "habitual actions becoming inherited," &c., Mr. Darwin means to allude to the principle of lapsing intelligence. This must be borne in mind while reading these quotations, where "habit" is always used in the sense of intelligent adjustment which has become partly automatic in the individual.

produced by the same unknown causes which produce slight deviation of bodily structure."

Again, in the "Descent of Man," he repeats substantially the same judgment (pp. 67–8); and among his MSS I find the following passage, which I shall quote because it serves to convey his opinion in a still more clear and emphatic manner.

"Although, as I have attempted to show, there is a striking and close parallelism between habits and instincts; and although habitual actions and states of mind do become hereditary, and may then, as far as I can see, most properly be called instinctive; yet it would be, I believe, the greatest error to look at the great majority of instincts as acquired through habit and become hereditary. I believe that most instincts are the accumulated result, through natural selection, of slight and profitable modifications of other instincts; which modifications I look at as due to the same causes which produce variations in corporeal structure. Indeed, I suppose that it will hardly be doubted, when an instinctive action is transmitted by inheritance in some slightly modified form, that this must be caused by some slight change in the organization of the brain. (Sir B. Brodie, 'Psychological Enquiries,' 1854, p. 199.) But in the case of the many instincts which, as I believe, have not at all originated in hereditary habit, I do not doubt that they have been strengthened and perfected by habit; just in the same manner as we may select corporeal structures conducing to fleetness of pace, but likewise improve this quality by training in each generation."

From these quotations it is evident that Mr. Darwin clearly recognized both the lapsing of intelligence and natural selection as operating causes in the formation of instinct; but that he regarded natural selection as the more important of the two. Although, however, he does not expressly say so, I cannot doubt—in fact I know—that he fully recognized the importance of intelligence in supplying adaptive as distinguished from fortuitous variations to be seized upon by natural selection. Viewed in this relation, natural selection may be deemed a promoting cause of the lapsing of intelligence into instinct, and the two principles working in conjunction must, I think, be more potent than either working alone. But if I were asked which of the two I deem more important, I should say that natural selection must be held to

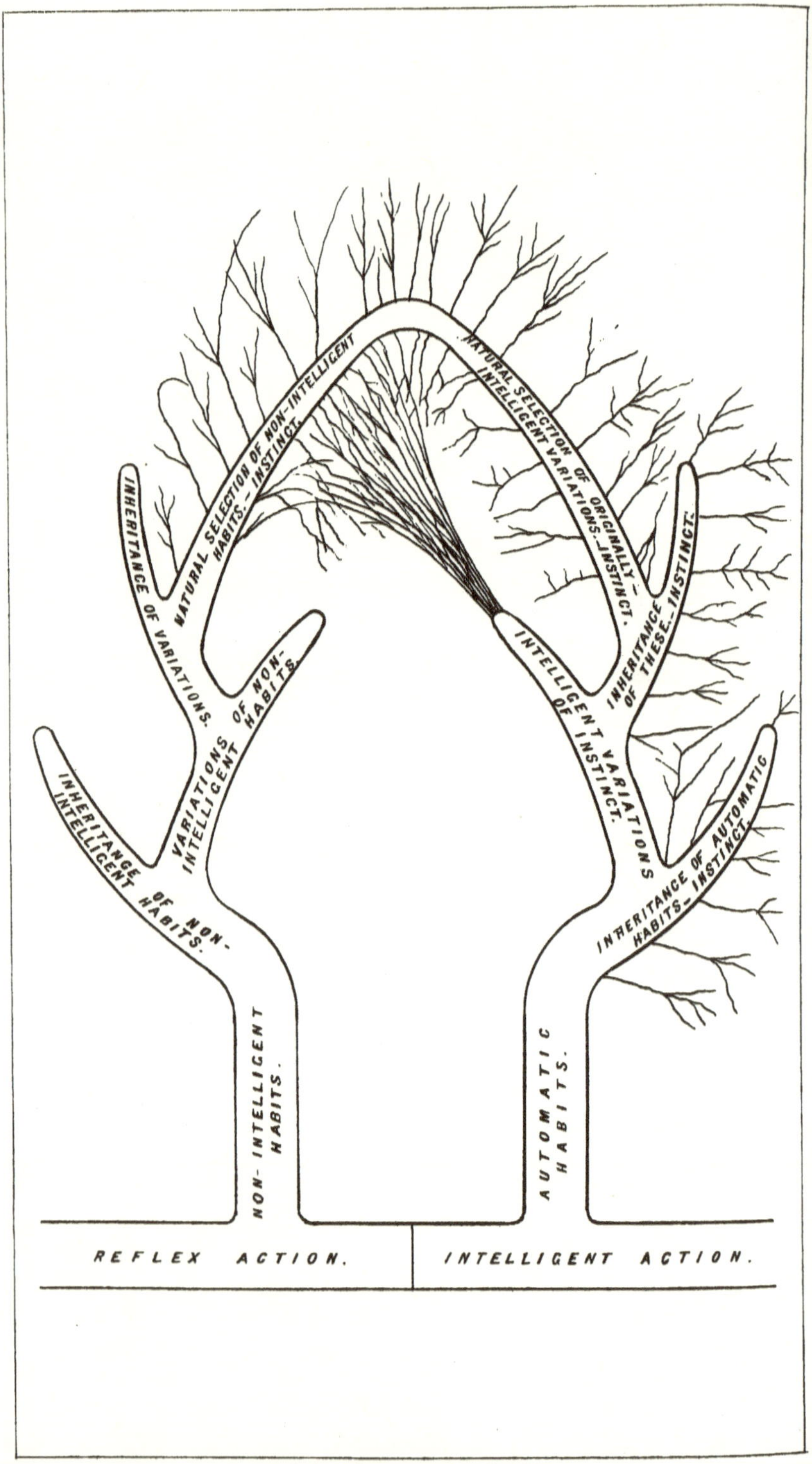

Harrison & Sons. Lith. St Martins Lane. W.C.

have the pre-eminence, inasmuch as the principle of lapsing intelligence can demonstrably have had no part at all in the formation of the "most complex and wonderful instincts" with which we are acquainted—viz., those of the social Hymenoptera.* And this, as we have seen, is the judgment of Mr. Darwin, which therefore appears to me, in consideration of all the reasons which I have now stated, to be the truest judgment—and this without reference to the unapproachable authority upon the subject with which he must be held to speak.

General Summary on Instinct.

For the sake of rendering clear the relations which the sundry principles that are concerned in the formation of instinct bear to one another, I append a diagram which is designed to show these relations in a graphic form. After what has now been said it is only needful, for the purpose of explaining the diagram, to observe the following points. The little twigs which are represented as growing out of the large branches or principles, are intended to represent instincts, and I have inserted them in order to mark the only principles from which instincts (in accordance with my definition of instincts) are able to spring. Here and there I have represented the branching structure of these instincts as inarching with one another—a device which is intended to display what I take to be an important additional principle, viz., that fully-formed instincts may occasionally blend, so giving rise to new instincts; this may be due either to novel circumstances leading to an intentionally adaptive blending

* It is demonstrable that lapsing intelligence can have played no part in the formation of these instincts, because the "workers," both among bees and ants, are sterile. Lewes can never have had this particular case presented to his mind, for it proves his theory of lapsing intelligence alone insufficient. It is likewise incompatible with Spencer's theory. Thus, for instance, he writes :—"The automatic actions of a bee building one of its wax cells, answer to outer relations so constantly experienced that they are, as it were, organically remembered" (*Principles of Psychology*, i, p. 445). But he forgets, as Lewes also forgot, that the insect which performs these automatic actions has *not* thus "constantly experienced" the "outer relations," for it begins by performing these actions before it has itself had any individual experience of cell-making, and without its parents ever having had any ancestral experience. In the whole range of instincts no more unfortunate illustration could have been chosen by Mr. Spencer. How the difficulty is met by Mr. Darwin's theory I shall consider at the beginning of the next chapter.

of instinctive habits, or to an originally conscious imitation by one species of the instinctive habits of another. Lastly, I have joined the two tree-like growths at their summits in order to represent the fact that intelligent and non-intelligent adaptation, or primary and secondary instincts, may fuse together and then possess a common sap or principle of further growth. I have also represented such union between the two sides of the diagram, primary and secondary, to take place at one other point—viz., between the branch Primary Instinct and the branch Intelligent Variation of Secondary Instinct. I do this to bring out into stronger prominence the fact that when once a non-intelligent or primary instinct has been formed, it is most ready to join with and become fertilised by the principle of intelligence at any point where this principle is, as it were, mobile, or not yet fixed and frozen into secondary instinct. But the most important thing to remember is that whether instincts have had an intelligent or a non-intelligent mode of origin, they may at any time after their full formation come into contact with intelligence at any point; so that the two sides of our diagram (being the embodiment of all the foregoing evidence upon the subject) illustrate at once the truth and the falsity of the common opinion which has been so neatly rendered by Pope, when he says of instinct and reason that they are things "for ever separate, yet for ever near."

I shall now proceed to give a general summary of all the preceding chapters on Instinct.

After defining the sense in which alone I use the word Instinct, I proceeded to give a few illustrations of the perfection of instinct as exhibited by very young animals, or by animals without individual experience of the circumstances to which their instinctive actions are adapted. Next I gave a few complementary illustrations of the imperfection of instinct, and pointed out that such imperfection might arise, either from a change in the conditions of the environment to which the ancestral instinct was adapted, or from the fact that the instinct is not yet completely formed. I also showed that imperfection of instinct might arise from internal or psychological changes throwing out of gear the delicate mechanism on which the perfect display of instinct depends. In this connection I gave instances to prove that such derangement of instinct is particularly apt to arise when the normal history of an animal's converse with its environment is interrupted for a time and again renewed. I also gave one case of such derangement where there had been no such interruption, and which, therefore, may most properly be regarded as a case of insanity.

If instincts are slowly evolved, we should expect to meet with some cases in which they are not yet fully evolved, and, as just observed, for this reason imperfect. Such cases we do find—as, for example, young turkeys pointing at flies, young chickens being half afraid of bees, rabbits only toddling instead of running away from weasels, &c., &c. We may also see instincts in course of development among young children learning to balance the head, to walk, to speak, &c. Moreover all cases of the education or improvement of instinct, whether in the individual or in the race, are so many cases of the original imperfection of instinct. But this brought us directly into the question as to the origin of instincts.

I have endeavoured to prove that the origin of instincts may be what I have called either primary or secondary. That is to say, I believe there is ample evidence to show that instincts may arise either by natural selection fixing on purposeless habits which chance to be profitable, so converting these habits into instincts without intelligence being ever concerned in the process; or by habits, originally intelligent, becoming by repetition automatic. As an example of a primary instinct I gave incubation; and as examples of secondary instincts I gave sundry cases of "practice making perfect." On *à priori* grounds we saw that instincts must arise by the processes thus explained, and then we proceeded to render *à posteriori* proof that they have. This proof undertook to show that purposeless habits occur in individuals, are inherited, vary, have their variations inherited, and then developed in beneficial lines by natural selection; also that habits originally intelligent by repetition become automatic, and, having lapsed from intelligence, are then inherited as instincts, which may then vary, have their variations inherited and developed in beneficial lines by natural selection, as in the previous and analogous case. These sundry propositions were substantiated by showing, first, that tricks of manner are displayed more or less by every one, and especially by idiots; also by animals, as in dogs barking round a carriage, differences of individual disposition and idiosyncracies, forming strange companionships, &c. Next, that automatic and useless or fortuitous habits are inherited, was amply proved by cases in which this has been observed of the tricks of manner displayed by men and animals; in disposition, as among the island races of monkeys described by Humboldt; in the paces of the horse in different parts of the world; in the remarkable and wholly useless habits of the tumbler and pouter pigeons, &c. Further, that such inherited, non-intelligent, or purposeless habits should vary, is a matter of certainty; seeing that, as was subsequently shown, useful habits may do so, and that even fully formed instincts are plastic; much more, then, must these fortuitous sports of habit be variable. Lastly, that when they vary in profitable directions the variations will be seized upon and fixed by natural selection is no less a matter of certainty, and will not be questioned by any one who believes in natural selection as a principle concerned in the evolution of organic structures. Thus only can we explain the instincts of many low animals (such as the caddis-worm), and certain instincts of the higher (such as that of incubation). Coming next to secondary instincts, it was first shown that intelligent adjustments when frequently performed become automatic in the individual, and next that they are inherited till they become automatic habits in the race. The former fact is familiar to every one; the latter was proved by such cases as those of hereditary handwriting, family aptitudes for particular pursuits, race characteristics of psychology in man, good breeding, and sense of modesty. In animals the same principle is seen in an hereditary tendency to "beg" in dogs, and even in cats; ponies from Norway not having "mouths;" Dr. Huggins's dog presenting an inherited antipathy to butchers; wild animals showing an instinctive fear of their particular enemies, such fear being lost as regards man in domesticated animals (notably in the rabbit and duck, where selection is not likely to have had any part in obliterating natural wildness); animals living on oceanic islands showing no fear of man for several generations after his first advent among them, then acquiring instinctive dread of him, and even learning what constitutes safe distance from fire-arms;

changed instincts of the woodcock; and the effects of blending instincts by crossing.

It having been fully shown by these selected examples that instincts may arise by natural selection alone, or by lapsing intelligence alone, the discussion went on to show that instincts in general are not necessarily confined to one or other of these two modes of origin; but, on the contrary, that these principles when working in cooperation have greater influence in evolving instincts than either of them can have when working singly. For, on the one hand, hereditary proclivities or habitual actions which, being useful though never intelligent, were originally fixed by natural selection, may come to furnish material for further improvement, or be put to better uses, by intelligence; and, conversely, adjustments originally due to lapsed intelligence may come to be greatly improved, or put to better uses, by natural selection. For, taking the latter case alone, if, as we have seen, intelligent actions may by repetition become automatic as secondary instincts, and if they may then vary and have their variations fixed in beneficial lines by natural selection, how much more scope may be given to natural selection in this further development of an instinct, if the variations of this instinct are not wholly fortuitous, but arise as intelligent adaptations of ancestral experience to the perceived requirements of individual experience. Clearly, natural selection must in such a case be working at a much greater advantage than it does when working alone in the formation of primary instincts, where it is supplied only with fortuitous variations, instead of with variations which, being determined by intelligence, are from the first adaptive. And no less clearly, the principle of lapsing intelligence must be working at a much greater advantage when thus in association with natural selection, than it is when working alone in the formation of secondary instincts; for natural selection in this case must always tend to favour the best of the intelligent adjustments, and by concentrating the power of heredity into them must tend the more speedily to render them automatic or instinctive.

It is of no moment, as regards instincts of blended origin, to determine in particular cases which of the two principles —natural selection or lapsing intelligence—has had the historic priority, even if from the first these two principles have not been in combination; the important fact to be shown is that even a fully formed instinct may prove itself, under the influence of intelligence, variable or plastic. I therefore demonstrated the plasticity of many existing instincts, dwelling especially upon the cell-making instinct of bees, and the incubating and maternal instinct of warm-blooded animals—choosing these instincts for special consideration because they must be of so ancient an origin, and are so strongly inherited.

Intelligence may operate in the modification of instinct, either by perceiving the need of a change in the dictates of heredity, by intelligent imitation of the habits of other animals, or by parents intentionally teaching their young. Copious facts on all these points were therefore given. But the best evidence of the extreme modification which instincts may be made to undergo by the combined effects of intelligence and selection, is that which is afforded by the facts of Domestication. These facts were therefore detailed at length, and they showed that domestication has not merely a negative influence in eradicating natural instincts (witness the loss of wildness in dogs, cats, horses, and cattle; dogs not attacking sheep, pigs, or poultry; the latter having lost their instinctive fear of dogs, so differing from pheasants; the incubating instinct being lost in the Spanish hen, and the maternal instinct in cows and sheep where the young have for generations been habitually removed from their mothers at birth; Polynesian dogs having lost their natural intelligence, together with their natural taste for flesh); but also a positive influence in developing new instincts. In the case of the Dog these new or "artificial" instincts were shown to be strikingly exhibited in the sheep-dog, pointer, and retriever; but perhaps still more remarkably in the instinctive love of man shown by nearly all the breeds; faithfulness to and sense of dependence upon man; inborn idea of protecting his master's property and of himself as constituting a part of that property; barking being an acquired instinct, and probably arising from this idea of protecting his master's property. Indeed so fundamental has been this psychological transformation in the dog, that the artificial instincts have frequently become stronger than even the strongest of the natural instincts, viz., the maternal—as is proved by cases in which the latter has given way when in conflict with the former. Lastly, I devoted a chapter to the consideration of local and specific variations of instinct, showing how these constituted a kind of palæontological evidence of the transmutation of instinct.

Such then is the *à posteriori* proof of the two ways which, either singly or in combination, must be regarded as those by which all properly so-called instincts have been developed. A diagram was given to show graphically how the sundry principles concerned are related and inter-related with one another. Here it was shown that when an instinct, whether of single or blended origin, was perfected, it might vary or ramify into modified forms, and even blend, or, as it were, inarch with other instincts to produce a new growth. It is difficult, or rather impossible, to trace the history of actual instincts in this respect, from the fact that instincts are not fossilized, and therefore leave no record of their transitional states. But from all the evidence together—and especially from what we may almost denominate the historical evidence supplied by the facts of domestication—there can be no reasonable doubt that instincts may not only have a double root—one in the principle of selection, and the other in that of lapsing intelligence—but also a more or less branching stem, which (or the branches of which) may in some cases become grafted with the stem or branches of other instincts.

In estimating the comparative importance of the two great factors in the formation of instinct, we had occasion to differ on the one hand from Mr. Spencer, who attributes the origin of all instincts to reflex action with little or no aid from natural selection, and on the other hand with Mr. Lewes, who goes to the opposite extreme of regarding all instincts as cases of lapsed intelligence. It was shown, however, that Mr. Spencer's view might be held to explain the rise of doubtfully instinctive actions displayed by very low animals, and that it is of much importance as an explanation of the origin of Consciousness. The view, however, which I adopt to explain the origin of instincts is substantially the same as that which has been propounded by Mr. Darwin, and which, while recognizing both the factors which I have now so repeatedly named—*i.e.*, natural selection and lapsing intelligence—whether singly or in combination, attributes most importance to the former, especially if it be remembered that in its work of organizing instincts, intelligent adjustment is always under the direction and control of natural selection, so that its chief function in the formative process is probably that of supplying to natural selection variations of ancestral instincts which are not merely fortuitous, but intentionally adapted to the conditions of the environment.

CHAPTER XVIII.

INSTINCT *(continued)*.

CASES OF SPECIAL DIFFICULTY WITH REGARD TO THE FOREGOING THEORY OF THE ORIGIN AND DEVELOPMENT OF INSTINCTS.

WE must not take leave of Instinct without looking into all the known cases of its exhibition which admit of being reasonably cited against the views here expressed on the rise and development of instincts generally. I shall therefore consider *seriatim* all such cases which I have met with in the writings of others, or which occur to me as admitting of being possibly cited in this connection.

Similar Instincts in Unallied Animals.

Mr. Darwin observes in the Appendix, "We occasionally meet with the same peculiar instinct in animals widely remote in the scale of nature, and which consequently cannot have derived the peculiarity from community of descent." The difficulty, of course, is to account for the parallelism, and the instances given by Mr. Darwin are those of the Molothrus having the same instinct of parasitism as the Cuckoo, the Termites having much the same instincts as the Ants, and a neuropterous and a dipterous larva having the same instinct of digging a pitfall for prey. He shows satisfactorily that the last-mentioned is the only case that offers any real difficulty; but even here, it seems to me, the difficulty is not one of any magnitude. For the instinct in question is not one of such complexity, or of such remote probability as to its formation where a larva habitually lives in sand, that we may not readily believe a similarity of environment should have determined its development independently in two lines of descent—just as for the same reason wings, for example, have been developed independently in at least four lines of descent.

Dissimilar Instincts in Allied Animals.

Mr. Darwin in the Appendix also alludes to this subject, and the few remarks which he makes upon it seem to me fully to dispose of the difficulty—which, indeed, with his characteristic candour, I cannot but think that he unduly magnifies. As I have observed in my chapter on Local and Specific Variations of Instinct, the theory of the formation of instincts by natural selection really leads us to anticipate the not infrequent occurrence of what we may term isolated instincts; for only if we were to suppose that all considerable variations of instinct (local or otherwise) are permanent, could we anticipate—in the absence of any palæontology of instinct—a graduated series of instincts in all cases, with the consequent absence of isolated instincts in every case. But to suppose this would be to run counter to the first principles of our theory. Of course if specific instincts were of very general occurrence, it might reasonably be objected that this theory would require to suppose too great a slaughter of intermediate species to be accepted as credible; but as matters actually stand I have felt that the occasional appearance of isolated instincts in about the proportion of cases that the theory would lead us to anticipate, really constitutes a corroboration of, rather than an objection to, the theory.

Trivial and Useless Instincts.

Mr. Darwin in the Appendix also refers to trivial and useless instincts, and says:—"I have not rarely felt that small and trifling instincts were a greater difficulty on our theory, than those which have so justly excited the wonder of mankind; for an instinct, if really of no considerable importance in the struggle for life, could not be modified or formed through natural selection."

This is no doubt an important point, and must be carefully considered. First of all it ought to be observed that if any such difficulty can be shown to stand against the theory of the formation of instinct by natural causes, much more must the difficulty stand against the older theory of the implanting of instincts by a supernatural cause. Next, we must be perfectly sure, in any given case, that the instinct which appears to be trivial or useless is really such. This point is mentioned by Mr. Darwin, and he cites some very good cases to show how the important utility, or even absolute necessity, of an instinct may readily escape observation. But even after due allowance is made on this score, some few instincts certainly do remain which it seems impossible to suppose of the smallest utility. How, then, are these to be explained?

I believe they admit of being satisfactorily explained by two considerations. The first of these is that our theory does not suppose natural selection to be the only influence at work in the formation of instincts. We have repeatedly insisted that the lapsing of intelligence is another influence of scarcely less importance; and we have also seen abundant evidence to show that non-adaptive habits occur in individuals and may be inherited in the race. Therefore, if from play, affection, curiosity, or even mere caprice, the intelligence of the animal should lead the animal to perform any useless kind of action habitually (as, for instance, in the case of the ratels tumbling head-over-heels),* and if this habit were to become hereditary in the similarly constituted progeny, we should have a trivial or useless instinct. The only condition, so far as I can see, that would require to be satisfied would be that the trivial or useless habit should not be actually detrimental to the species exhibiting it so that its growth into an instinct should not be *prevented* by natural selection.

The other consideration to which I have alluded as mitigating or dispelling the difficulty in question is this. In the analogous case of structures, as is well known, we meet with innumerable cases of useless organs; but here, so far from the fact being deemed a difficulty in the way of the theory of evolution by natural selection, it is justly deemed one of its strongest supports; and the reason is that in all such cases we have evidence of the useless and perhaps rudimentary organs being of use in other and allied animals. Now I see no reason to doubt that the same may be true of instincts, and therefore that what we now find to be apparently trivial and certainly useless exhibitions of hereditary

* See p. 189.

habit may, at an earlier period in the history of the species or of its allies, have been of real utility. We may, for example, readily imagine that the instinct displayed by many herbivorous animals of goring sick or wounded companions, is really of use in countries where the presence of weak members in a herd is a source of danger to the herd from the prevalence of wild beasts; and Mr. Darwin in the Appendix gives evidence that such is actually the case. Or, to take a more fanciful illustration, we may suppose the Megapodidæ mentioned in the Appendix, which incubate their eggs by placing them in a large heap of fermenting vegetable matter which they collect for this purpose, were to find, from a change of their habitat or of the Australian climate, that it was difficult to collect a sufficient quantity of vegetable matter, or that it would not ferment sufficiently for the purpose of incubation. The birds might then gradually revert to the usual mode of incubation, but might still retain a marked propensity to make tumuli of vegetable matter as nests. If so, the labour expended in making such tumuli would be obviously useless, and there being no analogy among the incubating habits of other birds to give us a clue as to the origin of such an instinct, we should be quite at a loss to account for it.

Instincts apparently Detrimental to the Species which exhibit them.

It constitutes no difficulty or objection to our general theory of instinct-formation to point to cases in which instincts are obviously detrimental to the individuals which manifest them; for it is of the essence of the theory of natural selection to suppose that the interests of the individual are, in the process of selection, subordinated to those of the species. It is, for example, manifestly to the detriment of an individual fly to procreate its kind, inasmuch as its own death is speedily induced by the act; but seeing that the act is essential to the continuance of the species, we perceive how natural selection must here have developed an instinct which virtually amounts to that of suicide. And the same remark applies to all similar cases, such as that alluded to in "Animal Intelligence" of soldier ants and termites sacrificing themselves for the benefit of the community—*i.e.*, the species.

But of course the case is entirely altered where we appear to meet with an instinct the operation of which is detrimental to the individual, without being attended with any compensating benefit to the species; for in such a case the detriment to the individual would also become a detriment to the species. Such apparent cases, in fact, are precisely analogous to those in which certain structures appear to be detrimental to their possessors, without seeming to confer any compensating benefit upon their species;* and, as Mr. Darwin observes, such an apparent case, if it could be shown to be a real one, would be incompatible with the theory of natural selection, inasmuch as "natural selection acts solely by and for the good of each." Further, as Mr. Darwin adds, "if it could be proved that any part of the structure of any one species had been formed for the exclusive good of another, it would annihilate my theory;" and it is obvious that the same remark would equally apply to the case of instincts.

It is therefore of the utmost importance to take a survey of all known instincts, in order to see whether there is any one case, either of an instinct which is detrimental to the species exhibiting it, or of one which has exclusive reference to the benefit of other species. For, on the one hand, if there is any one such case of an indisputable kind, we should clearly have to modify our whole theory in order to meet it; while, on the other hand, if there is no such case, the fact of all the innumerable multitude of animal instincts being of obvious use to the species which manifest them, and never of exclusive use to other species, must be taken as the strongest possible evidence of the theory that ascribes all instincts to the causes which we have assigned.

I may as well say at once that there is only one apparent case of an instinct in one species having exclusive reference to the benefit of another, although there are cases of instincts beneficial to the species presenting them being also beneficial to other species. With the latter cases we are not, of course, concerned. The former is the case of aphides yielding up their secretion to ants, and has already been considered by Mr. Darwin. His explanation is that, "as the excretion is extremely viscid, it is no doubt a convenience to the aphides

* See *Origin of Species*, 162–4, where the case of the rattle of the rattlesnake, &c., is considered.

to have it removed; therefore probably they do not excrete solely for the good of the ants."*

Coming now to the other branch of the subject, after due reflection I can only think of two or three instincts which could possibly be cited as presenting the appearance of being detrimental to the species which manifest them. I shall therefore consider these cases separately.

1. *Suicide of Scorpion.*—The state of the evidence on this subject will be found in my other work.† It will there be seen that two or three independent witnesses—including a friend of Dr. Allen Thomson on whose accuracy he says he can rely—bear testimony to the truth of the popular saying that when a scorpion is surrounded by fire, or otherwise exposed to undue heat, it will commit suicide by stinging itself to death. It will be seen, however, by referring to the correspondence in question, that the alleged facts are disputed by other observers, and also, as I have already indicated, that they were not observed by Dr. Thomson himself.

The effect of republishing this correspondence and of pointing out the desirability of obtaining further evidence upon the matter, has been to induce two very competent naturalists to make some observations upon the subject. One of these naturalists is Professor Lankester, who published his observations in the "Journal of the Linnean Society" (1882), and the other is Professor Lloyd Morgan, who published his results in "Nature" (vol. xxvii, p. 313). Both these observers agree that the scorpions never commit suicide, and as Mr. Morgan exposed the animals to a variety of dreadful tortures with a uniformly negative result, I think the question may now be considered as closed. Moreover Mr. G. Bidie, who started the previous correspondence in "Nature," has recently addressed another letter to that Journal‡ in which he makes the not improbable suggestion that, as in his experiments he applied heat by condensing the rays of the sun with a lens upon a small point of the scorpion's back, the animal in stinging itself "may have merely been trying to get rid of an imaginary enemy."

2. *Insects flying through Flame.*—The determination shown by many kinds of insects to fly towards and through a flame is unquestionably due to instinct, and as such might be ad-

* *Origin of Species*, p. 208. † *Animal Intelligence*, pp. 222–5.
‡ July 12, 1883.

duced as evidence of an instinct detrimental alike to the individual and to the species. But before this conclusion could be reached, several possibilities require to be attended to. In the first place, flame in Nature is an exceedingly rare phenomenon, so that we could scarcely expect that any instinct should have been developed for the express purpose of its avoidance. Therefore, if the general economy of night-flying insects is such that it is of advantage to approach and examine shining objects, there would be nothing anomalous in their failing to distinguish between flame and other shining objects—such as white flowers or, in the case of moths, pale coloured members of the opposite sex. But as the instinct of flying into flame is of such general occurrence among many species of insects, I think we certainly cannot attribute all the cases of it to a mistaking of flame for some other shining object; to meet all the cases some still more general explanation is required, and this, I think, is afforded by considering other and analogous cases. Thus many species of birds display an exactly similar propensity, as is proved by the experience of lighthouse keepers; and, according to Professor A. Newton, some species of birds are more readily attracted by light than others.* Here there can be no question about a possible mistaking of flame for white flowers, &c., and therefore the habit must be set down to mere curiosity, or desire to examine a new and striking object; and that the same explanation may be given in the case of insects seems not improbable, seeing that it must certainly be resorted to in the case of fish, which, as I pointed out in "Animal Intelligence," are likewise attracted by the light of lanterns, &c.; and the psychology of a fish is not much, if at all, in advance of that of many insects.

Thus, in any case, it seems certain that we have no reason to regard the propensity in question as an expression of any instinct specially formed with reference to flame, and this is really the only point with which we are directly concerned. But, as the subject is in itself an interesting one, I shall here add a few remarks with reference to other aspects of it.

Among Mr. Darwin's MSS I find the following note, which, however, is not in his hand-writing.

" Query. Why do moths and certain gnats fly into candles, and why are they not all on their way to the moon—at least

* *Yarrell's Brit. Birds*, 4th ed., II, 235

when the moon is in the horizon? I formerly observed that they fly very much less at candles on a moon-light night. Let a cloud pass over, and they are again attracted to the candle."

I do not know to whom this observation is due; but I quote it for the sake of the query. The answer, I think, must be, that as the moon is a familiar object, the insects regard it as a matter of course, and so have no desire to examine it. I have little doubt that if moonlight were concentrated to a point in a dark room, the moths and gnats would approach it.

In "Nature" (vol. xxv, p. 436), Mr. J. S. Gardener writes:—

"Whilst watching the great horse-shoe falls of the Skjalfandafljot near Sjosavan in Iceland, I saw moth after moth fly deliberately into the falling water and disappear. Some which I noticed arriving from a distance, fluttered at first deviously, but as they neared the water flew straight in. The gleaming falls seemed at least as attractive as artificial light." And doubtless the same explanation applies, inasmuch as a gleaming waterfall is not a sufficiently common object in Nature, either to fail in arresting the curiosity of the moths, or to ensure that a special instinct should be developed to warn the insects from approaching it.

3. Mr. Darwin in the Appendix points out two or three cases of instinct which are apparently at first sight detrimental to the species exhibiting them. Thus, the crowing of the cock-pheasant on going to roost reveals his presence to the poacher, the cackling of a hen after having laid an egg informs the natives of India where the nest is concealed, certain birds place their nests in very conspicuous situations, and a kind of Shrew-mouse betrays itself by screaming when approached. Now it seems to me that in all these cases—and many similar or analogous ones might be given—the difficulty is, if I may use the term, fictitious; for it only arises when we shut our eyes to some of the most important principles which in the previous chapters I have been endeavouring to explain. These principles do not imply that an instinct should ever be formed or modified with reference to a *prospective* change of environment, while they do imply that when such a change has taken place, time must be allowed for the compensating modification of the instinct—even suppos-

ing that any such modification is urgently required. Now it can scarcely be held probable on these principles that the instinct of crowing on the part of the pheasant should have been modified by natural selection during the short time that his ancestors have been naturalized in this country, and in consequence of one in a hundred having thus fallen a victim to poachers. The case of a wild hen cackling over its eggs may seem a stronger one; but here again the whole question really consists in the actual percentage of eggs thus discovered by the natives, and I should think this must be exceedingly small. Birds building in exposed situations only become an argument against the modificability of instinct by natural selection, when it is shown that the exposure has led to the destruction of nests by man or other animals for a great number of generations; and this has never been shown. Even in the most remarkable case—that of the Furnarius of La Plata—Mr. Darwin merely says that this bird "in a thickly peopled country, with mischievous boys, *would soon be* exterminated." And similarly it would require to be shown that the habit of the Shrew-mouse at the Mauritius has long led to the destruction of many individuals of each generation by man.

In all such cases we must remember how very insignificant the influence of man—and especially of savage man—usually is, as compared with the sum of other influences, organic and inorganic; we must remember the time which in any case is required for the modification of an instinct; and we must have proof that the instinct which is now injurious in some percentage of cases, has long been highly injurious in a large percentage of cases. I am not aware of any instance where all these conditions have been fulfilled, and where the species has not either been exterminated by man, or the required modification of instinct has not actually taken place.

4. Mr. Darwin in the Appendix also alludes to the injurious effects which frequently attend the exercise of the instinct of migration in certain animals. Thus, he says, the congregating of quadrupeds in Africa, and of the Passenger Pigeons in America is detrimental to the animals, in consequence of their being thus readily followed by beasts of prey as well as by man. But when we remember the enormous numbers of both kinds of animals which thus congregate, I cannot see that any difficulty remains; for not only is the

percentage of individuals destroyed in itself small, but I doubt whether it is much larger than would be the case if these multitudes of animals were segregated over a very much wider area. A stronger case, I think, is afforded by that of the Norwegian Lemming, and therefore I shall consider it at greater length.

Since Mr. Darwin wrote his remarks on this subject which are presented in the Appendix, further statements with reference to it have been published. These, therefore, I shall quote.

Mr. Crotch, who has had the opportunity of observing the phenomena for a number of years, thus briefly gives his account of the facts, so far as they concern us.

"The Lemmings (which are little rodents) certainly do not visit my part of Norway at any recurring period of years; but every third or fourth year they may be expected with tolerable regularity, though in variable numbers. Thus it is quite probable that some migrations may have so far escaped notice as to give rise to the old idea that they took place every tenth year.

"They are, however, always directed westwards; and thus the theory that they are caused by deficiency of food fails so far, that these migrations do not take place in a southerly direction by which a larger supply might be obtained. M. Guyne (*loc. cit.*) suggested that the course followed was merely that of the watershed. However, this runs east as well as west, and follows valleys which often run north and south for hundreds of miles, whereas the route pursued by the Lemming is due west. At all events this is the case in Norway, where they traverse the broadest lakes filled with water at an extremely low temperature, and cross alike the most rapid torrents and the deepest valleys.

"With no guiding pillar of fire, they pass on through a wilderness by night; they rear their families on their journey, and the three or four generations of a brief subarctic summer serve to swell the pilgrim caravan. They winter beneath more than six feet of snow during seven or eight weary months; and with the first days of summer (for in those regions there is no spring) the migration is renewed. At length the harassed crowd, thinned by the increasing attacks of the wolf, the fox, and even the reindeer, pursued by eagle, hawk and owl, and never spared by man himself, yet still a vast multitude, plunges into the Atlantic Ocean on the first calm day and perishes with its front still pointing westward. No faint heart lingers on the way, and no survivor returns to the mountains. Mr. R. Collett, a Norwegian naturalist, writes that in November, 1868 (quoted by Filleburg, *infrà*), a ship sailed for fifteen hours through a swarm of Lemmings, which extended as far over the Trondhjemsfiord as the eye could reach."*

Such, according to Mr. Crotch, are the facts, and the following are the hypotheses which have been propounded to explain them. Mr. Wallace suggests† that natural selection has played an important part in causing migration, by giving an advantage to those animals which enlarge their breeding area by travel. To this view, as applied to the lemming, Mr. Crotch objects that the animal, "it is true, always breeds during migration; but if none return or survive, it is difficult to say what becomes of the fittest." His own theory is a remarkable one. "There is," he says, "a solution of this difficulty, involving a subject of the deepest interest, and which led me to spend two years in the Canaries and adjacent islands. I allude to the island or continent of Atlantis. . . It is evident that land did exist in the North Atlantic Ocean at no very distant date. . . . Is it not then conçeivable, and even probable, that when a great part of Europe was submerged and dry land connected Norway with Greenland, the lemmings acquired the habit of migrating westward for the same reasons which govern more familiar migrations? . . . It appears to me quite as likely that the impetus of migration towards this continent should be retained as that a dog should turn round before lying down on a rug, merely because his ancestors found it necessary thus to hollow out a couch in the long grass."

In a later paper‡ he combats by the aid of charts the popular theory "that these migrations follow the natural declivities of the country," and then proceeds to add, "It is very remarkable that the average depth from Norway to Iceland does not exceed 250 fathoms, with the exception of a deep and narrow channel of 682 fathoms at 14° W. This probably represented the old Gulf Stream; and if this were so, the lemmings did wisely to migrate westwards in search

* *Linn. Soc. Jour.*, vol. xiii, p. 30; *et seq.* † *Nature*, vol. x, p. 459.

‡ *Linn. Soc. Jour.*, vol. xiii, p. 157, *et seq.*

of its genial influence. As little by little the ocean encroached on the land, the same advantages would remain, as in fact they do to this day."

To this ingenious theory dissent is expressed by another gentleman who has had a very large experience in observing these migrations, namely Mr. Robert Collett, of the University Museum, Christiania.* His view is that in years when reproduction is excessive, multitudes of individuals are led by hunger, as well as by "the natural desire to wander possessed by this species," to overflow the limits of their plateaux home, and spread out "over an area that is considerably larger than obtains in any other of the species under similar circumstances." As breeding continues throughout the wandering, in cases where in two or three succeeding years the production of young has been excessive, "the masses are incessantly pushed towards the sides of the fells; and the migration becomes an overrunning of the lower and far remote portions of the country, as the individuals gradually penetrate further in search of localities suitable to their habits (and which are capable of giving them a permanent subsistence), until they are stopped by the sea or destroyed in some other manner."

Looking to Mr. Collett's large experience on the subject, as well as to the intrinsically probable nature of his views, I think we may most safely lend countenance to the latter. The most important point of difference between Mr. Crotch and Mr. Collett has reference to a question of fact. For while Mr. Crotch states that the migrations are made westwards without reference to the declivities of the country, Mr. Collett is emphatic in saying that "the wanderings take place in *the direction of the valleys*, and therefore can branch out from the plateaux in any direction." If this is so, there is an end of Mr. Crotch's theory, and the only difficulty left to explain would be why, when the lemmings reach the sea, they still continue on their onward course to perish in their multitudes by drowning. The answer to this, however, is not far to seek. For their ordinary habits are such that when in their wanderings they come upon a stream or lake, they swim across it; and therefore when they come upon the coast line it is not surprising that they should behave in a similar manner, and, mistaking the sea for a large lake, swim per-

* *Linn. Soc. Jour.*, vol. xiii, p. 327, *et seq.*

sistently away from land with the view to reaching the opposite shore, till they succumb to fatigue and the waves. Therefore, pending further observations on the question of fact above alluded to, I cannot feel that the migration of the lemming furnishes any difficulty to the theory of evolution over and above that which is furnished by the larger and more important case of migration in general, to the consideration of which I shall now proceed.

Migration.

Taking the animal kingdom from below upwards, the first animals that can properly be said to present the instincts of migration are to be found in the group Articulata. I think it is sufficient to refer to "Animal Intelligence" for the facts concerning the migrations of Crabs (pp. 231–2)* and Caterpillars (238–40), though as regards the latter I may add the following remarkable account, which I quote from the "Colonies and India."

"To say that a train had been stopped by caterpillars would sound like a Yankee yarn, yet such a thing (according to the "Rangitikei Advocate") actually took place on the local railway a few days ago. In the neighbourhood of Turakina, New Zealand, an army of caterpillars, hundreds of thousands strong, was marching across the line, bound for a new field of oats, when the train came along. Thousands of the creeping vermin were crushed by the wheels of the engine, and suddenly the train came to a dead stop. On examination it was found that the wheels of the engine had become so greasy that they kept on revolving without advancing—they could not grip the rails. The guard and the engine-driver procured sand and strewed it on the rails, and the train made a fresh start, but it was found that during the stoppage caterpillars in thousands had crawled all over the engine, and over all the carriages inside and out."

With regard to Butterflies many instances of large migrations are on record. Thus, Madame de Meuron Wolff describes an immense swarm of the Painted Lady butterfly passing over Grandson, Canton de Vaud, flying closely together from south to north. The column, which was from ten to fifteen feet broad, flew low and equally, and took two hours to pass.

* See also Professor Moseley, *A Naturalist on the* Challenger, p. 561.

The caterpillar of this species is not gregarious. Professor Bonelli also describes a migration similar in all respects, including locality, except that it lasted longer—the insects covering the flowers at night and proceeding on the journey by day.

Immense swarms of migratory Dragon-flies have been at times observed, the most remarkable case being one that occurred in May, 1839, and which seems to have extended over a great part of Europe. The insects flew at a height of 100 to 150 feet, and seemed to follow the direction of the rivers.*

Many species of Fish are known to migrate regularly for purposes of spawning, such as the herring, salmon, &c., and also to find water; † while among Reptiles the most remarkable instance seems to be that which is furnished by the Turtles which visit Ascension Island to deposit their eggs. How the animals can find this comparatively small speck of land in the midst of a vast ocean is very unaccountable. I have recently written to Professor Moseley upon the subject, and in reply he says, "No man without proper modern means of finding latitude and longitude could reach either Tristan or Ascension; and it is especially difficult for animals whose eyes cannot be raised above the sea-level, and to whom, therefore, the islands are visible for a comparatively small radius only. Merchant skippers have several times been unable to find Bermuda, and on return baffled have reported the island gone down." But, as Professor Moseley adds, "It is *just* possible that the animals do not retire far from the land after all, but hang about unobserved," I think it is undesirable to enter into any discussion where the facts are still of an uncertain character.

Among Mammals, from whales to mice, we meet with many migratory species, but it is among Birds that the propensity is most prevalent. Indeed, a very competent authority on all matters pertaining to ornithology has said in the new "Encyclopædia Britannica:" "Every bird of the northern hemisphere is to a greater or less degree migratory in some part of its range. Such a conclusion brings us to a still more general inference—viz., that Migration, instead of

* For a full account see Weissenborne, *Loundoun's Mag. Nat. Hist.*, N.S., vol. iii.

† See *Animal Intelligence*, 248–50.

being the exceptional characteristic it used formerly to be thought, may really be almost universal."*

I have neither the occasion nor the space to discuss the large question of migration in general; and having now indicated the animals in which the instinct is most pronounced, I shall pass on to consider the theory of its formation. First I may allude to Mr. Darwin's remarks on Migration at the beginning of the Appendix. It will be seen from them that among others he establishes the following points:—

1. There is "in different breeds of birds a perfect series from those which occasionally or regularly shift their quarters within the same country, to those which periodically pass to far distant countries."

2. "The same species often migrates in one country and is stationary in another; or different individuals of the same species in the same country are migratory or stationary."

3. "The migratory instinct is made up of two very distinct factors—viz., an impulse to travel periodically, and a faculty of knowing the direction in which to travel."

4. "Savage man shows a sense of direction which may be analogous to that shown by migratory animals."

5. "Certain cases are on record of breeds of domesticated animals having truly migratory instincts."

Such being the data, the problem is to account for the origin of the instinct. Mr. Darwin's theory is that the ancestors of migratory animals were annually driven, by cold or want of food, slowly to travel southwards; "and in time we may well believe that this compulsory travelling would become an instinctive passion," as is the case with domesticated sheep in Spain. In the case of birds, the wings would be used, and if in the course of many successive generations the land over which they flew in their annual journeys were to become slowly submerged, the line of flight would remain unaltered, and thus we should have the state of things which we now perceive—viz., migratory birds flying over wide stretches of ocean.

Before I proceed to consider this theory, I should like to call prominent attention to the fact that it has been inde-

* Professor Newton, F.R.S., Art. *Birds*, where see for a good *résumé* of the main facts of migration as regards birds.

pendently arrived at by Mr. Wallace. It is only now that Mr. Darwin's views upon this subject are published, although they were committed to writing as they appear in the Appendix between twenty and thirty years ago. Mr. Wallace however enunciated substantially the same views in a letter to "Nature" in 1874 (Oct. 8),* from which I shall quote *in extenso*, not only for the purpose of showing the coincidence to which I have alluded, but also because I think that the additional element which Mr. Wallace mentions—*i.e.*, the separation of breeding and subsistence areas—is a most important one.

"Let us suppose that in any species of migratory bird, breeding can as a rule be only safely accomplished in a given area; and further, that during a great part of the rest of the year sufficient food cannot be obtained in that area. It will follow that these birds which do not leave the breeding area at the proper season will suffer, and ultimately become extinct; which will also be the fate of those which do not leave the feeding area at the proper time. Now if we suppose that the two areas were (for some remote ancestor of the existing species) coincident, but by geological and climatic changes gradually diverged from each other, we can easily understand how the habit of incipient and partial migration at the proper season would at last become hereditary, and so fixed as to be what we term an instinct. It will probably be found that every gradation still exists in various parts of the world, from a complete coincidence to a complete separation of the breeding and subsistence areas; and when the natural history of a sufficient number of species in all parts of the world is thoroughly worked out, we may find every link between species which never leave a restricted area in which they breed and live the whole year round, to those other cases in which the two areas are absolutely separated. The actual causes that determine the exact time, year by year, at which certain species migrate, will of course be difficult to ascertain. I would suggest, however, that they will be found to depend on those climatic changes which most affect the particular species. The change of colour, or the fall of certain leaves; the change to the pupa state of certain insects; prevalent winds or rains; or even the

* Captain Hutton also foreshadowed these views in 1872; see *Trans. New Zealand Inst.*, p. 235.

decreased temperature of the earth and water, may all have their influence."

It will be observed that this theory, besides being intrinsically probable, derives a good deal of support from the enquiries made by Mr. Darwin, which have shown that there is a general relationship between oceanic islands which there is independent reason to conclude have never been joined to the mainland, and an absence of migratory birds.*

It will also be observed this theory makes two important assumptions—first, that the birds have a very accurate sense of direction, and second, that a no less accurate knowledge of the particular direction to be pursued is inherited; for it is certain that the young Cuckoo (which leaves England after its parents) cannot be guided on its first journey by any other means, and it is asserted that the same is true of the young of many other species.† Taking then these assumptions separately, the first is no more than a statement of fact, unaccountable though the fact may be. That is to say, a very accurate sense of direction migratory birds unquestionably possess, and it is probably the same in kind as the so-called "homing" faculty which is shown by many domesticated animals, and also, as Mr. Darwin points out, by savage man. I could fill pages with letters which I have received from all parts of the world describing more or less remarkable cases of the display of this faculty by dogs, cats, horses,‡

* To be quite fair, however, I must here allude to the only fact I have met with which seems to me opposed to this theory. Mr. Hurdis in his work entitled *The Naturalist in Bermuda*, observes that the migratory golden plover (*Charadrius marmoratus*) passes over the islands in countless multitudes (but without ever alighting) on the journey south, while they are never seen passing over the islands on their return journey north. Now, if it is a fact that the two journeys are taken by different routes, a difficulty would be encountered by the above theory; but as Mr. Hurdis says that the birds fly at an enormous height while passing over the islands on their southern journey, it is not, I think, impossible that they may take the same route on their northern journey, although at a still higher elevation, and thus escape notice.

† See Temminck, *Man. d'Orn.*, ed. 2, iii, Introd., p. xliii, and Seebohm, *Siberia in Europe*. On the other hand Leroy says that in the case of swallows 'those who have had no instruction do not migrate, and the young birds are seen to be led by those whose age and experience give them knowledge and authority;" and adds that if a brood are hatched out too late to accompany the old birds in their migration, "it is in vain that they reach maturity . . . . they perish the victims of their ignorance, and of the tardy birth which made them unable to follow their parents" (*loc. cit.*, pp. 183–4).

‡ I have one instance of a cat returning in four days from London to

T

asses, cows, sheep, goats, and pigs; but as so many similar cases are already on record, I feel it is needless to add to the number. The remarkable fact is that the animals are able to find their way back over immense distances, even though the outgoing journey has been made at night, or in a closed box; so that it is truly upon some sense of direction, and not merely upon a memory of landmarks, that they must rely. Moreover, it is certain that in many cases, if not as a general rule, the animals on their return journey do not traverse the exact route which they had taken in the outgoing journey, but take the "bee-line";* so that, for instance, if the out-going journey has been made over two sides of a triangle, the return journey will most probably be made over the third side. One instance, the account of which I have received from a correspondent in Australia, is of sufficient interest in this connection to quote. "A pair of horses were sent many hundred miles round the Australian coast by ship; as they did not like their new quarters, they started back by land; but after returning 230 miles they were pulled up by a peninsula on the coast, where they were eventually recovered. They did not attempt to retrace their steps to clear this difficulty."†

Huddersfield, a distance of two hundred miles. A still more remarkable case, however, was published by Mr. J. B. Andrews in *Nature* several years ago (vol. viii, p. 6). The Archduchess Marie Régnier passed the winter of 1871–2 at the Hotel Victoria, in Mentone, and while there took a fancy to a spaniel belonging to the landlord, M. Milandri. In the spring of 1872 she brought the dog with her by rail to Vienna. Not long afterwards it reappeared at the hotel in Mentone, having thus run a distance of nearly a thousand miles. On arriving it died of fatigue and was buried in the hotel gardens, where a monument now commemorates the performance. Mr. A. W. Howitt writing to *Nature* from Victoria at about the same time (vol. viii, p. 322) gives a number of cases of horses and cattle finding their way home over greater or less distances, and I specially allude to his communication because he says that in some of the cases the return journey was made after a considerable lapse of time—months and even years.

* This is an American term which I employ because in itself showing the observed regularity of the fact as regards bees—it being the custom to find wild hives of honey by catching several bees, and letting them go again from different places. The insects under these circumstances make straight for their hive, so that by observing the point where several "bee-lines" intersect, the honey seekers are able to find the hive.

† I may here also quote an observation by Mr. Darwin to the same effect:—"I sent a riding-horse by railway from Kent *viâ* Yarmouth, to Freshwater Bay, in the Isle of Wight. On the first day that I rode eastward, my horse, when I turned to go home, was very unwilling to return towards his stable, and he several times turned round. This led me to make repeated trials, and every

Now it is evident that this fact alone—*i.e.*, of animals not requiring to return by the same route—is sufficient to dispose of the hypothesis advanced by Mr. Wallace* to the effect that the return journey is due to a memory of the odours perceived during the out-going journey, these odours thus serving as land-marks. Therefore it seems to me there are only two hypotheses open to us whereby to meet the facts. First, it has been thought possible that animals may be endowed with a special sense enabling them to perceive the magnetic currents of the earth, and so to guide themselves as by a compass. There is no inherent impossibility attaching to this hypothesis, but as it is wholly destitute of evidence, we may disregard it. The only other hypothesis is that animals are able to keep an unconscious register of the turns and curves taken in the outgoing journey, and so to retain a general impression of their bearings. This hypothesis is substantiated by the fact that, as Mr. Darwin observes, savage man is certainly endowed with some such faculty; and a friend of my own (Mr. Henry Forde quoted below), who has spent many years in the forests and prairies of America, informs me that even civilized man when long accustomed to such primitive habits of life, acquires this faculty in a degree of perfection quite comparable with that of savages. He also informs me that, occasionally, without any assignable reason, the sense of direction becomes confused, leading to a distressed sensation of bewilderment. He has seen a hunter thus reduced to a lamentable condition of nervousness, and when at last he abandoned himself to the leadership of his companions (who relied entirely on their own sense of direction), he felt persuaded that they were going the wrong way. But on approaching his dwelling-place he recognized one of the trees, and declared that a particular notch upon it had passed round to the other side of the trunk. Eventually he said that the whole world

time that I slackened the reins, he turned sharply round and began to trot to the eastward by a little north, which was nearly in the direction of his home in Kent. I had ridden this horse daily for several years, and he had never before behaved in this manner. My impression was that he somehow knew the direction whence he had been brought. I should state that the last stage from Yarmouth to Freshwater is almost due south, and along this road he had been ridden by my groom; but he never once showed any wish to return in this direction" (*Nature*, vol. vii, p. 360). See also *Nature*, viii, p. 322.

* *Nature, loc. cit.*

seemed to have turned round him as a centre. In this connection I may quote the following passage from a letter published some years ago by Mr. Darwin in "Nature" (vol. vii):—

"The manner in which the sense of direction is sometimes suddenly disarranged in very old and feeble persons, and the feeling of strong distress which, as I know, has been experienced by persons when they have suddenly found out that they have been proceeding in a wholly unexpected and wrong direction, leads to the suspicion that some part of the brain is specialized for the function of direction. Whether animals may not possess the faculty of keeping a dead-reckoning of their course in a much more perfect degree than man; or whether this faculty may not come into play on the commencement of a journey, when an animal is shut up in a basket, I will not attempt to discuss, as I have not sufficient data." He also alludes to the case of Audubon's pinioned wild goose, which showed a very determined impulse to migrate at the proper season, but mistook the direction and went due north instead of south.

Lastly, I may quote the following from Dr. Bastian's work on the Brain.*

"On this subject, G. C. Merrill, writing from Kansas, says:—

'I have learned from the hunters and guides who spend their lives on the plains and mountains west of us, that no matter how far, or with what turns, they may have been led, in chasing the bison or other game, they, on their return to camp, always take a straight line. In explanation, they say that, unconsciously to themselves, they have kept all the turns in their mind.'

"Referring to his travels in the State of Western Virginia, Mr. Henry Forde (Nature,' April 17, 1873, p. 463) writes as follows:—'It is said that even the most experienced hunters of the forest-covered mountains in that unsettled region are liable to a kind of seizure—that they 'lose their heads' all at once, and become convinced that they are going in quite the contrary direction to what they had intended, and that no reasoning nor pointing out of land-marks by their companions, nor observations of the position of the sun, can

* *Brain as an Organ of Mind*, p. 215, where see also for cases of way-finding in animals.

overcome their feeling; it is accompanied by great nervousness and a general sense of dismay and 'upset.' The nervousness comes after the seizure, and is not the cause of it. This is spoken of by the natives as 'getting turned round.' The feeling sometimes ceases suddenly, or it may wear away gradually. Colonel Lodge, in his 'Hunting Grounds of the Far West,' 1876, speaks of the same kind of feelings seizing upon, and occasionally demoralizing, old and experienced prairie travellers. Indian chiefs all concurred in assuring G. Catlin ('Life amongst the Indians,' p. 96) that 'whenever a man is lost on the prairies, he travels in a circle, and also that he invariably turns to the left; of which singular fact,' the author adds, 'I have become doubly convinced by subsequent proofs.'"

But it is evident that definite experiments on this homing faculty, both in men and in animals, are required before we can be in a position to say anything more with regard to it than admitting it as a matter of fact. The only experiments which have been made, so far as I am aware, are those of Sir John Lubbock, on the sense of direction in the Hymenoptera (to which I shall allude presently), and those which have more recently been published by M. Fabre,* who also experimented upon the Hymenoptera. As the last-named author believes that he has established a very definite conclusion by means of his experiments, it is necessary that I should make a few remarks upon them.

At the suggestion of Mr. Darwin, he placed some marked mason-bees in a closed paper box, carried them thus imprisoned for some distance in one direction, then rotated the box and carried them a much greater distance in the opposite direction, after which he released the insects. He found that when the distance to which the bees were taken was as much as three kilometres, and even when the rotation was very considerable (the box being placed in a sling and rotated in various planes at several points in the route) a certain percentage of the bees returned home. It made no difference whether the bees were released in an open space or in a thick wood; neither did it make any difference whether the outgoing journey were performed in a straight line or in a circuitous curve. From these experiments M. Fabre con-

* *Nouveaux Souvenirs Entomologiques*, 1882, pp. 99–123.

cludes that the sense of direction cannot depend upon any process of dead-reckoning. At the suggestion of Mr. Darwin he also tried the effect of attaching a magnetized needle to the thorax of a bee; but the bee having succeeded in getting rid of the encumbrance, he did not repeat the experiment.

Now, although the observations with the rotating box are no doubt very interesting, they do not appear to me to sustain the definite conclusion that the sense of direction is not due to a process of dead-reckoning. It is of course impossible to suppose that the bees could retain a register of all the turns to which they were submitted in the sling, and, therefore, if it were certain that they found their way home by means of their sense of direction, I should agree with M. Fabre in concluding, once for all, against the theory of dead-reckoning. But there is no evidence to show that the bees which found their way home did so by means of their sense of direction. It is quite possible that they found their way home simply from their knowledge of land-marks; for the distance to which they were taken was only three kilometres, and it is known that the hive-bee will go three times that distance in its ordinary foraging excursions.* Moreover, the fact that only a comparatively small number of the bees succeeded in returning (about 22 per cent.), is suggestive of the explanation that these were the only ones which, during the random flight of the whole number in sundry directions, happened to encounter land-marks with which they were familiar. I am therefore inclined to feel that any sense of direction which existed in these insects may very well have been rendered useless by these experiments, and yet that the results of the experiments might have been exactly those which M. Fabre describes.

Returning, however, to the case of migration, I think it is not very improbable that the sense of direction may be greatly assisted by observing the direction of the sun with reference to the appropriate line of flight. It is true that many migratory birds fly at night; but in this case, even if the moon is not available to steer by instead of the sun, during much of the night the directions of sun-set and sun-rise are clearly indicated by the light of the sky; and it appears that on very dark and cloudy nights migratory birds are apt to become

* See *Animal Intelligence*, p. 150.

confused.* The possibility thus suggested receives, I think, some countenance from the following fact. In "Animal Intelligence" I recorded a number of observations which had been made by Sir John Lubbock on the sense of direction as exhibited by ants. These experiments yielded results of a most definite nature, and thus led Sir John to conclude that ants are endowed with the sense of direction in a singular degree. Subsequently, however, he has found (accidentally in the first instance) that in all these experiments the ants found their way by observing the direction in which the light was falling; so that, as long as the source of light was stationary, no matter how many times he turns them round upon a rotating table, when the rotation ceased they knew their road to and from the hive as well as they did before the rotation; whereas, if the source of light were shifted, the insects at once became confused as to their bearings, even though not rotated at all.† Now if ants thus habitually guide themselves by observing the direction in which the light is falling (*i.e.*, the position of the sun), I do not see why migratory birds should not be assisted by similar means.

This, however, I only put forward as a conjecture. The fact that migratory birds, like many other animals, are in some way able to hold a true course in order to reach a particular locality, is a fact which confessedly we are not able to explain. But—and this is the most important point for us—our inability to explain this fact in the present state of our information, is no objection to the theory of instinct on which we are engaged. We cannot doubt that the fact admits of *some* explanation, and when we certainly know what this explanation is, we shall first be able to ascertain whether the faculty of way-finding is or is not compatible with the foregoing theory of the evolution of instinct.

Let us turn now to the second of the two assumptions above alluded to as necessary in order to embrace the facts of migration under the theory—viz., the assumption that some at least among migratory birds must possess, by inheritance alone, a very precise knowledge of the particular direction to

* See Professor Newton in *Nature*, vol. xi, p. 6, who says, "Dark cloudy nights seem to disconcert the travellers. On such nights the attention of others besides myself has often been directed to the cries of a mixed multitude of birds hovering over this (Cambridge) and other towns, apparently at a loss whither to proceed, and attracted by the light of the street lamps."

† See *Journ. Linn. Soc.*, 1883.

be pursued. It is, without question, an astonishing fact that a young cuckoo should be prompted to leave its foster-parents at a particular season of the year, and without any guide to show the course previously taken by its own parents; but this is a fact which must be met by any theory of instinct which aims at being complete. Now upon our own theory it can only be met by taking it to be due to inherited memory. I confess to me it seems incredible that many hundred miles of landscape scenery should constitute an object of inherited memory,* to say nothing of long stretches of ocean; but the case is not quite so hopeless as to require so extreme a hypothesis. When we say that upon our theory the young cuckoo must be supposed to find its way on its first journey by inherited memory, we need not necessarily affirm that this is the memory of a landscape. As I have said in the previous paragraphs, we do not yet know what it is that guides the course of migratory birds in general; but whatever this may be, it can scarcely be the appearance of the country over which they pass, seeing not only that the distances are so great and that two hundred or three hundred miles of ocean may separate one piece of country over which they travel from another, but also that the journeys may be taken by night. Of what, then, is the inherited memory on which the young cuckoo (if not also other migratory birds) depends? We can only answer, Of the same (whatever this may be) as that upon which the old birds depend. When we certainly know what this is, we shall first be able to ascertain whether it is incompatible with the theory of evolution to suppose that it can be an object of hereditary memory. Thus, for the sake of example, let us suppose that the old birds in their outgoing journey guide their way by flying towards the south wind (as has been suggested to me by Mr. William Black, who believes that swallows always start against the south wind); heredity would in this case have an easy task in associating the warm moist breath of this wind with a desire to fly against it. Of course I only adduce this suggestion in order to show how simple the mere question of heredity might become, if once we knew the means whereby migratory birds in general find their way. The only difference between the faculty of homing and the instinct of migration, so far as

* This theory was first advanced by Canon Kingsley (*Nature*, Jan. 18, 1867), and has since been independently suggested by several writers.

the matter of way-finding is concerned, seems to me to be this—that in the case of the young cuckoo, and perhaps also in that of certain other migratory birds, the animals know their way instinctively, or without even one lesson. But if we could ascertain upon what it is that the faculty of homing depends (which, be it observed, is not an instinct, seeing that its occurrence is the exception and not the rule, even in the species which exhibit it), we might very probably get a clue to explain the manner in which heredity has been able to work up this faculty into the instinct of migration.

No doubt this discussion is most unsatisfactory, and the reason is that we are so much in the dark about the facts. All, therefore, that I have attempted to do is to show that, in the present state of our information, the migratory instinct cannot fairly be quoted as a difficulty in the way of our theory of the formation of instincts in general. And, in order to give emphasis to this statement, I may allude to the general facts already mentioned—viz., that the migratory instinct is both variable and graduated, that it is occasionally exhibited by domesticated animals, and that the sense of direction on which it depends is a very general one among animals, if not also in savage man; for all these facts tend to show that whatever the causation of the migratory instinct may be, it has probably been proceeding upon the lines of evolution in general.

Instincts of Neuter Insects.

Mr. Darwin has pointed out a serious difficulty lying against his theory of the origin of instincts by natural selection, and one which, as he justly remarks, it is surprising that no one should have previously advanced against the well-known doctrine of inherited habit, as taught by Lamarck. The difficulty is that among various species of social insects, such as bees and ants, there occur "neuter," or asexual individuals, which manifest entirely different instincts from the other or sexual individuals, and, as the neuters cannot breed, it is difficult to understand how their peculiar and distinctive instincts can be formed by natural selection, which, as we have seen, requires for its operation the transmission of mental faculties by heredity. The difficulty is increased by the fact that among the termites and many species of ants several varieties or "castes" of neuters occur in the same nest, which differ widely from one another both in structure and in instincts. The only possible way in which this difficulty can be met is the way in which it has been met by Mr. Darwin, viz., by supposing "that selection may be applied to the family as to the individual." "Such faith may be placed in the power of selection that a breed of cattle always yielding oxen with extraordinarily long horns could, it is probable, be formed by carefully watching which individual bulls and cows, when matched, produced oxen with the longest horns; and yet no one ox would ever have propagated its kind;" and similarly, of course, with regard to the instincts of neuters. Otherwise stated, we may regard the nest or hive as itself an organism of which the sexual insects and the several castes of neuters constitute the organs; and we may then suppose natural relation to operate upon this organism as a whole, somewhat in the same way as we habitually suppose it to operate upon the "social organisms" or communities of mankind. No doubt, when carefully considered, the analogy between a hive and an organism, or even between a hive and a social community, is not a close analogy so far as the *modus operandi* of natural selection is concerned; for in the one case the analogue of organs is a variety of separate individuals, while in the other case there is no such great contrast between different classes of a human community as that which obtains among the different castes of an insect community. The root of the question really consists in whether or not it is possible to suppose that natural selection may operate upon specific types as distinguished from individual members of species. During his life-time I had the advantage of discussing this question with Mr. Darwin, and I ascertained from him that it had greatly occupied his thoughts while writing the "Origin of Species;" but that, finding it to be a question of so much intricacy, he deemed it unadvisable to enter upon its exposition. It would occupy too much space were I to attempt such an exposition here, and I have alluded to the subject only because I desire to show that it is really this general question which is involved in the case of the special difficulty with which we are now concerned. On some future occasion I intend to argue this general question, and then I shall hope to mitigate the force of this special difficulty. I may, however, point to one fact which Mr. Darwin has observed, and which is of much importance as indicating that the different castes of neuters have arisen by degrees, and therefore presumably under the influence of natural selection. This fact is that, when carefully searched for, neuters presenting more or less well-marked gradations of structure between one caste and another may be occasionally found in the same nest.* On the whole, therefore, I conclude, with regard to this particular case of difficulty, that it is not so formidable as to exclude the explanation furnished by the hypothesis of natural selection, supposing that we have already accepted this hypothesis as explanatory of other and less difficult cases.

Instincts of the Sphex.

Several species of this division of the Hymenoptera display what I think may be justly deemed the most remarkable instincts in the world. These consist in stinging spiders, insects, and caterpillars in their chief nerve-centres, in consequence of which the victims are not killed outright, but rendered motionless; they are then conveyed to a burrow previously formed by the Sphex, and, continuing to live in their paralyzed condition for several weeks, are at last available as food for the larvæ when these are hatched. Of course the extraordinary fact which stands to be explained is that of the precise anatomical, not to say also physiological knowledge which appears to be displayed by the insect in stinging only the nerve-centres of its prey. The following, so far as is at present known, are the main features of this very surprising case.

The same species of Sphex always preys upon the same species of victim. When the victim is a spider, the instinct of its assailant dictates that a single sting shall be given in the large ganglion where, in the case of the spider, most of the central nervous matter is aggregated. When the victim is a beetle, the Sphex which preys upon it—there are eight species which prey upon two genera—first throws the insect upon its back, then embraces it and plunges the sting through the membranes between the first and the second pairs of legs; the sting thus strikes the main nerve-centre, which is unusually agglomerated in beetles of this genus. When the prey is a

* See *Origin of Species*, 231-2.

cricket, the insect is thrown, as in the previous case, upon its back, and while holding it down with her mandibles firmly fastened upon the last segment of its abdomen, her feet on the sides holding down the body of the cricket—the anterior feet holding down the long posterior legs of the prey, and the hind feet holding back the mandibles, so as to prevent these from biting, and at the same time making tense the membranous junction of the head with the body—the Sphex darts her sting successively into three nerve-centres; first into the one below the neck which she has stretched back for the purpose, next into the one behind the prothorax, and lastly into the one lower down. A cricket thus paralyzed will live for six weeks or more. When the prey is a caterpillar, a series of six to nine stings are given, one between each of the segments of the body beginning from the anterior end; the brain is then partially crushed by a bite with the mandibles.*

Now so far as the spider and the beetle are concerned, I do not see much difficulty presented by the facts to our theory of the formation of instincts. For as both the large nerve-centres of the Spider and the sting of the Sphex occur upon the median line of their respective possessors, if the stinging of the ganglion were in the first instance accidentally favoured by this coincidence—which appears to me not improbable, seeing that the nerve-centre is thus the most likely place for the sting to strike,—it is evident that natural selection would have had excellent material on which to work for the purpose of developing such an instinct as we now observe. Again, in the case of the beetle, M. Fabre expressly notices that the only vulnerable point in the hard casing of the animal is in the articulation where the Sphex thrusts her sting; so that there is nothing very remarkable in natural selection having developed an instinct to sting at the only place in the body of the prey where stinging is mechanically possible.

But the case is certainly very different with the cricket and the caterpillar; for here—or at least in the latter case—we encounter the extraordinary and unavoidable fact of an insect, without any accidental guiding or mechanically

* All the above facts are taken from the works of M. J. H. Fabre (*Souvenirs Entomologiques*, 1879 and 1883), who was the first to observe and describe them.

imposed necessity, instinctively choosing a number of minute points in the uniformly soft body of its prey, with an apparently very precise knowledge that it is only at these particular points that the peculiar paralyzing influence of its sting can be exercised. After duly considering this case, I must candidly say that I feel it to be the most perplexing which has yet been brought to light, and the one which is most difficult of explanation upon the principles of the foregoing theory. It is, however, most desirable that the facts should be more thoroughly investigated, for it might then appear that some clue would be given as to the origin and development of this instinct. So far as our information at present extends, I can only suggest that this origin must have been of a purely secondary kind, although its subsequent development may probably have been assisted by natural selection. In other words, so far as we have any means of judging, I can see no alternative but to conclude that these wasp-like animals owe their present instincts to the high intelligence of their ancestors, who found from experience the effects of stinging caterpillars between the segments of their bodies, and consequently practised the art of so stinging them till it became an instinct.

During the last year of his life I had some conversation with Mr. Darwin upon this matter, and, after deliberating upon it for some time, he eventually came to the conclusion which I have just stated—as will be at once apparent from the following letter which he wrote to me, and which will serve in a few words to indicate what appears, I think, to be the most probable steps by which these singular instincts were acquired.

"I have been thinking about Pompilius and its allies.—Please take the trouble to read on perforation of the corolla by Bees, p. 425 of my "Cross-fertilization," to end of chapter. Bees show so much *intelligence* in their acts, that it seems not improbable to me that the progenitors of Pompilius originally stung caterpillars and spiders, &c., in any part of their bodies, and then observed by their intelligence that if they stung them in one particular place, as between certain segments on the lower side, their prey was at once paralyzed. It does not seem to me at all incredible that this action should then become instinctive, *i.e.*, memory transmitted from one generation to another. It does not seem necessary to suppose that when Pompilius stung its prey in the ganglion it intended or knew that their prey would keep long alive. The development of the larvæ may have been subsequently modified in relation to their half-dead, instead of wholly dead prey; supposing that the prey was at first quite killed, which would have required much stinging. Turn this over in your mind," &c.

Now in Chapter XIV I have already given a short epitome of the facts concerning the boring by humble-bees of holes in corollas, and the subsequent utilization of the holes by hive-bees. It will be remembered that the connection in which I there alluded to the facts was that of the power of imitation by one species of the habits of another—the hive-bees observing that the humble-bees were saving time by sucking through the holes instead of entering the flowers. But the point which is of importance in the present connection is the intelligence displayed by the humble-bees in originating the idea, so to speak, of boring the holes. For close observation shows that they bore the holes with as precise an appreciation of the morphology of the flowers, as is shown by the Sphex of the morphology of spiders, insects, or caterpillars. Thus in the case of leguminous flowers they bite only through the standard petal, and always on the left side just over the passage to the nectar, which is larger than the corresponding passage on the right side. Therefore, as Mr. Francis Darwin observes, "it is difficult to say how the bees could have acquired this habit. Whether they discovered the inequality in the size of the nectar-holes in sucking the flowers in the proper way, and then utilized this knowledge in determining where to gnaw the hole; or whether they found out the best situation by biting through the standard at various points, and afterwards remembered its situation in visiting other flowers. But in either case they show a remarkable power of making use of what they have learnt by experience."*

Seeing, then, that Hymenopterous insects are certainly proved by these observations to be capable of marvellously intelligent appreciation of morphological structure, I think with Mr. Darwin that these observations are most apposite to the case of the Sphex. There is not, after all, so very much more of this kind of appreciation required to observe the effects of stinging a caterpillar between its segments, than to hit upon the idea of going outside a flower and biting a hole on the

* *Nature*, Jan. 8, 1874, p. 189.

left side of a particular petal, just over the spot where the larger passage to the nectar is to be found. But, as I have said, I feel that further observation—especially in the way of experiment—of the facts is required before we should be justified in giving a very definite opinion upon the theoretical interpretation of them. All I can say is that, in the present state of our information upon the subject, Mr. Darwin's view, as above stated, appears to be the most probable one that can be taken. We are not much surprised at the instinct of a Ferret in attacking the medulla oblongata of a rabbit, or at that of a Pole-cat in paralyzing frogs and toads by injuring the cerebral hemispheres;* and in both these cases—so analogous to that which we are now considering—the instinct must have originated by intelligent observation of the effects of biting these particular parts of the prey. But neither a ferret nor a pole-cat is a particularly intelligent animal, so that we are perhaps too ready to feel surprise at the possibly similar degree of intelligence displayed by insects which belong to the most intelligent group of invertebrated animals.

Feigning Death.

It is a matter of common knowledge and wonder that sundry species of animals belonging to different orders and even classes, manifest the instinct, when in danger, of feigning death. As it is clearly impossible to attribute this fact to any idea of death and of its conscious simulation on the part of the animals, the subject becomes one of importance for us to consider. I shall first give all the facts that I have been able to collect with regard to it, and then proceed to discuss their explanation.

The most familiar example of the instinct in question is furnished by sundry species of insects and spiders, many of which allow themselves to be slowly dismembered, or gradually roasted to death, without betraying the slightest movement. "Among fish, the captured sturgeon remains quiet and passive in the net, while the perch feigns death and floats on its back."† According to Wrangle,‡ the wild geese of Siberia, if alarmed during the moulting season when

* See *Animal Intelligence*, p. 347.

† Couch, *Illustrations of Instinct*, p. 199, *et seq.*

‡ *Travels in Siberia*, p. 312, Eng. Transl.

they are unable to fly, stretch themselves at length upon the ground with their heads concealed, so as to feign death and deceive the sportsman. According to Couch, the habit is common to the landrail, the skylark, and other birds.* Among mammals, the same author says, "The opossum of North America is so famous for feigning death, that its name has become proverbial as an expression of this deceit;"* and he narrates instances of the same fact with regard to mice, squirrels, and weasles. The testimony on the subject with regard to wolves and foxes is so abundant that I do not think there can be any reasonable question concerning its accuracy. Thus Captain Lyon, in the account of his Polar Expedition, says that a wolf caught in a trap which was set by Mr. Griffiths, was apparently killed and then dragged on board. "The eyes, however, as it lay on deck, were observed to wink, whenever any object was placed near them; some precautions were therefore considered necessary; and the legs being tied, the animal was hoisted up with his head downwards. He then, to our surprise, made a vigorous spring at those near him; and afterwards repeatedly turned himself upwards, so as to reach the rope by which he was suspended, endeavouring to gnaw it asunder," &c.

The testimony is abundant on the subject of foxes shamming dead. As Mr. Blyth observes,† "a fox has been known to personate a defunct carcase, when surprised in a hen-house, and it has even suffered itself to be carried out by the brush and thrown out on a dungheap, whereupon it instantly rose and took to its heels, to the astounding dismay of its human dupe. In like manner this animal has submitted to be carried for more than a mile, swinging over the shoulder with its head hanging downwards, till at length it has very speedily effected its release by biting."

Similarly Couch, who gives a number of instances of the fact, summarizes them thus:—"When suddenly surprised by man, he has been known to assume the appearance of being dead, and has suffered himself to be handled, and even ill-treated, without betraying any signs of sensibility. This high degree of simulation and dissimulation has been ascribed to consummate wisdom, which, when a better means of escape did not offer itself, prompted him to the stratagem of feigning

* *Loc. cit.*

† *Loundoun's Mag. Nat. Hist.*, N.S., vol. i, p. 5.

to be incapable of defence or flight until he had disarmed suspicion, and so escaped hostility.*

According to Jesse, "Snakes, too, will pretend to be dead, and lie motionless, as long as they think they are observed, and in danger, but, when they believe that all foes have withdrawn, and they are out of peril, they will glide away with the greatest speed into the nearest hole or covert.

"Among birds, the corncrake has been most remarked for this species of art. The author of 'The Natural History of the Corncrake' relates that one of these birds was brought to a gentleman by his dog, to all appearance quite dead. The gentleman turned it over with his foot, as it lay upon the ground, and was convinced there was no more life in it. But after a while he saw it open one eye; and he then took it up again, when its head fell, its legs hung loose, and it once more appeared to be certainly dead. He next put it into his pocket, and before long felt it struggling to escape; he took it out, and it seemed lifeless as before. He then laid it on the ground and retired to a little distance to watch it, and saw it in about five minutes raise its head warily, look round, and decamp at full speed."

Bingley observes, "This stratagem is also said to be practised by the common crab, which, when it apprehends danger, will lie as if dead, waiting for an opportunity to sink itself into the sand, keeping only its eyes above it."

Hence, it appears that from insects upwards, the instinct

* *Illustrations of Instinct*, p. 197. Sir E. Tennent, in his *Natural History of Ceylon*, gives the case of a wild elephant apparently feigning death; but as under the circumstances mentioned elephants very often actually do die (see *Animal Intelligence*, p. 396), this case is probably not to be attributed to intentional deception on the part of the animal. The case is as follows:—"Mr. Cripps has related to me an instance in which a recently captured elephant was either rendered senseless from fear, or, as the native attendants asserted, *feigned death* in order to regain its freedom. It was led from the corral as usual between two tame ones, and had already proceeded far towards its destination; when night closing in, and the torches being lighted, it refused to go on, and finally sank to the ground, apparently lifeless. Mr. Cripps ordered the fastenings to be removed from its legs, and when attempts to raise it had failed, so convinced was he that it was dead, that he ordered the ropes to be taken off and the carcase abandoned. While this was being done he and a gentleman by whom he was accompanied leaned against the body to rest. They had scarcely taken their departure and proceeded a few yards, when, to their astonishment, the elephant rose with the utmost alacrity, and fled towards the jungle, screaming at the top of its voice, its cries being audible long after it had disappeared in the shades of the forest."

of shamming dead occurs in most if not all the classes of the animal kingdom. The subject therefore demands from us serious attention, because on the one hand, as previously remarked, it is obvious that the idea of death and of its conscious simulation would involve abstraction of a higher order than we could readily ascribe to any animal, and on the other hand it is not very easy otherwise to explain the facts.

I shall first of all quote what Couch says upon the subject, as he is the first author, so far as I am aware, who did not at once take it for granted that animals consciously feign death, and who also supplied a reasonable hypothesis to account for the facts. He says:—

"But a more probable explanation is, that the suddenness of the encounter, at a time when the creature thought of no such thing, had the effect of stupefying his senses, so that an effort of escape was out of his power, and the appearance of death was not the fictitious contrivance of cunning, but the consequence of terror. And that this explanation is the true one appears, among other proofs, from the conduct of a bolder and more ferocious animal, the Wolf, under similar circumstances. If taken in a pitfall it is said that it is so subdued by surprise, that a man may safely descend and bind and lead it away or knock it on the head; and it is also said that when it has wandered into a country to which it is a stranger, it loses much of its courage and may be assailed almost with impunity."*

"A similar action to that of the Fox has been observed in a little animal to which it is not common to ascribe more than an ordinary degree of cunning or confidence in its own resources. In a bookcase of wainscot, impervious to light, certain articles were kept more agreeable to the taste of mice than books, and, at midday, when the doors were suddenly opened, a mouse was seen on one of the shelves; and so rivetted was the little creature to the spot that it showed all the signs of death, not even moving a limb when taken into the hand. On another occasion, on opening a parlour-door, in broad daylight, a mouse was seen fixed and motionless in the middle of the room; and, on advancing towards it, its appearance in no way differed from that of a dead animal, excepting that it had not fallen over on its side. Neither of these

* *Mag. Nat. Hist.*, New Ser., vol. ii, p. 124.

creatures made an effort to escape, and were taken up at leisure; nor had they received any hurt or injury, for they soon displayed every mark of being alive and well.

"It would be as easy to catch a Weasel asleep as off its guard; but it seems still more unlikely that, in the disguise of death it should suffer itself to be cuffed, pawed, and handled with impunity by a cat: yet it so happened that, while Puss was reclining at ease, seemingly inattentive to all the world around her, a Weasel came unexpectedly up, was seized in a moment, and dangling from her teeth as if dead, was thus carried to the house at no great distance. The door being shut, Puss, deceived by its apparent lifelessness, laid her victim on the step, while she gave her usual mewing cry as for admittance. But by this time the active little creature had recovered its recollection, and in a moment struck its teeth into its enemy's nose. It is probable that, besides the sudden surprise of the capture, the firm grasp which the cat had of it round the body had prevented any earlier effort at resistance from the Weasel; for in this manner our smaller quadrupeds which bite so fiercely may be held without injury; but the Weasel can hardly be supposed to have been practising a deception all the while it was in the Cat's mouth."*

This hypothesis would require to be substantiated by special experiments before it would merit unreserved acceptance. These experiments would consist in allowing an animal, immediately that it is observed shamming dead, to regain its liberty, and to watch it without the animal knowing that it is being observed. If it were then to continue for any considerable time motionless, the fact would tend to prove Couch's hypothesis, whereas if it were quickly to recover, the inference would lie in the direction of supposing the passiveness in the presence of danger due to conscious purpose. I once thought that I had myself the opportunity of trying this experiment; for having caught a wild squirrel in a cloth I observed that the animal immediately became motionless. Turning it out upon the ground and concealing myself from its view, I waited a long time for it to recover; but as it did not do so I went up to examine it, and found it was not shamming, but really dead. This incident I mention here, because it has an important bearing on Couch's hypothesis; it shows that the terror of a

* *Illustrations of Instinct*, p. 125.

wild animal on being captured may be sufficient to cause actual death, and the researches of Professor Preyer on the hypnotism of animals (conducted long after Couch's book was published and having no special reference to the present question), showed that fright is a strong predisposing cause of " Kataplexy," or mesmeric sleep in animals.

This allusion to Professor Preyer's researches leads me next to remark that he ascribes the shamming dead of insects to the exclusive influence of kataplexy. Having observed the potency of this influence in producing analogous conditions in the neuro-muscular systems of higher animals—even as far down the series as the cray-fish, which were made to stand upon their heads while in the hypnotic state—it was perfectly logical in him to attribute the shamming dead of insects to the same cause. And his reasoning might have been greatly strengthened had he been aware of the important facts which had been observed by Mr. Darwin, and which are now given in the Appendix. These facts, it will be noted, are, that there is no species of spider or insect of which it can be said that the attitude assumed when shamming death at all closely resembles the one which the animal assumes when really dead; that in many cases this attitude is very dissimilar; and therefore that all " shamming dead" amounts to in these animals is an instinct to remain motionless, and thus inconspicuous, in the presence of enemies. And it is easy to see that this instinct may have been developed by natural selection without ever having been of an intelligent nature—those individuals which were least inclined to run away from enemies being preserved rather than those which rendered themselves conspicuous by movement.

That is to say, it is easy to see how the instinct may have become developed by primary means; for if it were of more advantage to an animal when in danger to become motionless, and therefore inconspicuous or unattractive to enemies, than it would be to seek safety in flight, of course it is obvious that in such cases natural selection would always have operated in the direction of producing quiescence, no less than in other cases it would have operated in the direction of producing activity. Now, I think it is not at all improbable that " kataplexy " may have been of much assistance in originating, and possibly also in developing, this instinct.

For if this peculiar physiological condition is apt to occur among insects and spiders—as it certainly occurs in an animal belonging to the same class, the cray-fish—there would be supplied to natural selection the material, as it were, out of which to form this instinct. And if such were the origin of the instinct, we may presume that its development to its present state of perfection would most likely be continued along the same lines—natural selection always improving the kataplectic susceptibility, so as to make the state occur with great suddenness under the influence of a certain class of stimuli, and to prevent it from lasting for an unnecessary time after such stimuli had ceased to operate. Thus we might arrive at the existing state of things, in such an animal as a wood-louse or death-watch, which fall into a kataplectic state immediately on being alarmed (when, on the present hypothesis, they are quite insensitive to pain), but quickly recover as soon as the source of alarming stimulation is removed.*

We have here, then, a rather interesting speculation of a not improbable kind as to the strange, and, so to speak, far-off peculiarities of organization on which natural selection may seize for the developing of a beneficial instinct. But I desire it to be particularly noted that I only adduce this speculation, as it were, parenthetically. I think with Preyer that the shamming dead of insects is a phenomenon in which the principles of hypnotism are probably concerned. But if so, I regard these principles only as furnishing the materials out of which natural selection has constructed this particular instinct. Therefore, whether or not these principles are really concerned in the phenomenon, is only a side question; the important consideration for us is, that the instinct, whether or not developed from materials supplied by kataplexy, must certainly have been developed by natural selection. Mr. Darwin's observations place this conclusion beyond the reach

* An objection to this view may here be disposed of: Duncan, "On Instinct," after observing that spiders while shamming dead, "will suffer themselves to be pierced with pins and torn to pieces without discovering the smallest signs of terror," adds that if the cause were, as often supposed, "a kind of stupor occasioned by terror," the animal ought not so soon to recover when the object of terror is removed. But as a matter of fact the "stupor" does not pass off immediately upon the cessation of the stimulus; it lasts as long as the kataplectic state does in certain birds, such as the owl when held on its back.

of doubt, and even if the phenomena of kataplexy were not available for natural selection to seize upon for the purpose in question, there can be no doubt that other materials might have been so; for, *à priori*, there seems to be at least not more difficulty in developing an instinct to remain motionless under certain circumstances, than in developing one to run away; and as a matter of fact, all animals which are protectively coloured have, either as cause or consequence, developed their instincts in the former direction. Therefore we must suppose that an animal which was not sufficiently locomotive to find safety in flight, would be most closely attended to by natural selection in the direction of encouraging quiescence—and this whether or not natural selection were provided with kataplectic susceptibilities on which to operate; kataplexy alone could not form the instinct.

So far, then, the subject is sufficiently clear. But now, we have obviously some important distinctions to draw. For the shamming dead of a highly intelligent animal like a fox is a widely different matter, psychologically considered, from the shamming dead of insects; so that an explanation which might be held fully adequate to account for the latter might not be so to account for the former. Thus while I have no hesitation in regarding the fact in insects as due to a non-intelligent instinct developed by natural selection in the way just explained, I cannot see how this could well be the case in vertebrated animals. A fox would never have so good a chance of escape from an enemy by remaining motionless as it would by the use of its legs, which it requires a fox-hound to overtake. Moreover the shamming dead is here far from invariable, and so is not, as in the case of insects, instinctive. Therefore, although I did not fully agree with Preyer in assigning the universal (instinctive) quiescence of certain insects when alarmed to the unassisted influence of kataplexy, I think that the occasional (accidental) display of quiescence by wild vertebrated animals under similar circumstances tends much more unequivocally in favour of his view. For here the action is not universal, or even usual; and when it does take place it must, as a rule, be rather detrimental to the animal than otherwise—seeing that the whole economy of the animal is here adapted to rapid movement. Therefore I think that in the case of Birds and Mammals the hypothesis of Couch already quoted is the most reasonable—especially if we supplement it with our knowledge concerning the recently discovered facts of kataplexy.*

On the other hand, not to shirk a difficulty, I have some remarkable evidence which tends to show that certain monkeys sham dead with the deliberate purpose, not of escaping from enemies, but of deceiving intended victims. Here, of course there can be no terror and no kataplexy, so that if we accept the evidence of the fact we must seek for some other explanation.

Thompson gives in his "Passions of Animals" (pp. 455-7), the case of a captive monkey which was tied to a long upright pole of bamboo in the jungles of Tillicherry. The ring at the end of its chain fitting loosely to the slippery pole, the animal was able to ascend and descend the latter at pleasure. He was in the habit of sitting on the top of the pole, and the crows taking advantage of his elevated position, used to steal his food which was placed every morning and evening at the foot of the pole. "To this he had vainly expressed his dislike by chattering, and other indications of his displeasure equally ineffectual; but they continued their periodical depredations. Finding that he was perfectly unheeded, he adopted a plan of retribution as effectual as it was ingenious. One morning when his tormenters had been particularly troublesome, he appeared as if seriously indisposed; he closed his eyes, dropped his head and exhibited various other symptoms of severe suffering. No sooner were his ordinary rations placed at the foot of the bamboo, than the crows watching their opportunity, descended in great numbers, and according to their usual custom, began to demolish his provisions. The monkey began now to descend the pole by slow degrees as if the effort were painful to him, and as if so overcome by indisposition that his remaining strength was scarcely equal to such an exertion. When he reached the ground he rolled about for some time, seeming in great agony, until he found himself close to the vessel employed to contain his food which the crows had by this time well nigh devoured. There was still, however, some remaining, which a solitary bird, emboldened by the apparent indisposition of the monkey, advanced to seize. The wily creature was at this time lying in a state of apparent insensibility at the

* The winking of the wolf's eye, mentioned by Captain Lyon, would be quite compatible with a certain phase of the hypnotic state.

foot of the pole and close by the pan. The moment the crow stretched out his head, and ere it could secure a mouthful of the interdicted food, the watchful avenger seized the depredator by the neck with the rapidity of thought and secured it from doing further mischief. He now began to chatter and grin with every expression of gratified triumph, while the crows flew round, cawing, as if deprecating the chastisement about to be inflicted on their captive companion. The monkey continued for a while to chatter and grin in triumph; he then deliberately placed the crow between his knees and began to pluck it with the most humorous gravity. When he had completely stripped it, except of the large feathers in the pinions and tail, he flung it into the air as high as his strength would permit, and after flapping its wings for a few seconds, it fell to the ground with a stunning shock. The other crows, which had been fortunate enough to escape a similar castigation, now surrounded it and immediately pecked it to death. The animal then ascended its pole, and the next time his food was brought, not a single crow approached it."

I have quoted this case although it sounds well nigh incredible, not merely because Thompson is a good authority, but because in all its essential details it has been unconsciously corroborated by the observations of a friend of my own, viz., the late Dr. W. Bryden, C.B. This gentleman, without being cognizant of the above anecdote, told me that he had himself repeatedly witnessed a tame monkey (I forget the species) in India lying on its back perfectly motionless for long periods of time, till the crows in the neighbourhood, supposing him to be dead, approached within grasping distance, when he used to make a sudden spring at one of them, and proceed slowly to pluck it alive, apparently for the mere love of gratifying his passion of cruelty—although, however, he used to suck the juicy ends of the larger feathers. As I can quite rely on Dr. Bryden's veracity and cannot imagine how in such a case there can have been any room for malobservation, I am inclined to lend a credence to the above anecdote which I should otherwise have regarded with distrust.

Now if, as I can scarcely doubt from Dr. Bryden's account, some monkeys display the remarkable trick of really and of set purpose shamming death, the only possible explanation of the fact is that, having observed crows to congregate round motionless carcasses, they infer that by remaining motionless they may induce these animals to come within grasping distance. No doubt this displays an astonishing amount of deliberative inference; but it is to be observed that the fact, if it is a fact, does not imply any abstract idea of death; it implies only the idea of imitating a previously observed quiescence with the purpose of bringing about the same result—approach of birds—which that quiescence had previously been observed to produce. Seeing that monkeys are highly imitative as well as highly observant animals, this interpretation is not so antecedently incredible as at first sight it no doubt appears.

But now it follows that if monkeys are able consciously and with deliberate intent to remain motionless for the purpose of gaining a particular object, other and almost as intelligent animals may do the same. Thus, notwithstanding the probability previously pointed out that the shamming dead of wolves and foxes may be due to kataplexy, there here arises a possibility of its being due to intelligent purpose. As bearing on this possibility, I will quote two cases which appear to have been sufficiently well observed.

The first is one which has been recently published by Brigade Surgeon G. Bidie in "Nature" (vol. xviii, p. 244). He says:—

"Some years ago, while living in Western Mysore, I occupied a house surrounded by several acres of fine pasture land. The superior grass in this preserve was a great temptation to the village cattle, and whenever the gates were open trespass was common. My servants did their best to drive off intruders, but one day they came to me rather troubled, stating that a Brahmin-bull which they had beaten had fallen down dead. It may be remarked that these bulls are sacred and privileged animals—being allowed to roam at large and eat whatever they may fancy in the open shops of the bazaar-men. On hearing that the trespasser was dead, I immediately went to view the body, and there sure enough it was lying exactly as if life were extinct. Being rather vexed about the occurrence in case of getting into trouble with the natives, I did not stay to make any minute examination, but at once returned to the house with the view of reporting the affair to the district authorities. I had only just gone for a short time when a man, with joy in his face, came running to tell me that the bull was on his legs again and quietly grazing. Suffice it to say that the brute had acquired the trick of feigning death which practically rendered its expulsion impossible, when it found itself in a desirable situation which it did not wish to quit. The ruse was practised frequently with the object of enjoying my excellent grass, and although for a time amusing, it at length became troublesome, and resolving to get rid of it the sooner, I one day, when he had fallen down, sent to the kitchen for a supply of hot cinders, which we placed on his rump. At first he did not seem to mind this much, but as the application waxed hot, he gradually raised his head, took a steady look at the site of the cinders, and finally getting on his legs went off at a racing pace and cleared the fence like a deer. This was the last occasion on which we were favoured with a visit from our friend."

Now here we have a case of apparent simulation of death frequently repeated with an intelligent purpose, and as the narrator is a medical man, we must suppose that the simulation was well done. Nevertheless, the idea which the animal had may only have been that of remaining inert, and trusting to his weight in preventing his removal. The case, however, is unquestionably a remarkable one, and the interpretation which I have suggested becomes perhaps less probable in view of the other case to which I have alluded, and which I shall now proceed to give. This case is published in the late Mr. Morgan's book on the Beaver (p. 269), and he says it "was communicated to the author by Mr. Coral C. White of Aurora, New York, who carried out the fox. His veracity is unimpeachable."

"A fox one night entered the hen-house of a farmer, and after destroying a large number of fowls, gorged himself to such repletion that he could not pass out through the small aperture by which he had entered. The proprietor found him in the morning sprawled out upon the floor, apparently dead from surfeit; and taking him up by the legs carried him out unsuspectingly, and for some distance to the side of his house, where he dropped him upon the grass. No sooner did Reynard find himself free than he sprang to his feet and made his escape. He seemed to know that it was only as a dead fox that he would be allowed to leave the scene of his spoliations; and yet to devise this plan of escape required no ordinary effort of intelligence," &c.

If the facts are here correctly recorded (and in all the points upon which I am about to dwell they agree closely with some of the cases given by Couch), one would scarcely suppose that the mere approach of a man in opening the door of the hen-house could have caused either the kind or degree of alarm which is known to produce kataplexy; while it is somewhat doubtful whether the stimulus occasioned by dropping the fox upon the grass would have been sufficient suddenly to dispel the kataplectic state. Therefore, in such a case as this it seems to me that the probability rather inclines to the shamming dead having been due to an intelligent purpose, even although we may not suppose the animal to have had any idea either of death as such, or of its conscious simulation. Thus the case with respect to the higher animals —if we have due regard to all the evidence which has now been presented—seems to me one of no small difficulty. The truth simply is that there is a lack of sufficient observation, by experimental means, to determine whether wolves, and more especially foxes, simulate death—*i.e.*, remain motionless in certain circumstances of danger with the conscious purpose of furthering their escape; or, perhaps almost as probably, whether the motionless condition of these animals under such circumstances is due to the occurrence of the hypnotic state. With regard to these animals, therefore, as with regard to the Brahmin-bull, I have thought it best not to express a definite opinion either way; but rather to present all the evidence on both sides with the view of stimulating experimental enquiry of the kind that I have suggested by any one who may have the opportunity of conducting it.* Such an enquiry having been conducted by Mr. Darwin in the case of insects and spiders has closed the question as far as they are concerned, by leaving no room to suppose that their behaviour is due to conscious purpose. The evidence with regard to the higher Mammalia, on the other hand, points to a different conclusion, for the full establishment of which further and corroborative evidence is doubtless necessary.

Be it observed, however, that in these cases the difficulty

* If Mr. C. C. White, after having read the above and so having understood the nature of the question, had laid down his fox upon the grass with extreme gentleness, and immediately concealed himself, he might greatly have furthered the solution of the question.

has no reference to any question of instinct—for, unlike the case of insects, the habit is much too exceptional to be regarded as instinctive—but to determining whether the facts are due to intelligent purpose or to some purely physiological effects of fear. In the more remarkable of the above-quoted cases, no doubt, the latter hypothesis is not available; but it may be so in some of the others, and even where this hypothesis is not available, it becomes most desirable to understand the class of ideas which induce the animal to behave in a manner so closely simulating death. Here, however, I am only concerned with showing that the difficulty of arriving at such an understanding has nothing to do with the present theory as to the formation of instinct.

Feigning Injury.

In the "Contemporary Review" (July 1875) the Duke of Argyll, in an article on "Animal Instinct," argues that the female duck could hardly have consciously learnt to imitate the movements of a wounded bird; and that the young mergansers, which squat on the mud when alarmed and are thus made inconspicuous while the old ones fly away, are in the same case. Mr. Darwin, in some MS notes on this article, observes that he agrees with the Duke in not ascribing the deceptive movements of the female duck, &c., to conscious imitation of wounded birds; but thinks that a female bird which, from solicitude for her nestlings, would endeavour to fight a threatening quadruped as a hen does a dog, might, by alternately attacking and retreating, inadvertently draw the enemy away from the nest. Natural selection, acting on this primitive habit, might then develop the running away from the nest as an instinct; and if, as is probable, carnivorous quadrupeds would be more likely to follow birds apparently unable to fly than birds apparently well, the action of drooping the wing, &c., might have been slowly developed.

The instinct of squatting shown by young birds, which are thus rendered inconspicuous, was no doubt acquired in the same way and for the same reason as the instinct of shamming dead in insects. The instinct, however, in the case of young birds may have originally been acquired by older animals (due in the first instance to being partly paralyzed by fright), and then, in accordance with the general principles of heredity, being inherited at an earlier age by the progeny.

It will thus be seen that Mr. Darwin was disposed to attribute the instinct, both of the mother and young, to an exclusively primary origin; but I confess that the case does appear to my mind one of difficulty, and I am rather inclined to think that the instinct of the mother in the case of the duck, peeweet, partridge, and all birds which present it, must have originally been assisted by intelligence. It must be admitted, from what we know of hens, that the maternal feelings may be so strong as to lead to a readiness to incur danger or death rather than that the brood should do so. Therefore, when in the presence of a four-footed enemy the mother bird begins alternately attacking and retreating in the manner alluded to by Mr. Darwin, if she were intelligent enough to *observe* that on retreating without taking wing she was followed up, there can be no doubt that she might with intentional purpose thus lure away the enemy from her young. If so, those parents which had sense enough to adopt this device would no doubt be able to rear a greater number of broods than could the less observant parents; and the young broods of such intelligent parents would inherit a tendency to adopt this device when they themselves became mothers. Thus the originally intelligent device would slowly become organized into an instinct, and so be now performed with mechanical promptitude by every individual partridge, plover, and duck. The greatest difficulty is to account for the drooping of the wing, and this, I think, can only be done by regarding it, with Mr. Darwin, as of an unblended primary origin. The case, however, is unquestionably very remarkable.

Such are the only instincts which have occurred to me as likely to present any special difficulty to the foregoing theory of the origin and development of instincts in general. Mr. Darwin in his chapter on Instinct in the "Origin of Species," has fully discussed several other instincts in this connection (viz., the parasitic instinct of the cuckoo, the cell-making instinct of bees, and the slave-making instinct of ants); but as these do not present any real difficulty, I shall not wait to go over the ground already so thoroughly traversed by him.

CHAPTER XIX.

REASON.

I SHALL begin this chapter by quoting from "Animal Intelligence" my definition of the word "Reason," in order that my use of the word may be clearly understood.

"Reason is the faculty which is concerned in the intentional adaptation of means to ends. It therefore implies the conscious knowledge of the relation between means employed and ends attained, and may be exercised in adaptation to circumstances novel alike to the experience of the individual and to that of the species." In other words, "it implies the power of perceiving analogies or ratios, and is in this sense equivalent to the term 'ratiocination,' or the faculty of deducing inferences from a perceived equivalency of relations. This latter is the only use of the word that is strictly legitimate, and it is thus that I shall use it throughout the present treatise. This faculty, however, of balancing relations, drawing inferences, and so of forecasting probabilities, admits of numberless degrees."

The object of the present chapter will be that of tracing the probable genesis of this faculty, and, in order to give clearness to the discussion, I desire it to be remembered that I reserve the terms Reason and Ratiocination to designate the faculty above defined. I shall use the term Inference to designate the less highly developed mental antecedents out of which, as I shall show, I conceive Reason to have been evolved. No doubt every act of reason is also an act of inference, but we shall find that it is absolutely necessary to have some word to signify indifferently the lowest and the highest stages of that whole class of mental processes which culminates in symbolic calculation. The word Inference is the best that I can find, and therefore it will be understood that in my usage, while all acts of reason are likewise acts of inference, all acts of inference need not be acts of reason.

Thus much as to terminology being premised, I may pass to the subject of the present chapter. I have already, in earlier chapters, endeavoured to show how it is probable that consciousness arises out of reflex action (or that the mind-element becomes attached to nervous processes of adjustment), when the latter arrives at such a degree of complexity, or has reference to external circumstances having such a degree of inconstancy, that the nerve-centre becomes a seat of comparative turmoil among molecular forces. Whenever this stage is reached, and a nerve-centre begins to become conscious of its own working, we pass, according to my classification, from the domain of reflex action into that of instinct —instinct being, in my terminology, reflex action into which there is imported the element of consciousness. But now, as during the course of evolution the lower forms of life are required progressively to adjust their actions to circumstances of growing complexity and inconstancy, or to occasions of growing infrequency, it follows that the organized instincts with which they are endowed must at some point begin to become inadequate; a greater flexibility in the power of adjustive response is needed, and if any such flexibility is possible under the conditions of ganglionic action, those individuals which best attain to it will survive, and so the improvement will become general to the species. Now we know that such an increase of flexibility is possible under the conditions of ganglionic action, and this increase of flexibility on its subjective side we know as the faculty of reason. It is here needful to consider in what this faculty consists.

While treating of the genesis of Perception I pointed out that the faculty admits of numberless degrees of elaboration. These we found to depend largely, or even chiefly, upon the degree of complexity presented by the objects or relations perceived. Now when a perception reaches a certain degree of elaboration, so that it is able to take cognizance of the relation between relations, it begins to pass into reason, or ratiocination. Contrariwise, in its highest stages of development, ratiocination is merely a highly complex process of perception—*i.e.*, a perception of the equivalency of perceived ratios, which are themselves more or less elaborated percepts formed out of simpler percepts, or percepts lying nearer to the immediate data of sensation. Thus, universally ratiocination may be considered as the higher development of perception; for at no point can we draw the line and say that the two are distinct. In other words, a perception is always in its essential nature what logicians term a *conclusion*, whether it has reference to the simplest memory of a past sensation or to the highest product of abstract thought. For when the highest product of abstract thought is analyzed, the ultimate elements must always be found to consist in material given directly by the senses; and every stage in the symbolic construction of ideas in which the process of abstraction consists, depends upon acts of perception taking place in the lower stages. True it is that these acts of perception here have reference to the symbols of ideas, which may themselves be far removed from the simple and immediate memories of past sensations; but as we can nowhere draw the line between perception of the one order and perception of the other, we ought to recognize that in the case of this faculty there is nowhere any difference in kind, although everywhere a difference in degree: or, otherwise stated, intellectual processes which culminate in symbolic reasoning are everywhere processes of cognition, and of these processes the term perception is a generic name.

But having thus shown that in my opinion there is no real break between cognition of the lowest and of the highest order, I must next show at what places I think it is convenient, for the sake of historical description, to mark off what I may term conventional stages in the development of cognition. This I have already done for the lower stages of such development in my chapter on Perception, where it was shown that the first stage consists in merely perceiving an external object as an external object, the next stage in recognizing the simplest qualities of an object, the third stage in mentally grouping objects with reference to their perceived qualities or relations, and the fourth stage in inferring unperceived qualities or relations from perceived ones—as when on hearing a growl I immediately infer the presence of a dangerous dog.

Now from this it is apparent that the process of Inference, with which we are in this chapter concerned, is never in its earlier or least developed, stages a process of conscious comparison. The inference is formed out of the perception, as it were, immediately, and does not require to pass through any such process of reflection as the term ratiocination is apt, and indeed ought, to imply; the ratios at this stage are perceived, compared, and the inference from them drawn, without the need of deliberate thought. For instance, I am hurrying to catch a train, and meet a man in the street hurrying in the opposite direction; we both begin rapidly to dance from side to side in our endeavour to pass one another, and each time we do so it is evident that we have each inferred that the other will pass on the opposite side: yet these successive acts of inference are made with such rapidity, that not only has there been no deliberate thought in the matter, but it is only by such thought that I can afterwards find that I must have performed so many separate acts of inference.

Clearly, then, it is in these lower stages of perception that we have to look for the first germ of reason: for this purpose, let us first interrogate our own perceptions. The large measure in which inference enters into the very structure even of our most habitual perceptions is easily shown. Sir David Brewster has noticed the fact, which must have been observed by every one, that when looking through a window on the pane of which there is a fly or a gnat, if the eyes are adjusted for a greater distance, so that the gnat is not clearly focussed, the mind at once infers that it is a bird, or some other much larger object seen at a greater distance.* Now this shows that in the case of all our visual perceptions mental inference is perpetually at work, compensating for the effects of distance in diminishing apparent size. No less constant must be the work of mental inference in compensating for the effects of the "blind spot" upon the retina. For if the vision be directed to a coloured surface, the part of the surface which, on account of the blind spot, is not really seen, yet appears to be seen; and not only so, but it appears to be coloured the same tint as the rest of the surface, whatever this may happen to be: unconscious inference supplies the colour. Mr. Sully has devoted a large part of his work on "Illusions" to a survey and classification of "The Illusions of Perception;" and in most of the instances which he gives it is apparent, as he observes, that the illusion arises through the mental "application of a rule, valid for the majority of

* *Letters on Natural Magic*, VII.

cases, to an exceptional case"—*i.e.*, the illusion arises from an erroneous inference. It therefore seems needless for me to occupy space with an enumeration of instances.

The first or earliest stage of inference, then, is that in which the inference arises *in* or *together with* the perception, as when we infer that a gnat is a bird, or that the portion of a surface corresponding to the blind spot of the retina is coloured like the surrounding portions of the surface; inference may here be said to be a constituent part of perception.* In other words, we do not in such cases really *sensate* all that we perceive, and the residue of the perception is supplied by inference which is unconscious only because it is so instantaneous. The reason why in such cases it is so instantaneous, is because the part furnished by inference has been so habitually associated with the part furnished by sensation, that the instant the sensation is perceived the mental addition is supplied. That this is the true explanation of the matter is rendered evident, not only from the deductive considerations just stated, but also from the inductive verification which they receive from the facts that arise when a man who has been born blind is suddenly made to see. A good case of this kind is the celebrated one of the youth (about twelve years of age) whom Mr. Cheselden couched for removing congenital cataracts from both eyes. I shall therefore quote a few passages from Mr. Cheselden's account of the case.

"When he first saw he was so far from making any judgment about distances, that he thought all objects whatever touched his eyes (as he expressed) as what he felt touched his skin, and thought no objects so agreeable as those which were smooth and regular, though he could form no judgment of their shape, or guess what it was in any object that was pleasing to him. He knew not the shape of anything, nor any one thing from another, however different in shape and magnitude; but upon being told what things were, whose form he before knew from feeling, he would carefully observe, that he might know them again; but having too many objects to learn at once, he forgot many of them; and (as he said) at first learnt to know, and again forgot a thousand things in a day. One particular only (though it may appear trifling) I will relate. Having often forgotten which was the

* Just in the same way as we found perception to form an integral part of Memory and of the Association of Ideas.

cat and which was the dog, he was ashamed to ask; but catching the cat (which he knew by feeling) he was observed to look at her steadfastly, and then setting her down said, 'So puss! I shall know you another time.' . . . We thought he soon knew what pictures were which were showed to him, but we found afterwards we were mistaken; for about two months after he was couched, he discovered at once they represented solid bodies, when to that time he considered them as only party-coloured planes, or surfaces diversified with variety of paints; but even then he was no less surprised, expecting the pictures would feel like the things they represented; and was amazed when he found those parts, which by their light and shadow appeared round and uneven, felt only flat like the rest; and asked which was the lying sense, feeling or seeing."

Dr. W. B. Carpenter gives a somewhat similar case which fell within his own observation;* but taking the above as sufficient for our purposes, it is evident that the youth, upon being first able to see, was not able to supply any of those mental inferences from his visual perceptions which alone could make these sensations of any practical use as guides or stimuli to action: that is to say, in the absence of these inferences, these perceptions were imperfect. But he immediately set about establishing consciously, or with deliberate intention, those numberless associations between sight and touch which are usually acquired in early infancy, and which are required to constitute the data of the mental inferences which we are considering. The number of such special associations required being so great and varied, we may wonder that even within the space of three months he should have been able to have made so much progress as to feel his visual perception deceived by the arts of shading and perspective; but on this point I shall have more to say presently. Meanwhile it is enough to remember that the case proves the utility of all our visual perceptions to depend upon the ingredient of mental inference which is supplied by habitual association; and, of course, we cannot doubt that the same is true of perceptions yielded by the other senses.†

* *Human Physiology*, 7th ed., p. 103, and in more detail, *Contemp. Rev.*, vol. xxi, pp. 781–2.

† As Adam Smith observes, in his comments upon this case, "When the young gentleman said that objects which he saw touched his eyes, he cer-

Such, then, I conceive to be the first or most rudimentary stage of inference, where, in virtue of constant association, the act is organically bound up with a sensuous perception, so as in fact to constitute an integral part of such perception, and therefore to be precluded from ever emerging into consciousness as a separate act of mind. The next stage in the process of inference I take to be the one which Mr. Spencer regards as the earliest stage. This, in his words, is "that reasoning through which the great mass of surrounding co-existences and sequences are known."* That is to say, when habitually co-existing groups of external objects, attributes, and relations recognized become too numerous and too complex to be all recognized simultaneously, or when the first in a series of habitually successive groups occurs, the unperceived objects, attributes, or relations are inferred. For instance, if a sportsman while shooting woodcock in cover sees a bird about the size and colour of a woodcock get up and fly through the foliage, not having time to see more than that it is a bird of such a size and colour, he immediately supplies by inference the other qualities of a woodcock, and is afterwards disgusted to find that he has shot a thrush. I have done so myself, and could hardly believe that the thrush was the bird I had fired at, so complete was my mental supplement to my visual perception. And, without waiting to give illustrations, it is evident that the same principles apply to the case of habitual sequences.

The second stage of inference, then, is reached when, owing to a constant association of objects, qualities, or relations in the environment, a correspondingly constant association of ideas is produced in the mind, such that when some members of the external group are perceived, the other members of it are inferred. Inference at this stage resembles inference at the earlier stage which we have considered in one respect, and differs from it in another. The resemblance consists in the act of inference being too rapid to admit of its

tainly could not mean that they pressed upon or resisted his eyes . . . . He could mean no more than that they were close upon his eyes, or to speak more properly, perhaps, that they were in his eyes. A deaf man, who was made all at once to hear, might in the same manner naturally enough say, that the sounds which he heard touched his ears, meaning that he felt them as close upon his ears, or, to speak perhaps more properly, as in his ears." (*Essay on External Senses.*)

* *Principles of Psychology*, vol. i, p. 458.

being consciously recognized as an act of mind separate or distinct from the perception. The difference consists in subsequent reflection being able to show that the act of inference *was* distinct from the act of perception, and *must have been* separated from it by a short interval of time; the inference did not, as in the previous cases, constitute an integral part of the perception.

The next stage which we are able to distinguish in the faculty of inference is, I think, that of the conscious comparison of objects, qualities, or relations. Here we arrive at ratiocination strictly so called; but still not necessarily at self-conscious thought. At this stage we make what Mr. Mivart calls "practical inferences;" that is to say, we compare one group of ratios with another, but without thinking of them *as* ratios. Thus, for instance, if I meet a cut-throat looking man upon a lonely road in Ireland, I may begin consciously to determine the probabilities whether he is one of a "brotherhood," and if so whether he is waiting for me; but I cast the matter over in my mind while we are approaching one another, without waiting to think about my thoughts. If I do wait to think about them, I know that I have been carrying on a process of reasoning; but I have equally carried on that process whether or not I ever think about it afterwards as a process.

The last or highest stage of reasoning is attained when the process admits of being consciously recognized as a process, or itself becomes an object of knowledge. This is the stage at which it first becomes possible intentionally to abstract qualities or relations for the purposes of inference. Here, therefore, it first becomes possible to use symbols of ideas instead of the actual ideas themselves, and so it is here that the "Logic of Signs" first emerges from the "Logic of Feelings." In my next work I shall have a great deal to say touching this final stage; but as it only occurs in Man, I have nothing more to say about it at present.

Turning now to animals, it is evident that they must present the first, or, as we may call it, the perceptive stage of inference; for otherwise their whole mechanism of perception would need to be supposed different from our own. But there is only one respect in which this mechanism can be shown to be different, and this consists in the fact already mentioned in former chapters—viz., that newly-hatched birds and newly-born mammals are able, without such individual experience as is required in the case of man, immediately and correctly to supply all the mental inferences which are needed to complete their sensuous perceptions. Of course the explanation of this must be that heredity in these cases has already done the work, so that the young animal comes into the world with its mental endowments of perceptive inference as fully elaborated and as completely efficient as its bodily endowments of perceptive sensations. But the question arises, Why is not this also the case with Man? That it is not the case is sufficiently proved by the results of couching for cataract previously quoted; but why it should not be the case is not quite so clear, and hitherto has not been sufficiently considered; for it is only since the experiments of Mr. Spalding that the facts with regard to animals have become known.* I think the answer to this question is as follows.

First of all, there is no evidence to show that even in the case of man heredity has not played a very important part (though not so important as with animals) in supplying the machinery of perceptive inference. Indeed I think we have some evidence to show that it has; for only by supposing this are we able to explain why the youth whose case was so well described by Mr. Cheselden was able, after so short an interval as three months, to perceive the illusory effects of shading and perspective in a picture. But, even if it be allowed that heredity here played an important part, there is still, no doubt, a great discrepancy to be explained in the degree of its influence as compared with its absolute perfection in the case of the lower animals. But I think there are two considerations which, taken together, are sufficient to explain the discrepancy. In the first place we have already seen, when treating of the hereditary instinctive endowments of animals, that the machinery of these endowments is apt to be thrown out of gear if they are not allowed full play at the time of life when they ought normally to have first come into operation. Therefore in the case of this youth it seems highly probable that during the twelve years

* Or rather, I should say, so well known. Houzeau had pointed out that while young infants are unable to localize a pain or other sensation, newly-born calves are able to do so with precision (see *Fac. Mém. des Animaux*, tom. i, p. 52).

of his congenital blindness, whatever hereditary endowments he may have had in the way of forming perceptive inferences relating to sight, were largely aborted by disuse, if not also thrown out of gear. The other consideration is that, during these twelve years his faculties of perceptive inference were not lying idle, but were thrown with all the greater strength into his perceptions arising from touch and hearing. It is therefore abundantly probable that, even upon this lowest plane of inference, the strong organization which had been formed between this faculty and the perceptions of touch and hearing, made it all the more difficult for this faculty to form a new organization with the perceptions of sight. Further than this, I think it is not improbable that the human mind, in being so habitually concerned with processes of inference on higher planes, would not be so ready to build up by unconscious association a mechanism of perceptive or automatic inference, as would the less highly elaborated mind of an animal similarly situated. Still, notwithstanding these considerations, I feel that it would be well worth while to try the experiment of keeping an animal blindfolded from the time of its birth till it is a year or two old, and then to see, when the blindfolding was removed, whether or not its faculties of perceptive inference resemble those of a similar animal soon after its birth.

That inference of what I have called the second stage also occurs in animals no one will dispute, although, of course, some psychologists may object to my calling this particular case of the association of ideas by the name of inference. I have already said in the chapter which deals with Memory and the Association of Ideas, that it is impossible to say which are really the lowest animals that possess these faculties; and therefore it is still more impossible to say where in the animal kingdom inference of the first or of the second stage begins: we can only say that wherever there is visual or other sensuous perception which, as a perception, requires to form an estimate of distance or other simple relation not immediately given by sensation, but mentally deduced from sensation—there we must suppose that inference of the first stage obtains; and that wherever there is an association of ideas, such that the occurrence of one perception arouses an inferred knowledge of a complement of that perception, or an inferred anticipation of a future event

—there we must suppose that inference of the second stage obtains. And, although we are not able to draw the lines with precision, we know that both these conditions occur low down among the Invertebrata.

The next stage of inference is the highest that obtains among animals. This is the stage in which objects, qualities, and relations are deliberately compared with the intention of perceiving likenesses and unlikenesses (analogies); the action which follows is therefore undertaken with a knowledge, or perception, of the relation between the means employed and the ends attained. This, as I have before said, is the stage of the process of inference at which the term Reason or Ratiocination first becomes appropriate, and therefore it is with reference to this stage that I first use the word. That this stage of the process of inference is reached by nearly all the warm-blooded animals, and even by some of the Invertebrata, no one, I think, can possibly question. If, however, any one should do so, I must refer him to my previous work; for the instances there given are so numerous that it would be tedious here to reproduce even the more striking among them.* To my mind the most remarkable of these instances are those which have reference to the Hymenoptera; for although the faculty does not attain to so high a level of development among them as it does among some of the warm-blooded Vertebrata, it certainly has attained to much more than a

* For the sake at once of giving a striking example of reason in an animal most nearly approaching man, and of supplementing a deficiency in my former treatise, I shall here quote a passage from Dr. Bastian's work on *The Brain as an Organ of Mind* (p. 329). "In regard to the high degree of Intelligence of the Orang, we have the following, on the best of testimony, from Leuret, who says (*Anat. Comp. du Syst. Nerv.*, tom. i, p. 540):—

"'One of the Orangs, which recently died at the Ménagerie of the Musée, was accustomed, when the dinner-hour had come, to open the door of the room where he took his meals in company with several persons. As he was not sufficeintly tall to reach as far as the key of the door, he hung on to a rope, balanced himself, and after a few oscillations very quickly reached the key. His keeper, who was rather worried by so much exactitude, one day took occasion to make three knots in the rope, which, having thus been made too short, no longer permitted the Orang-outang to seize the key. The animal, after an ineffectual attempt, *recognizing the nature of the obstacle which opposed his desire, climbed up the rope, placed himself above the knots, and untied all three,* in the presence of M. Geoffroy Saint-Hilaire, who related the fact to me. The same ape wishing to open a door, his keeper gave him a bunch of fifteen keys; the ape tried them in turn till he had found the one which he wanted. Another time a bar of iron was put into his hands, and he made use of it as a lever."'

proportional development in them; and this whether we consider their position in the zoological scale, or the general structure of their psychology as compared with that of other animals—so that if the whole structure of their psychology were correspondingly advanced, these insects would deserve to be placed on a psychological level with Birds, if not with some of the more intelligent of the Mammalia. But looking to their psychology as a whole, I think that its *status* may most fairly be assigned to the level on which I have placed it in the diagram. However, I do not conceal that the peculiar nature of ant and bee intelligence makes it most difficult to compare with the intelligence of higher animals.

Another special difficulty with reference to reason in animals meets us in the case of the Beaver. For, as remarked in "Animal Intelligence," "on the one hand it seems incredible that the beaver should attain to such a level of abstract thought as would be implied by his forming his various structures with the calculated purpose of achieving the ends which they undoubtedly subserve. On the other hand, as we have seen, it seems little less than impossible that the formation of these structures can be due to instinct." The structures specially alluded to in this connection were the beaver canals, and my information concerning them was derived exclusively from the work of the late Mr. Lewis H. Morgan. Since the publication of "Animal Intelligence," however, I am informed from private sources that the intelligence of the beaver has been greatly over-estimated. My correspondents have undoubtedly seen much of the habits of American beavers; but as I place confidence in the observations of Mr. Morgan, I do not feet entitled to allow the counter-statements of my correspondents to nullify them. Still, I must allow such counter-statements to carry a considerable degree of weight, and therefore I feel that at present it is most judicious to say that, pending further and trustworthy observations, I am not really in a position to discuss the quality of reason as it occurs in this animal. On this account I should not here have referred to the subject at all, were it not that in my previous work I promised to discuss it in the present one. Finding, however, since then, that the facts do not appear to be so certain as I supposed, I prefer, with this explanation, to allow the matter drop.

Recurring now to my views on the origin and development of Reason, it will have been noticed that they differ materially from those of Mr. Herbert Spencer, and, therefore, looking to the influence which he justly exerts upon all matters relating to psychological analysis, I feel that it is desirable to enter at some length into an explanation of the ground on which I have been here reluctantly compelled to disagree with him. Possibly the divergence between us may not be so important as at present I am led to suppose; but if it should hereafter admit of being shown that such is not the case, I need scarcely say that the fact would be a matter of sincere gratification to me.

According to Mr. Spencer, Reason arises out of "compound reflex action" or "Instinct," when this reaches a certain level of compounding or complexity.* Now I have already given the considerations which induce me to differ from Mr. Spencer in regarding Instinct as compound reflex action, and therefore it is only in a general way that I am able to agree with him in his theory of the origin and development of Reason. Nevertheless, in a general way I am able to agree with him, and therefore I shall begin by stating the points in which I do so.

First he says:—"The impossibility of establishing any line of demarkation between the two [Instinct and Reason] may be clearly demonstrated. If every instinctive action is an adjustment of inner to outer relations, and if every rational action is also an adjustment of inner relations to outer relations; then, any alleged distinction can have no other basis than some difference in the characters of the relations to which the adjustments are made. It must be that while, in Instinct, the correspondence is between inner and outer relations that are very simple or general; in Reason, the correspondence is between inner and outer relations that are complex, or special, or abstract, or infrequent. But the complexity, speciality, abstractness, and infrequency of relations, are entirely matters of degree. . . . How then can any particular phase of complexity or infrequency be fixed upon as that at which Instinct ends and Reason begins?"†

With this statement I quite agree, provided I am allowed

* See *Principles of Psychology*, i, pp. 253-71.
† *Loc. cit.*, pp. 453-4.

to make the important addition that it must be strictly limited to the objective aspect, as distinguished both from the subjective and ejective aspects of the phenomena. In other words, if we have regard only to the physical aspect of the phenomena (*i.e.*, the physiology of ganglionic processes as expressed in the adjustive movements of organisms), this statement of the case is unexceptionable. But if we pass from physiology to psychology, the statement ceases to be adequate; for both in the region of subjective and of ejective psychology it would fail to express the important distinction between two very different acts of mind—viz., one in which there is no knowledge of the relation between means employed and ends attained, and one in which there is such knowledge.*

But, passing over this point, we arrive at a lucid statement of the view that "when the correspondence has advanced to those environing objects and acts which present groups of attributes and relations of considerable complexity, and which occur with comparative infrequency—when, consequently, the repetition of experiences has been insufficient to make the sensory changes produced by such groups cohere perfectly with the adaptive motor changes—when such motor changes and the impressions that accompany them simply become nascent: then, by implication, there result *ideas* of such motor changes and impressions, or, as already explained, *memories* of the motor changes before performed under like circumstances, and of the concomitant impressions." Still there is not yet any manifestation of rationality. But now, "when the confusion of a complex impression with some allied one causes a confusion among the nascent motor excitations, there is entailed a certain hesitation, and . . . . ultimately some one set of motor excitations will prevail over the rest." The strongest set will eventually pass into action, and as this set will usually have reference to the circumstances which have recurred most frequently in experience, "the action will, on the average of cases, be the one best adapted to the circumstances. But an action thus produced is nothing

* It will be observed that if we adopt Mr. Spencer's definition of Instinct, the breach on the mental side is still further widened—the distinction between Instinct and Reason being then equivalent to the distinction between nervous actions having no mental counterparts at all, and nervous actions which on their subjective side are intentionally adaptive.

else than a rational one. . . . . This, however, is just the process which we saw must arise whenever, from increasing complexity and decreasing frequency, the automatic adjustment of inner to outer relations becomes uncertain and hesitating. Hence it is clear that the actions we call instinctive pass gradually into the actions we call rational."

Now in an earlier part of this treatise I have stated my belief that consciousness arises when a nerve-centre is subjected to a comparative turmoil of molecular forces, which finds its physiological expression in delay of response, or, as Mr. Spencer says, in "hesitation." But I do not believe that in all such cases Reason, as distinguished from Consciousness, must arise. Therefore I should say that, although there cannot be Reason without such ganglionic friction, there may be such ganglionic friction without Reason. There may, for example, be a large, and even a distressing amount of such friction produced in the case of a conflict of instincts; there may in such cases be prolonged delay ending in "the strongest group of antagonistic tendencies at length passing into action;" and yet no act of reason need arise.

In what respect, then, do I differ from Mr. Spencer touching the genesis of Reason? I differ from him, firstly, in not deeming an act of reason as such a constant or invariable index of ganglionic disturbance greater than that which may arise under other circumstances of psychical activity (and therefore in not deeming that reason must *necessarily* arise out of such disturbance); and, secondly, in not deeming that Reason can *only* arise out of Instinct.

Taking these two points of difference separately, it will be enough to say of the first that it has reference only to the earliest origin of Reason, or to acts of reason of the simplest kind; in the case of more elaborate processes of reasoning I have no doubt that the ganglionic disturbance must be great, and that without such disturbance these more elaborate processes would not be possible. But this, of course, is a widely different matter from concluding that wherever ganglionic disturbance reaches a certain degree of complexity, leading to a consequent delay of response, there Reason (as distinguished from vividness of consciousness) must necessarily arise. On the contrary, I hold that in the lower stages of what I have defined as Reason (and, *à fortiori*, in all the stages of what I have defined as Inference), there may not be more, and there may not be even so much ganglionic disturbance or consequent delay of response, as there may be where no act of rationality is concerned—as, *e.g.*, in a conflict of instincts.

Turning now to my second point of difference with Mr. Spencer, I can see no adequate ground for concluding with him that Reason can only arise out of Instinct. On the contrary, holding, as I have explained, that Reason has its antecedents in the habitual inferences of sensuous perception, that Instinct (as distinguished from Reflex Action) likewise has its antecedents in sensuous perception, and that neither Reason nor Instinct can advance beyond this first origin without an always corresponding advance in the powers of perception; holding these views, I am forced to conclude that Perception is the common stem out of which Instinct and Reason arise as independent branches. In so far as Perception involves Inference, Instinct involves Perception, and Reason involves Inference, there arises, of course, a genetic connection between Instinct and Reason; but this connection is clearly not of the kind which Mr. Spencer indicates: it is organic, and not historic.

This important divergence in my views from those of Mr. Spencer I take to be due to his manner of regarding the relations that subsist between nervous changes which are accompanied by Consciousness, and nervous changes which are not so accompanied. Thus the divergence between our views on this matter began so far back as in our respective analyses of Memory, where I observed, "I cannot agree that if *psychical* changes (as distinguished from physiological changes) are completely automatic, they are on this account to be precluded from being regarded as mnemonic. . . . . In so far as they involve the presence of *conscious recognition*, as distinguished from reflex action, so far, I think, no line of demarcation should be drawn between them and any less perfect memories."* Again, the divergence was manifested when I came to treat of Perception, and I there gave my reasons for regarding it as "very questionable whether the only factors which lead to the differentiating of psychical processes from reflex nervous processes are (as Mr. Spencer alleges) complexity of operation combined with infrequency of occurrence."† And the divergence in question became still more

* See p. 130. † See p. 140.

pronounced when I arrived at my analysis of Instinct; for by identifying Instinct with compound reflex action, we found it to be evident that Mr. Spencer wholly disregards what I take to be the essentially distinguishing feature of Instinct, viz., the presence of Perception as distinguished from Sensation. Thus, lastly, when we now come to the province of Reason, the same divergence recurs. Whether for the special purpose in hand I accept Mr. Spencer's definition of Instinct as compound reflex action, or adhere to my own definition of it as reflex action into which there is imported the element of consciousness, I alike find it impossible to agree with him that Reason necessarily and only arises out of Instinct.

For, taking first Mr. Spencer's definition of Instinct, I cannot agree that Reason necessarily and only arises out of compound reflex action, because I see it to be a fact that in the higher organisms we meet with numerous cases of enormously compound reflex actions which present no indications of rationality. And some of these cases, it may be parenthetically observed, can never at any period of their developmental history have been rational, and afterwards have become automatic by frequency of repetition. Such, for example, cannot have been the case with the compound reflex actions which are concerned in parturition, nor with those more obscure reflex actions which now baffle our rationality to comprehend—I mean the changes set up by an impregnated ovum in the walls of the uterus. These are instances of immensely compound reflex actions which must always have occurred with great rarity in the life-history of individuals, and can never at any time have been either the cause or the effect of rationality.

Again, taking my own definition of Instinct, I cannot agree that Reason necessarily and only arises out of reflex action into which there is imported the element of consciousness. For this element is merely the element of Perception, and I do not know of any evidence to justify the conclusion that Perception can only arise out of the growing complexity and infrequency of reflex actions. As I said in my chapter on Perception, "the truth is that, so far as definite knowledge entitles us to say anything, the only constant physiological difference between a nervous process accompanied by consciousness and a nervous process not so accompanied, is that of time. In very many cases no doubt this difference may be caused by the intricacy or the novelty of the nervous process which is accompanied by consciousness;"* but seeing that in ourselves, as just observed, highly intricate and very infrequent nervous processes may take place mechanically, I do not think we are justified in concluding that complexity and infrequency of ganglionic action are the only factors in determining the rise of consciousness. But even supposing, for the sake of argument, that they are, still it would not follow that the only road to Reason lies through Instinct. Perception being the element common both to Instinct and to Reason, it may very well happen (and indeed I think actually does happen) that Reason arises directly out of those automatic inferences which, as we have seen, are given in Perception, and which, as we have also seen, furnish the conditions to the origin of Instinct.

From this statement, however, I hope it will be manifest that I do not dispute that Reason may, and probably does in many cases arise out of Instinct, in that the perceptive basis of Instinct is so apt to yield material for the higher perceptions of Reason. I merely object to the doctrine that Reason can arise in no other way. And, as further showing the untruth of this doctrine, I may in conclusion point to the numberless instances given in my chapters on Instinct of the reciprocal action between Instinct and Reason—the development of the former sometimes leading to the higher development of the latter, and sometimes, as in all cases of the formation of Instinct by lapsing intelligence, the development of the latter leading to the higher development of the former. Such reciprocal action could not take place were it true that Instinct is always and necessarily the precursor of Reason.

I must not take leave of this discussion on Reason without briefly alluding to the very prevalent view—with which of course I do not agree—that the faculty in question is the special prerogative of Man. As the most enlightened and best informed writer who of late years has espoused this doctrine is Mr. Mivart, I shall take him as its exponent, and in examining his arguments on the subject I shall consider that I am examining the best arguments which can be adduced in support of the view in question.

Mr. Darwin, in his "Descent of Man," gives the follow-

* See p. 140.

ing account of the exhibition of Reason on the part of a Crab:—" Mr. Gardner, whilst watching a shore crab (*gelasimus*) making its burrow, threw some shells towards the hole. One rolled in, and three other shells remained within a few inches of the mouth. In about five minutes the crab brought out the shell which had fallen in, and carried it away to the distance of a foot; it then saw the three other shells lying near, and evidently thinking that they might likewise roll in, carried them to the spot where it had laid the first. It would, I think, be difficult to distinguish this act from one performed by man by the aid of reason."*

Mr. Mivart, after quoting the above, calls the concluding sentence an "astonishing remark."† I shall, therefore, proceed to consider the very prevalent opinion to which such a commentary introduces us, and which consists, as I have said, in regarding the faculty of Reason as the special prerogative of Man.

I must begin by again observing that the faculty of Reason, in the sense of a "knowledge of the relation between means employed and ends attained, ... admits of numberless degrees;" and I hold it to be a mistake, greater than any other that has been committed in psychological science, to suppose that there is any difference of *kind* whether this faculty is exercised with reference to the highest abstractions of introspective thought, or to the lowest products of sensuous perception; whether the ideas involved are general or special, complex or simple, *wherever* there is a process of inference from them which results in establishing a proportional conclusion among them, there we have something more than the mere association of ideas; and this something is Reason. If I were to see a large stone falling through the roof of my conservatory, and on climbing to the wall above saw three or four other stones just upon the edge, I should infer that the stones which fell previously stood in a similar relation to my conservatory, and therefore that it would be desirable to remove the others from their threatening position. This would not be an act of association, but an act of reason (though a simple one), and it is psychologically identical with the act which was performed by the crab.

Further, according to J. S. Mill, "all inference is from particulars to particulars: General propositions are merely

* *Descent of Man*, p. 270. † *Lessons from Nature*, p. 213.

registers of such inferences already made, and short formulæ for making more." Now although this doctrine is not universally accepted by logicians—Whately, for instance, having maintained the exact converse, and many minor writers more or less agreeing with him,—I feel compelled to fall in with it on purely logical grounds, or without reference to any considerations drawn from the theory of evolution. For it appears to me that Mill is completely successful in showing that only by this doctrine can the syllogism be shown to have any functions or any value. "It must be granted that in every syllogism, considered as an argument to prove the conclusion, there is a *petitio principii*. When we say, All men are mortal; Socrates is a man; therefore Socrates is mortal; it is unanswerably urged by the adversaries of the syllogistic theory that the proposition, Socrates is mortal, is presupposed in the more general assumption, All men are mortal." Therefore, "no reasoning from generals to particulars can, as such, prove anything: since from a general proposition we cannot infer any particulars, but those which the principle itself assumes as known." This is not a suitable place in which to discuss such a question of logic at length, and therefore I shall merely refer to Mill's exposition of it.* But as I can see no escape from the view which he enforces that the major premiss of a syllogism is merely a generalized memorandum of past particular experiences, and therefore that all reasoning is, in the last resort, an inference from particulars to particulars; I think that this conclusion (arrived at without reference to the theory of evolution) is available to argue that there is no difference *in kind* between the act of reason performed by the crab and any act of reason performed by a man.

It must be remembered that I am not now discussing the larger question as to whether there is any distinction in kind between the whole mental organization of an animal and the whole mental organization of a man. This larger question I shall fully discuss in my subsequent work. Here I am only endeavouring to show that so far as the particular faculty of mind is concerned which falls under my definition of reason, there is no such distinction. A process of conscious inference, considered merely as a process of conscious inference,

* *Logic*, vol. i, Chap. III.

is the same in kind wherever it occurs and whatever degree of elaboration it presents.

But here I must meet an assertion which is often made, and which has been presented by Mr. Mivart with his accustomed adherence to logical form, and therefore with much apparent cogency. He says:—"Two faculties are distinct *in kind* if we may possess the one in perfection without thereby implying that we possess the other; and still more so if the two faculties tend to increase in an inverse ratio, the perfection of the one being accompanied by a degradation of the other. Yet this is just the distinction between the instinctive and rational parts of man's nature. His instinctive actions are, as all admit, not rational ones; his rational actions are not instinctive. Even more than this, we may say the *more* instinctive a man's actions the *less* are they rational, and *vice versâ;* and this amounts to a demonstration that reason has not, and by no possibility could have been, developed from instinct. In man we have this inverse ratio between sensation and perception, and in brutes it is just where the absence of reason is most generally admitted (*e.g.*, in insects) that we have the very summit and perfection of instinct made known to us by the ant and the bee. . . . Sir William Hamilton long ago called attention to this inverse relation; but when two faculties tend to increase in an inverse ratio, it becomes unquestionable that the difference between them is one of kind."*

Now I meet this argument by denying the alleged fact on which it reposes. It is simply not true that there is any constant inverse ratio of the kind stated. It is no doubt true in a general way (as the principles of evolution would lead us to anticipate), that as animals advance in the scale of mental development their powers of intelligent adjustment are apt to become added in larger measure to their less elaborate powers of instinctive adjustment; but that there is no inverse proportion between the two must be evident to any one who has directed his attention to the mental endowments of animals. Thus, so far is it from being the case that "the absence of reason is most generally admitted" among the ants and bees, that all the observers with whose writings I am acquainted are unanimous in their opinion that there are no animals among the Invertebrata which can be said to

* *Lessons from Nature*, pp. 230–1.

equal the ants and bees in respect of drawing intelligent inferences. Furthermore, looking to the animal kingdom as a whole, I should say that while there is no very constant relationship between the powers of instinct and those of intelligent inference, such relationship as there is points rather to the view that the complexity of mental organization which finds expression in a high development of the instinctive faculties, is favourable to the development of the more intelligent faculties.* And that there should be such a general correspondence is no more than the theory of evolution might lead us to expect; for the progressive complication of instincts tends to diminish, as Mr. Spencer observes, their purely automatic character. But, on the other hand, that this correspondence should be *general* and not *constant* might also be anticipated, seeing that instincts may arise either without the precedence of intelligence, or by means of the lapsing of intelligence.

In the next place, as regards Man, I do not think that Mr. Mivart's argument is any more satisfactorily established by fact. It is no doubt true that "the *more* instinctive are a man's actions the *less* are they rational, and *vice versâ;*" but this, again, is no more than we should expect, on the hypothesis of human instincts being due to hereditary experience, while processes of conscious inference are chiefly due to individual experience. It thus happens that the instinctive actions preponderate over the intelligent actions during infancy, and that the scale begins to turn during childhood. But in all this there is nothing to show that the two are distinct in kind; and in subsequent life their generic identity is shown by the fact that the principle of lapsing intelligence may cause, even in the experience of the individual, actions which are at first consciously adaptive or rational to become by repetition automatic or instinctive.

To what misconception, then, are we to ascribe the very prevalent doctrine that Reason is the special prerogative of Man? I think the misconception arises from an erroneous meaning which is attached to the word Reason. Mr. Mivart, for instance, habitually follows the traditional usage and invests the word with the meaning that belongs to self-conscious Thought. Thus he expressly says that in denying Reason to

* Cf. *Pouchet*, "*L'Instinct chez les Insectes,*" in *Rev. des Deux Mondes*, Feb. 1870, p. 690.

brutes all he maintains is that "they have not the power of forming judgments;"* that is, in his own definition of a judgment, the power of reflective or self-conscious Thought. In my subsequent work I shall have much to say upon the psychology of Judgment; but here it is enough to observe that I hold the power of reflective thought, which the formation of a judgment implies, to constitute no essential part of a process of reason as such, although when present it unquestionably affords that process much new material with which to be concerned. As I have said, I regard reasoning to be a process of consciously inferring, and therefore conclude that it should make no difference to our classification of the rational faculty whether the subject matter on which it may happen to be exercised has reference to the sphere of feeling or to that of thought. And, as Mr. Mivart allows that animals perform "practical inferences," I further conclude that my difference with the school which he represents has reference, so far, only to a matter of terminology. There is, without question, some enormous distinction between the psychology of man and that of the lower animals, and hereafter I shall have to consider at much length what this distinction is. Here I am only concerned with showing that it does not consist in animals having no vestige of the faculty of Reason in the sense above defined. And, in order to show this, I feel, as I have already remarked, that it would be superfluous to render specific instances of the display of animal reason; for they have already been given in such abundance in my former work.

"Is not the earth
With various living creatures, and the air
Replenished? . . . . know'st thou not
Their language and their ways? They also know
And reason not contemptibly."—MILTON.

* *Lessons from Nature*, p. 217.

CHAPTER XX.

ANIMAL EMOTIONS, AND SUMMARY OF INTELLECTUAL FACULTIES.

IT will be observed on turning to the diagram that I attribute to animals the following emotions, which I name in the probable order of their historical development:—Surprise, Fear, Sexual and Parental Affection, Social Feelings, Pugnacity, Industry, Curiosity, Jealousy, Anger, Play, Affection, Sympathy, Emulation, Pride, Resentment, Æsthetic Love of Ornament, Terror, Grief, Hate, Cruelty, Benevolence, Revenge, Rage, Shame, Remorse, Deceit, Ludicrous. This list, which leaves many of the human emotions without mention, exhausts all the emotions of which I have found any evidence in the psychology of animals. Before presenting this evidence in detail, perhaps it will not be thought superfluous again to insist that in attributing this and that emotion to such and such an animal, we can depend only upon inference drawn from actions, and that this inference necessarily becomes of less and less validity as we pass through the animal kingdom to organisms less and less like our own; so that, for instance, "when we get as low down as the insects, I think the most we can confidently assert is, that the known facts of human psychology furnish the best available pattern of the probable facts of insect psychology."* Still, as the known facts of human psychology do furnish the best available pattern, we must here, while treating of the emotional faculties, follow the same method which we have hitherto followed while treating of the intellectual faculties—viz., while having full regard to the progressive weakening of the analogy from human to brute psychology as we recede through the animal kingdom downwards from man, nevertheless using the analogy so far as it goes as the only instrument of analysis that we possess.

* *Animal Intelligence*, pp. 9–10, where see for a more full discussion of this point.

I shall now proceed, as briefly as possible, to render the evidence which has induced me to ascribe each of the above-named emotions to animals, and remembering that I have in each case written the emotion upon the diagram at the level of mental evolution where I have found the earliest evidence of its occurrence, it follows that in the majority of cases the emotion is present in the higher levels of mental evolution in a more highly-developed form.

It will be observed that in the diagram I represent the Emotions as a class to take their origin from the growing structure of mind at the same level as that at which the faculty of Perception takes its origin. I do this because I think that as soon as an animal or a young child is able to perceive its sensations, it must be able to perceive pleasures and pains; hence, when the antecedents of a painful perception recur in consciousness, the animal or child must anticipate the recurrence of that perception—must suffer an ideal representation of the pains, and such suffering is Fear. And that, as a matter of fact, Fear of this low or vague order is manifested at about the second or third week of infancy, is the general opinion of those who have most carefully observed the development of infant psychology.* To specify the class in the animal kingdom where a true emotion of Fear first arises is clearly a more difficult matter, and indeed it is impossible to do so in the absence of any definite knowledge as to the class in which Perception first arises. But while, as previously explained, I am not able to say whether or not the Cœlenterata, and still less the Echinodermata, are able to perceive their sensations, I think the evidence becomes very strong in the case of insect Larvæ and Worms. And that both the one and the other manifest striking symptoms of alarm in the presence of danger may be easily shown. For instance, a few months ago I had an opportunity of observing the habits of the processional caterpillar mentioned in "Animal Intelligence."† Wishing to ascertain whether I could artifi-

* See Preyer, *loc. cit.*

‡ Pp. 238–40. It will be seen on referring to this passage that De Villiers' account differs materially from that of Mr. Davis. For he says that, on removing one of the chain of caterpillars, the whole chain stopped immediately with one consent, like a single organism. Mr. Davis on the other hand said that the information was communicated from caterpillar to caterpillar at the rate of somewhat less than a second per caterpillar. On repeating this observation a great number of times, I could obtain no corroboration at all of De Villiers' statement, while I found that of Mr. Davis to be correct

cially imitate the stimulus which the head of one caterpillar supplied to the tail of the next in the series (and which serves to let the latter known that the series is not interrupted), I removed the last member of the series. As always happens when this is done, the next member stopped, then the next, and the next, and so on, till the whole series were at a halt. If I had now replaced the last member with its head touching the tail of the penultimate member, the latter would again have begun to move, then the next, and the next, and so on, till the whole series would again have been in motion. Instead of doing this, however, I took a camel-hair brush and gently brushed the tail of the then last member. Immediately this member again began to move, and so set the whole train again upon the march. But in order that the march should continue, it was necessary that I should continue brushing the tail of the last member. Now I found that if I brushed in the least degree too hard, so as not sufficiently well to imitate the stimulus supplied by the hairy head of a caterpillar, the animal became alarmed and threw itself upon its side in the form of a coil. I therefore tried the experiment of puzzling the animal, by first brushing its tail gently for a considerable time—so that it should have no reason to doubt, as it were, that I was a caterpillar—and then beginning by degrees to brush it more and more strongly. I could then see that a point came at which the animal was puzzled, so that it hesitated whether to go on or to throw itself upon its side. It appeared to me that at this point the animal began to become alarmed; for the brushing was still exceedingly gentle, so that if the animal were actuated only by a pure reflex mechanism, I should not have expected so infinitesimally small a difference in the amount of stimulation to produce so great a difference in the nature of the response.

in all particulars. I am likewise able to confirm all the other points in his account of the remarkable habits of these larvæ. I may add that as soon as a member of a moving chain is removed, the next member in advance not only stops, but begins to wag its head in a peculiar manner from side to side. This perhaps may serve as a signal to the next member to stop; but, however this may be, as soon as the next one does stop, it also begins to wag its head in the same manner, and so on till all the caterpillars in advance of the interruption are standing still and wagging their heads. And they all continue without interruption thus to wag their heads until the procession again begins to move. I have never seen this peculiar movement performed except under these circumstances.

Again as regards Worms, Mr. Darwin has shown in his work on the Earth-worm that this animal is of a "timid" disposition, darting into its burrow "like a rabbit" when alarmed. Probably other kinds of worms, which are better provided with organs of special sense and consequently have more intelligence, may have more emotion.

With reference to young children, Preyer is of the opinion that the earliest emotion is one of surprise or astonishment upon perceiving any change, or strikingly novel feature, in the environment. In deference to his opinion, therefore, I have placed Surprise upon the same level of emotional development as Fear; but of course in both cases this level is so low that it is but the germs of such emotions that are here supposed to be present.

This earliest stage of emotional development (18) I have made to correspond with "Emotions preservative of Self." The next stage (19) I make to coincide with the origin of "Emotions preservative of Species;" and of these the first to appear are the Sexual. In the animal kingdom—or rather let us say in the psychological scale—these emotions are first unequivocally exhibited by the Mollusca,* which on this account, as well as for the reasons given while treating of the association of ideas, I have made to fill the corresponding level on the other side of the diagram.

The next level (20) is occupied by Parental Affection, Social Feelings, Pugnacity, Emotions conducing to Sexual Selection, Industry, and Curiosity. The level, therefore, corresponds with the origin of the branch marked Social Emotions in the central psychological tree, and with the earliest Recognition of Offspring on the side of the intellectual faculties. The animals which first satisfy all these conditions are the Insects and Spiders.† For here, even if we exclude the Hymenoptera, we have evidence of parental affection in the care which spiders, earwigs, and sundry other insects take of their eggs and broods.† Again, numberless species of insects are highly social in their habits; others are highly pugnacious; some are conspicuously industrious;‡ most flying insects (as we have already seen in Chapter XVIII) display curiosity; and, according to Mr. Darwin's elaborate enquiries, it is also in

* See *Animal Intelligence*, p. 26.

† For remarkable instances of this see *ibid.*, p. 205 and p. 229.

‡ *Ibid.*, pp. 226–8.

this class that we find the earliest evidence of sexual selection.

Coming now to level 21, I have assigned to it the first appearance of the emotions of Jealousy, Anger, and Play, which unquestionably occur in Fish.* On level 23 I have placed the dawn of Affection other than sexual, in view of the evidence of the emotional attachment of a python which was exhibited towards those who had kept it as a pet.†

On level 24 I have placed the dawn of Sympathy, seeing that this emotion appears to be unquestionably, though very fitfully, displayed by the Hymenoptera,‡ which for other reasons I have felt obliged to assign to this comparatively high stage of psychological development.

On the next level (25) I have given Emulation, Pride, Resentment, Æsthetic Love of Ornament, and Terror as distinguished from Fear. All these emotions, so far as I have been able to ascertain, first occur in Birds; and in this class some of the emotions which I have already named as occurring in lower classes, are much more highly developed.§

Next we arrive at Grief, Hate, Cruelty, and Benevolence, as first displayed in some of the more intelligent of the Mammalia. Grief is shown by pining, even to death, upon the removal of a favourite master or companion; Hate by persisting resentment; Cruelty by a cat's treatment of a mouse;‖ and Benevolence by the following instances which I have met with since the publication of "Animal Intelligence." Writing of a domestic cat, Mr. Oswald Fitch says that it "was observed to take out some fish-bones from the house to the garden, and, being followed, was seen to have placed them in front of a miserably thin and evidently hungry stranger cat, who was devouring them; not satisfied with that, our cat returned, procured a fresh supply, and repeated its charitable offer, which was apparently as gratefully accepted. This act of benevolence over, our cat returned to its customary dining-place, the scullery, and ate its own dinner off the remainder

* See *Animal Intelligence*, pp. 242–47.

† *Ibid.*, pp. 261–2.

‡ *Ibid.*, pp. 48-9 and p. 156.

§ *Ibid.*, pp. 270–82. Birds are the lowest animals which I have myself seen, or have heard of others having seen, to die of fright.

‖ For instances of all these facts in Mammals other than Elephants, Dogs or Monkeys, see *Animal Intelligence.*

of the bones."* An almost precisely similar case has been independently communicated to me by Dr. Allen Thomson, F.R.S. The only difference was that Dr. Thomson's cat drew the attention of the cook to the famishing stranger outside by pulling her dress and leading her to the place. When the cook supplied the hungry cat with some food, the other one paraded round and round while the meal was being discussed, purring loudly. One further instance of the display of benevolent feeling by a cat will suffice. Mr. H. A. Macpherson writes me that in 1876 he had an old male cat and a kitten aged a few months. The cat, who had long been a favourite, was jealous of the kitten and "showed considerable aversion to it." One day the floor of a room in the basement of the house was partly taken up in order to repair some pipes. The day after the boards had been replaced, the cat "entered the kitchen (he lived almost wholly on the drawing-room floor above), rubbed against the cook and mewed without ceasing until he had engaged her attention. He then, by running to and fro, drew her to the room in which the work had taken place. The servant was puzzled until she heard a faint mew from beneath her feet. On the boards being lifted the kitten emerged safe and sound, though half-starved. The cat watched the proceedings with the greatest interest until the kitten was released; but on ascertaining that it was safe he at once left the room, without evincing any pleasure at its return. Nor did he subsequently become really friendly with it."

On the next level I have placed Revenge as distinguished from Resentment, and Rage, as distinguished from Anger. In "Animal Intelligence" I give some cases of apparent vindictiveness occurring in birds;† but as the exact nature of the emotions in these cases appears to me somewhat doubtful, I here disregard them, and place Revenge on the psychological level which is occupied by the Elephant and Monkeys, in which animals this passion is very conspicuous.‡ The same remarks apply to Rage, as distinguished from the less violent display of hostile feeling which is suitably expressed by the term Anger.

Lastly, at level 28 we arrive at the highest products of emotional development which are manifested in animal psychology, and therefore at the highest of those products

* *Nature*, April 19, 1883, p. 580. † Pp. 277–8.
‡ *Animal Intelligence*, pp. 387–8 and 478.

with which the present treatise is concerned. These are Shame, Remorse, Deceit, and the Emotion of the Ludicrous. For instances of the display of these emotions by Dogs and Apes, I need merely again refer to "Animal Intelligence."*

In this brief sketch of the emotional faculties as they occur in the animal kingdom, my aim has been to give a generic rather than a specific representation. I have therefore omitted all details of the emotional character of this and that particular animal, as well as the narration of particular instances of the display of emotions. Such details and particular instances will be found in sufficient abundance in my previous work, and it seems undesirable, for the larger purpose now in hand, either to repeat what I have said before, or to burden the discussion with additional facts serving only to corroborate the general assignment of levels which I have now given.

Before concluding the present chapter, and with it the present work, I shall give a similar outline sketch of the assignment of levels on the other and corresponding side of the diagram, which serves to show the probable history of mental evolution so far as the faculties of intellect are concerned. This, of course, has already been done throughout the course of all the preceding pages; but I think it desirable to terminate our analysis of the psychology of animals, by briefly stating in a serial form the reasons which have induced me to assign the various classes of animals to the levels of psychological development in which I have respectively placed them. It is only needful to premise that in considering this side of the diagram I shall not at present wait to treat of the column which has to do with the psychogenesis of the child, for this will require to be treated *ab initio* in my work on Mental Evolution in Man. I may further observe that the sundry psychological faculties which I have written on one of the vertical columns are intended as so many indices of mental evolution, and not as exhausting all the distinctions between one level of such evolution and another. Indeed, looking to the fact that our classification of faculties is conventional rather than natural, we cannot expect that any diagrammatic representation of the order in which they have been developed should admit of being made very

* Pp. 438–45, and 471–78; also 484–98.

precise; for in some existing animals certain faculties are more highly developed than they are in other existing animals, which nevertheless with regard to their general psychology occupy a higher level of mental evolution. Therefore the faculties which I have named in the vertical column have been chosen only because they serve as convenient indices to mark the general upward progress of mental evolution in the animal kingdom.

I have already sufficiently expressed my doubt as to the levels at which all animals below the Articulata should be placed, and I have explained that this doubt arises from the difficulty, or rather the impossibility of ascertaining at what grade of psychological evolution consciousness first occurs. The positions, therefore, which I have assigned to the Cœlenterata and Echinodermata are confessedly arbitrary, and have been determined only because I have not been able to observe that these animals give any unmistakable evidence of perception as distinguished from sensation. This remark applies especially to the Cœlenterata, which in my opinion present no semblance of evidence that any of their responsive movements are of a perceptive, or even of a conscious nature. My judgment with respect to the Echinodermata is less confident, for although I am sure that I am right in placing them on a higher level of sensuous capability than the Cœlenterata, I am not at all sure that I ought not to have placed them one stage higher (*i.e.*, on 18 instead of 17), so as to have brought them within the first rise of perception. For the "acrobatic" and "righting" movements which are performed by these animals, and which I have described elsewhere, are, to say the least, strongly suggestive of true powers of perception. It is, therefore, on the principle of preferring to err on the side of safety that I have placed the Echinodermata on level 17 and not on level 18. That I am justified in attributing to these animals faint powers of memory (as distinguished from the association of ideas) may, I think, be shown by the fact that when a star-fish is crawling along the perpendicular wall of a tank at the level of the surface of the water, it every now and then throws back its rays to feel for other surfaces of attachment, and if it does not succeed in finding such a surface, it again applies its rays to crawling along the side of the tank in the same direction as before, in order that it may again and again repeat the manœuvre in different localities. Now, as this manœuvre requires a long time to execute, I think the fact that after it has been executed the animal continues its advance in the same direction as that in which it was crawling before the manœuvre began, constitutes tolerable evidence in favour of an abiding impression upon the nerve-centres concerned, and one which assuredly is not due to any organically imposed conditions, seeing that on no two occasions is the manœuvre performed in exactly the same way, or even at the same intervals of time.

On the next level I have placed Larvæ of Insects and Annelida. My reason for doing so is that both these classes of organisms unquestionably exhibit instincts of the primary kind,* the origin of which is also assigned to this level. In both cases, however, we meet with certain facts which may justly lead us to question whether in these animals intelligence of a higher order may not be present;† but here again I think it is better to err on the safer side.

It is in the Mollusca that we first undoubtedly meet with a demonstrable power of learning by individual experience,‡ and therefore I have placed this class of animals upon the next level, which is occupied by the first appearance of the power of association by contiguity. Of course, if the account given by Mr. Lonsdale to Mr. Darwin of the pair of land-snails§ were ever to be corroborated by further observations, the Gasteropoda would require to be separated from the other Mollusca and placed on a higher level in the diagram, as I have done in the case of the Cephalopoda.

Next we come to Insects and Spiders on a level with the first Recognition of Offspring and the rise of Secondary Instincts. The evidence that both these faculties occur in both these divisions of the Articulata is unquestionable—and this even when the Hymenoptera are removed for separate psychological classification.‖

Fish and Batrachia are assigned to the next level which corresponds with the rise of Association by Similarity, which I think we are justified in first ascribing to these animals.¶

On level 22 I have written the higher Crustacea. I

* See *Animal Intelligence*, 234–40, and 24.

† *Ibid.*, and Mr. Darwin's work on *Worms*.

‡ *Ibid.*, pp. 25–9.

§ *Ibid.*, p. 27.

‖ *Ibid.*, pp. 207–222, and 226–31.

¶ *Ibid.*, pp. 250–1, and 255.

have done so because this is the stage where, from independent considerations already explained, I have assigned the dawn of Reason (as distinguished from Inference), and the lowest animal psychologically considered in which I have found any evidence of this faculty is the crab.*

Next we come to level 23 where I have placed the Reptiles and Cephalopoda. My reason for so doing is that this is the level where I have represented psychological development to have advanced sufficiently far to admit of the recognition of persons, and this degree of advance has undoubtedly been attained by the Reptiles and the Cephalopoda.† It will be observed that I have bracketed this and the two preceding levels together. My reason for doing so is that the animals and the faculties named upon these levels in some degree overlap. Thus the Batrachia are able to recognize persons,‡ and it is possible that Fish may be able to reason,§ while, on the other hand, the Reptiles and Cephalopoda are not in their general psychology so far above the Batrachia and Fish as would be implied without the bracket; yet I should not be justified in placing them all upon the same level, because I have no such definite evidence that Batrachia and Fish are able to reason as I have in the case of Crustacea, Cephalopoda, and Reptiles. On the whole, therefore, I think that the fairest mode of expressing these various cross relations is the one which I have adopted. It is not to be expected that our essentially artificial mode of distinguishing between psychological faculties should so far agree with nature, that when applied to the animal kingdom our classification of faculties should always be found exactly to fit with our classification of organisms, so that every branch in our psychological tree should precisely correspond with some branch in our zoological tree. Some amount of overlapping must be expected, and in thus comparing the one classification with the other my only surprise has been how, in a general way, the two so closely coincide.

On level 24 I have placed the Hymenoptera, together with the distinction which I think most sharply marks off this stage of mental evolution, *i.e.*, the power of communicating ideas—a power which ants and bees undoubtedly possess.||

* See *Animal Intelligence*, p. 233.
† *Ibid.*, pp. 259, 260–1, and 30.
‡ *Ibid.*, p. 255.
§ *Ibid.*, pp. 250–1.
|| *Ibid.*, pp. 49–57, and 156–60.

Next we arrive at Birds with the psychological distinction of recognizing pictorial representations, understanding words, and dreaming.* If any of these faculties occur in any of the lower vertebrata, I have not found evidence of the fact.

To the next level I have assigned the Rodents and Carnivora, with the exception of the dog. The most marked psychological distinction which I take to mark this level is the understanding of mechanisms. For, although I have found one instance of such understanding to occur in Birds,† and although it likewise unquestionably occurs in Ruminants,‡ in neither case does the understanding appear to extend further than to the simplest order of mechanisms, and therefore is only comparable in kind with the much greater aptitude in this respect which is shown by rats,§ foxes,‖ cats,¶ and the wolverine.**

* *Animal Intelligence*, pp. 311–12. † *Ibid.*, p. 316. ‡ *Ibid.*, pp. 338–9. § *Ibid.*, p. 361. ‖ *Ibid.*, pp. 428–31. ¶ *Ibid.*, pp. 420–22.

** *Ibid.*, pp. 348–50.—Sir James Paget has told me of a parrot which by attentive study learned how to open a lock; but although such cases may occasionally occur in birds, they are so comparatively rare that I have thought it best to place the faculty of appreciating simple mechanical appliances on the next level, for it is here only that we may first be sure that the actions are not due to mere association. A cat which jumps at a thumb-latch, and while holding on to the curved handle beneath with one fore-leg, depresses the thumb-piece with the other and pushes the door-posts with the hind-leg, clearly shows that she has an intelligent appreciation of the facts that the latch fastens the door, that when it is depressed the door will be liberated and that when then pushed the door will open. And if it can still be supposed that all this knowledge can be obtained by simple association, there is the yet more remarkable case of the monkey described in *Animal Intelligence*, which by patient investigation discovered for himself, and without ever having observed any one perform a similar action, the mechanical principle of the screw, not to say also of the lever.

It is remarkable, as I observed in *Animal Intelligence*, that this faculty of appreciating simple mechanical appliances does not seem always to stand in any very precise or quantitative relation to the general mental development of the species which exhibit it. Thus the dog is, as to his general intelligence, unquestionably superior to the cat, and yet his ability in the particular we are considering is certainly not so high; while bovine animals and horses seem to show more cleverness in this respect than in any other. Probably the explanation of this apparent disproportion in the development of the psychical faculties is to be found in the corporeal members which minister to them; the monkey, which shows the highest power of appreciating mechanical appliances, is the animal which is best endowed with the organs of tactual examination; the fore-paws of a cat are better instruments in this respect than those of the dog; while the trunk of the elephant, the lips of the horse, and horns of ruminants give them in the same respects an advantage over most other mammals of a comparable grade of intelligence.

Next we arrive at Monkeys and the Elephant, which, with the exception of the Anthropoid Apes, are the only animals that, so far as I have been able to ascertain, make use of tools.*

Lastly on level 28 we arrive at the highest development of the psychical powers which are to be met with in existing animals, and to this level I have assigned the Anthropoid Apes and Dogs. The meaning of the term "Indefinite Morality," which I give as distinctive of this grade of mental evolution, I shall explain in my next work, when I shall have to discuss the question touching the probable genesis of the moral sense. It is, I think, undesirable to divide this discussion, and therefore I prefer to postpone the consideration of this which I take to be the earliest phase in the development of the faculty of Conscience. And for the same reason I shall postpone my analysis of the lower stages of Abstraction and Volition, both of which are crossed by the level which we have now reached, where our enquiry into the Mental Evolution of Animals comes to an end.

* *Animal Intelligence*, pp. 408-9 and 480-94.

THE END

APPENDIX.

A POSTHUMOUS ESSAY ON INSTINCT,

BY

CHARLES DARWIN, M.A., LL.D., F.R.S.

APPENDIX.

[The full text of a part of Mr. Darwin's chapter on Instinct written for the "Origin of Species," but afterwards suppressed for the sake of condensation.]

Migration.—The migration of young birds across broad tracts of the sea, and the migration of young salmon from fresh into salt water, and the return of both to their birth-places, have often been justly advanced as surprising instincts. With respect to the two main points which concern us, we have, firstly, in different breeds of birds a perfect series from those which occasionally or regularly shift their quarters within the same country to those which periodically pass to far distant countries, traversing, often by night, the open sea over spaces of from 240 to 300 miles, as from the north-eastern shores of Britain to Southern Scandinavia. Secondly, in regard to the variability of the migratory instinct, the very same species often migrates in one country and is stationary in another; or different individuals of the same species in the same country are migratory or stationary, and these can sometimes be distinguished from one another by slight differences.* Dr. Andrew Smith has often remarked to me how inveterate is the instinct of migration in some quadrupeds of S. Africa, notwithstanding the persecution to which they are in consequence subjected: in N. America, however, persecution has driven the Buffalo within

* Mr. Gould has observed this fact in Malta, and in Tasmania in the southern hemisphere. Bechstein (*Stubenvögel*, 1840, s. 293) says that in Germany the migratory and non-migratory Thrushes can be distinguished by the yellow tinge of the soles of their feet. The Quail is migratory in S. Africa, but stationary in Robin Island, only two leagues from the continent (*Le Vaillant's Travels*, vol. i, p. 105): Dr. Andrew Smith confirms this. In Ireland the Quail has lately taken to remain in numbers to breed there (W. Thompson, *Nat. Hist. of Ireland, vide* "Birds," vol. ii, p. 70).

a late period* to cross in its migrations the Rocky Mountains; and those "great highways, continuous for a hundred miles, always several inches, sometimes several feet in depth," worn by migrating buffaloes on the eastern plains, are never found westward of the Rocky Mountains. In the United States, swallows and other birds have largely extended, within quite a late period, the range of their migrations.†

The migratory instinct in Birds is occasionally lost; as in the case of the Woodcock, some of which have totally, without any assignable cause, taken to breed and become stationary in Scotland.‡ In Madeira the first arrival of the Woodcock is known,§ and it is not there migratory; nor is our common Swift, though belonging to a group of birds almost emblematical of migration. A Brent Goose, which had been wounded, lived for nineteen years in confinement; and for about the first twelve years, every spring at the migratory period it became uneasy, and would, like other confined individuals of the species, wander as far northwards as possible; but after this period "it ceased to exhibit any particular feeling at this season."‖ So that we have seen the migratory impulse at last worn out.

In the migration of animals, the instinct which impels them to proceed in a certain direction ought, I think, to be distinguished from the unknown means by which they can tell one direction from another, and by which, after starting, they are enabled to keep their course in a dark night over the open sea; and likewise from the means—whether some instinctive association with changing temperature, or with want of food, &c.—which leads them to start at the proper period. In this, and other cases, the several parts of the

* Col. Frémont, *Report of Exploring Expedition*, 1845, p. 144.

† See Dr. Bachman's excellent memoir on this subject in *Silliman's Philosoph. Journ.*, vol. 30, p. 81.

‡ Mr. W. Thompson has given an excellent and full account of this whole subject (see *Nat. Hist. of Ireland*, "Birds," vol. ii, pp. 247–57), where he discusses the cause. There seems reason to believe (p. 254) that the migratory and non-migratory individuals can be distinguished. For Scotland see St. John's *Wild Sports of the Highlands*, 1846, p. 220.

§ Dr. Heineken in *Zoological Journal*, vol. v, p. 75. See also Mr. E. V. Harcourt's *Sketch of Madeira*, 1851, p. 120.

‖ W. Thompson, *loc. cit.*, vol. iii, p. 63. In Dr. Bachman's paper just referred to cases of Canada geese in confinement periodically trying to escape northward are given.

problem have often been confused together under the word instinct.* With respect to the period of starting, it cannot of course be memory in the young cuckoos' start for the first time two months after their parents have departed: yet it deserves notice that animals somehow acquire a surprisingly accurate idea of time. A. d'Orbigny shows that a lame hawk in S. America knew the period of three weeks, and used at this interval to visit monasteries when food was distributed to the poor. Difficult though it may be to conceive how animals either intelligently or instinctively come to know a given period, yet we shall immediately see that in some cases our domestic animals have acquired an annual recurring impulse to travel, extremely like, if not identical with, a true migratory instinct, and which can hardly be due to mere memory.

It is a true instinct which leads the Brent Goose to try to escape northwards; but how the bird distinguishes north and south we know not. Nor do we know how a bird which starts in the night, as many do, to traverse the ocean, keeps its course as if provided with a compass. But we should be very cautious in attributing to migratory animals any capacity in this respect which we do not ourselves possess;† though certainly in them carried to a wonderful perfection. To give one instance, the experienced navigator Wrangel‡ expatiates with astonishment on the "unerring instinct" of the natives of N. Siberia, by which they guided him through an intricate labyrinth of hummocks of ice with incessant changes of direction; while Wrangel "was watching the different turns compass in hand and trying to reason the true route, the native had always a perfect knowledge of it instinctively." Moreover, the power in migratory animals of keeping their course is not unerring, as may be inferred from

* See E. P. Thompson on the *Passions of Animals*, 1851, p. 9; and Alison's remarks on this head in the *Cyclopædia of Anatomy and Physiology*, article "Instinct," p. 23.

† [I cannot refrain from drawing attention to the superiority of scientific method and philosophical caution here displayed as contrasted with Professor Hæckel's views on the same subject, which in presence of this difficulty at once conclude in favour of some mysterious additional sense (see p. 95).—G. J. R.]

‡ *Wrangel's Travels*, Eng. trans., p. 146. See also Sir G. Grey's *Expedition to Australia*, vol. ii, p. 72, for interesting account of the powers of the Australians in this same respect. The old French missionaries used to believe that the N. American Indians were actually guided by instinct in finding their way.

the numbers of lost swallows often met with by ships in the Atlantic: the migratory salmon, also, often fails in returning to its own river, "many Tweed salmon being found in the Forth." But how a small and tender bird coming from Africa or Spain, after traversing the sea, finds the very same hedge-row in the middle of England, where it made its nest last season, is truly marvellous.*

Let us now turn to our domesticated animals. Many cases are on record of animals finding their way home in a mysterious manner, and it is asserted that Highland sheep have actually swum over the Frith of Forth to their home a hundred miles distant;† when bred for three or four generations in the lowlands, they retain their restless disposition. I know of no reason to doubt the minute account given by Hogg of a family of sheep which had a *hereditary propensity* to return at the breeding season to a place ten miles off, whence the first of the lot had been brought; and, after their lambs were old enough, they returned by themselves to the place where they usually lived; so troublesome was this inherited propensity, associated with the period of parturition, that the owner was compelled to sell the lot.‡ Still more interesting is the account given by several authors of certain sheep in Spain, which from ancient times have annually migrated during May from one part of the country to another distant four hundred miles: all the authors§ agree that "as soon as April comes the sheep express, by curious uneasy motions, a strong desire to return to their summer habita-

* The number of birds which by chance visit the Azores (Consul C. Hunt, in *Journ. Geograph. Soc.*, vol. xv, Pt. 2, p. 282), so distant from Europe, is probably in part due to lost directions during migration: W. Thompson (*Nat. Hist. of Ireland*, "Birds," vol. ii, p. 172) shows that N. American birds, which occasionally wander to Ireland, generally arrive at the period when they are migrating in N. America. In regard to Salmon, see *Scope's Days of Salmon Fishing*, p. 47.

† *Gardener's Chronicle*, 1852, p. 798: other cases are given by *Youatt on Sheep*, p. 377.

‡ Quoted by Youatt in *Veterinary Journal*, vol. v, p. 282.

§ Bourgoanne's *Travels in Spain* (Eng. trans.), 1789, vol. i, pp. 38–54. In Mills' *Treatise on Cattle*, 1776, p. 342, there is an extract of a letter from a gentleman in Spain from which I have made extract. *Youatt on the Sheep*, p. 153, gives references to three other publications with similar accounts. I may add that von Tschudi (*Sketches of Nature in the Alps*, Eng. trans., 1856, p. 160) states that annually in the spring the cattle are greatly excited, when they hear the great bell which is carried with them; well knowing that this is the signal for their "approaching migration" to the higher Alps.

tions." "The unquietude," says another author, "which they manifest might in case of need serve as an almanack." "The shepherds must then exert all their vigilance to prevent them escaping," "for it is a known truth that they would go to the very place where they had been born." Many cases have occurred of three or four sheep having started and performed the journey by themselves, though generally these wanderers are destroyed by the wolves. It is very doubtful whether these migratory sheep are aborigines of the country; and it is certain that within a comparatively recent period their migrations have been widely extended: this being the case, I think there can hardly be a doubt that this "natural instinct," as one author calls it, to migrate at one particular season in one direction has been acquired during domestication, based no doubt on that passionate desire to return to their birth-place which, as we have seen, is common to many breeds of sheep. The whole case seems to me strictly parallel to the migrations of wild animals.

Let us now consider how the more remarkable migrations could possibly have originated. Take the case of a bird being driven each year, by cold or want of food, slowly to travel northward, as is the case with some birds; and in time we may well believe that this compulsory travelling would become an instinctive passion, as with the sheep in Spain. Now during the long course of ages, let valleys become converted into estuaries, and then into wider and wider arms of the sea; and still I can well believe that the impulse which leads the pinioned goose to scramble northward would lead our bird over the trackless waters; and that, by the aid of the unknown power by which many animals (and savage men) can retain a true course, it would safely cross the sea now covering the submerged path of its ancient land journey.*

* I do not suppose that the line of migration of birds always marks the line of formerly continuous land. It is possible that a bird accidentally blown to a distant land or island, after staying some time and breeding there, might be induced by its innate instinct to fly away, and again to return there in the breeding season. But I know of no facts to countenance the idea; and I have been much struck in the case of oceanic islands, lying at no excessive distance from the mainland, but which from reasons to be given in a future chapter I do not believe have ever been joined to the mainland, with the fact that they seem most rarely to have any migratory birds. Mr. E. V. Harcourt, who has written on the birds of Madeira, informs me that there are none in that island; so, I am informed by Mr. Carew Hunt, it is in the

[I will give one case of migration which seemed to me at first to offer especial difficulty. It is asserted that in the extreme north of America, Elk and Reindeer annually cross, as if they could smell the herbage at the distance of a hundred miles, a tract of *absolute* desert, to visit certain islands where there is a better (but still scanty) supply of food. How could their migration have been first established? If the climate formerly had been a little more favourable, the desert a hundred miles in width might then have been clothed with vegetation sufficient to have just tempted the quadrupeds over it, and so to have found out the more fertile northern islet. But the intense Glacial preceded our present climate, and therefore the idea of a former better climate seemed quite untenable; but if those American geologists are right who believe, from the range of recent shells, that subsequently to the Glacial period there was one slightly warmer than the present period, then perhaps we have a key to the migration across the desert of the Elk and Reindeer.*]

Instinctive Fear.—I have already discussed the hereditary tameness of our domestic animals; from what follows I have no doubt that the fear of man has always first to be acquired in a state of nature, and that under domestication it is merely lost. In all the few archipelagoes and islands inhabited by man, of which I have been able to find an early account, the native animals were entirely void of fear of man: I have ascertained this in six cases in the most distant parts of the world, and with birds and mammals of the most different kinds.† At the Galapagos Islands I pushed a hawk off a

Azores, though he thinks that perhaps the Quail, which migrates from island to island, may leave the Archipelago. [In pencil it is added "Canaries none."—G. J. R.]

In the Falkland Islands, so far as I can find, no *land-bird* is migratory. From enquiries which I have made, I find there is no migratory bird in Mauritius or Bourbon. Colenso asserts (*Tasmanian Journal*, vol. ii, p. 227) that a cuckoo, *C. lucidus*, is migratory, remaining only three or four months in New Zealand; but New Zealand is so large an island that it may very easily migrate to the south and remain there quite unknown to the natives of the north. Faröe, situated about 180 miles from the north of Scotland, have several migratory birds (Graber, *Tagebuch*, 1830, s. 205); Iceland seems to be the strongest exception to the general rule, but it lies only   miles from the   line of   100 fathoms. [The last ten words are added in pencil with the blanks left for subsequent filling in.—G. J. R.]

* [The paragraph which I have enclosed in square brackets is faintly struck out in pencil.—G. J. R.].

† I have given in my *Journal of Researches* (1845), p. 378, details on the Falkland and Galapagos. Mr. Cada Mosto (Kerr's *Collection of Voyages*,

tree with the muzzle of my gun, and the little birds drank water out of a vessel which I held in my hand. But I have in my "Journal" given details on this subject, and I will here only remark that the tameness is not general, but special towards man; for at the Falklands the geese build on the outlying islands on account of the foxes. These wolf-like foxes were here as fearless of man as were the birds, and the sailors in Byron's voyage, mistaking their curiosity for fierceness, ran into the water to avoid them. In all old civilized countries the wariness and fear of even young foxes and wolves are well known.* At the Galapagos Islands the great land lizards (*Amblyrhynchus*) were extremely tame, so that I could pull them by the tail; whereas in other parts of the world *large* lizards are wary enough. The aquatic lizard of the same genus lives on the coast, is adapted to swim and dive perfectly, and feeds on submerged algæ: no doubt it must be exposed to danger from the sharks, and consequently, though quite tame on the land, I could not drive them into the water, and when I threw them in they always swam directly back to the shore. See what a contrast with all amphibious animals in Europe, which when disturbed by the most dangerous animal, man, instinctively and instantly take to the water.

The tameness of the birds at the Falklands is particularly interesting, because most of the very same species, more especially the larger birds, are excessively wild in Tierra del Fuego, where for generations they have been persecuted by the savages. Both at these islands and at the Galapagos it is particularly noteworthy, as I have shown in my "Journal" by the comparison of the several accounts up to the time when we visited these islands, that the birds are gradually getting less and less tame; and it is surprising, considering the degree of persecution which they have occasionally suf-

vol. ii, p. 246) says that at the C. de Verde Islands the pigeons were so tame as readily to be caught. These, then, are the only large groups of islands, with the exception of the oceanic (of which I can find no early account) which were uninhabited when discovered. Thos. Herbert in 1626 in his *Travels* (p. 349) describes the tameness of the birds at Mauritius, and Du Bois in 1669-72 enters into details on this head with respect to all the birds at Bourbon. Capt. Moresby lent me a MS account of his survey of St. Pierre and Providence Islands, north of Madagascar, in which he describes the extreme tameness of the pigeons. Capt. Carmichael has described the tameness of the birds at Tristan d'Acunha.

* Le Roy, *Lettres Philosoph.*, p. 86.

fered during the last one or two centuries, that they have not become wilder; it shows that the fear of man is not soon acquired.

In old inhabited countries, where the animals have acquired much general and instinctive suspicion and fear, they seem very soon to learn from each other, and perhaps even from other species, caution directed towards any particular object. It is notorious that rats and mice cannot long be caught by the same sort of trap,* however tempting the bait may be; yet, as it is rare that one which has actually been caught escapes, the others must have learnt the danger from seeing their companions suffer. Even the most terrific object, if never causing danger, and if not *instinctively* dreaded, is immediately viewed with indifference, as we see in our railway trains. What bird is so difficult to approach as the heron, and how many generations would it not require to make herons fearless of man? Yet Mr. Thompson says† that these birds, after a few days' experience, would fearlessly allow a train to pass within half gun-shot distance.‡ Although it cannot be doubted that the fear of man in old inhabited countries is partly acquired, yet it also certainly is instinctive; for nesting birds are generally terrified at the first sight of man, and certainly far more so than most of the old birds at the Falklands and Galapagos Archipelago after years of persecution.

We have in England excellent evidence of the fear of man being acquired and inherited in proportion to the danger incurred; for, as was long ago remarked by the Hon. Daines Barrington,§ all our *large* birds, young and old, are extremely wild. Yet there can be no relation between size and fear;

* E. P. Thompson, *Passions of Animals*, p. 29.

† *Nat. Hist. of Ireland*, "Birds," vol. ii, p. 133.

‡ [I may here refer to the corroboration which this statement has recently received in a correspondence between Dr. Rae and Mr. Goodsir (*Nature*, July 3rd, 12th, and 19th, 1883). The former says that the wild duck, teal, &c., which frequent certain districts through which the Pacific Railway has been carried in Canada, became quite fearless of the trains the first few days after traffic was opened, and the latter gives similar testimony concerning the wild fowl of Australia, adding, "The constant roar of a great passing traffic, as well as the unceasing turmoil and unearthly noises of a large railway station within a stone's throw of their haunts, is now quite unnoticed by these usually most watchful and wary of all birds. [*i.e.*, wild ducks.] But for fear of trespassing on your space, I could give many more illustrations of the truth of Dr. Rae's remarks."—G. J. R.]

§ *Phil. Trans.*, 1773, p. 264.

for on unfrequented islands, when first visited, the large birds were as tame as the small. How exceedingly wary is our magpie; yet it fears not horses or cows, and sometimes alights on their backs, just like the doves at the Galapagos did in 1684 on Cowley. In Norway, where the magpie is not persecuted, it picks up food "close about the doors, sometimes walking inside the houses."* The hooded crow (*C. cornix*), again, is one of our wildest birds; yet in Egypt† is perfectly tame. Every single young magpie and crow cannot have been frightened in England, and yet all are fearful of man in the extreme: on the other hand, in the Falkland and Galapagos Islands many old birds, and their parents before them, must have been frightened and seen others killed; yet they have not acquired a salutary dread of the most destructive animal, man.‡

Animals feigning, as it is said, Death—an unknown state to each living creature—seemed to me a remarkable instinct. I agree with those authors§ who think that there has been much exaggeration on this subject: I do not doubt that fainting (I have had a Robin faint in my hands) and the paralyzing effects of excessive fear have sometimes been mistaken for the simulation of death.‖ Insects are most notori-

* Mr. C. Hewitson in *Magazine of Zoology and Botany*, vol. ii, p. 311.

† Geoffry St. Hilaire, *Anns. des Mus.*, tome ix, p. 471.

‡ [I have already pointed out the refined degree to which such instinctive dread of man is developed when it is able accurately to discriminate what constitutes safe distance from fire-arms. Since writing the passage to which I allude (see p. 197), I have met with the following observation in the letters recently published by Dr. Rae in *Nature*, which is of interest as showing how rapidly such refinement of discrimination is attained:—"I may perhaps be permitted to give one of many instances known to me of the quickness of birds in acquiring a knowledge of danger. Golden plover, when coming from their breeding-places in high latitudes, visit the islands north of Scotland in large numbers, and keep together in great packs. At first they are easily approached, but after a very few shots being fired at them, they become not only much more shy, but seem to measure with great accuracy the distance at which they are safe from harm."—G. J. R.]

§ Couch, *Illustrations of Instinct*, p. 201.

‖ The most curious case of apparently true simulation of death is that given by Wrangel (*Travels in Siberia*, p. 312, Eng. trans.) of the geese which migrate to the Tundras to moult, and are then quite incapable of flight. He says they feigned death so well "with their legs and necks stretched out quite stiff, that I passed them by, thinking they were dead." But the natives were not thus taken in. This simulation would not save them from foxes or wolves, &c., which I presume inhabit the Tundras: would it save them from hawks? The case seems a strange one. A lizard in Patagonia (*Journal of Researches*, p. 97), which lives on the sand near the coast, and is speckled like it, when frightened feigned death with outstretched legs, depressed body,

ous in this respect. We have amongst them a most perfect series, even within the same genus (as I have observed in Curculio and Chrysomela), from species which feign only for a second and sometimes imperfectly, still moving their antennæ (as with some Histers), and which will not feign a second time however much irritated, to other species which, according to De Geer, may be cruelly roasted at a slow fire, without the slightest movement—to others, again, which will long remain motionless as much as twenty-three minutes, as I find with *Chrysomela spartii*. Some individuals of the same species of Ptinus assumed a different position from that of others. Now it will not be disputed that the manner and duration of the feint is useful to each species, according to the kind of danger which it has to escape; therefore there is no more real difficulty in its acquirement, through natural selection, of this hereditary attitude than of any other. Nevertheless, it struck me as a strange coincidence that the insects should thus have come to exactly simulate the state which they took when dead. Hence I carefully noted the simulated positions of seventeen different kinds of insects (including an Iulus, Spider, and Oniscus) belonging to the most distinct genera, both poor and first-rate shammers; afterwards I procured naturally dead specimens of some of these insects, others I killed with camphor by an easy slow death; the result was that in no one instance was the attitude exactly the same, and in several instances the attitude of the feigners and of the really dead were as unlike as they possibly could be.

Nidification and Habitation.—We come now to more complex instincts. The nests of Birds have been carefully attended to, at least in Europe and the United States; so that we have a good and rare opportunity of seeing whether there is any variation in an important instinct, and we shall find that this is the case. We shall further find that favourable opportunities and intelligence sometimes slightly modify the constructive instinct. In the nests of birds, also, we have an unusually perfect series, from those which build none, but lay on the bare ground, to others which make a most imperfect and simple nest, to others more perfect, and

and closed eyes; if further disturbed, it buried itself quickly in the sand. If the Hare had been a small insignificant animal, and if she had closed her eyes when on her form, should we not perhaps have said that she was feigning death? In regard to Insects, see Kirby and Spence, *Introduction to Entomology*, vol. ii, p. 234.

so on, till we arrive at marvellous structures, rivalling the weavers' art.

Even in so singular a nest as that of the Hirundo (*Collocalia esculenta*), eaten by the Chinese, we can, I think, trace the stages by which the necessary instinct has been acquired. The nest is composed of a brittle white translucent substance, very like pure gum arabic, or even glass, lined with adherent feather-down. The nest of an allied species in the British Museum consists of irregularly reticulated fibres, some as fine as * of the same substance; in another species bits of sea-weed are agglutinated together with a similar substance. This dry mucilaginous matter soon absorbs water and softens: examined under the microscope it exhibits no structure, except traces of lamination, and very generally pear-shaped bubbles of various sizes; these, indeed, are very conspicuous in small dry fragments, and some bits looked almost like vesicular lava. A small pure piece put into flame crackles, swells, does not readily burn, and smells strongly of animal matter. The genus Collocalia, according to Mr. G. R. Gray, to whom I am much obliged for allowing me to examine all the specimens in the British Museum, ranks in the same sub-family with our common Swift. The latter bird generally seizes on the nest of a sparrow, but Mr. Macgillivray has carefully described two nests in which the confusedly fitted materials were agglutinated together by extremely thin shreds of a substance which crackles but does not readily burn when put into a flame. In N. America† another species of Swift causes its nest to adhere against the vertical wall of a chimney, and builds it of small sticks placed parallel and agglutinated together

* [In the MS a blank is here intentionally left for the subsequent filling in of an appropriate word.—G. J. R.]

† For *Cypselus murarius* see Macgillivray, *British Birds*, vol. iii, 1840, p. 625. For *C. pelasgius*, see Mr. Peabody's excellent paper on the Birds of Massachussetts in the *Boston Journal of Nat. Hist.*, vol. iii, p. 187. M. E. Robert (*Comptes Rendus*, quoted in *Anns. and Mag. of Nat. Hist.*, vol. viii, 1842, p. 476) found that the nests of the *Hirundo riparia*, made in the gravelly banks of the Volga, had their upper surfaces plastered with a yellow animal substance, which he imagined to be fishes' spawn. Could he have mistaken the species, for there is no reason to suppose our bank-martin has any such habit? This would be a very remarkable variation of instinct, if it could be proved; and the more remarkable that this bird belongs to a different sub-family from the Swifts and Collocalia. Yet I am inclined to believe it, for it has been affirmed with apparent truth that the House-martin moistens the mud, with which it builds its nest, with adhesive saliva.

with cakes of a brittle mucilage which, like that of the esculent swallow, swells and softens in water; in flame it crackles, swells, does not readily burn, and emits a strong animal odour: it differs only in being yellowish-brown, in not having so many large air-bubbles, in being more plainly laminated, and in having even a striated appearance, caused by innumerable elliptical excessively minute points, which I believe to be drawn-out minute air-bubbles.

Most authors believe that the nest of the esculent swallow is formed of either a Fucus or of the roe of a fish; others, I believe, have suspected that it is formed of a secretion from the salivary glands of the bird. The latter view I cannot doubt, from the preceding observations, is the correct one. The inland habits of the Swifts and the manner in which the substance behaves in flame almost disposes of the supposition of Fucus. Nor can I believe, after having examined the dried roe of fishes, that we should find no trace of cellular matter in the nests, had they been thus formed. How could our Swifts, the habits of which are so well known, obtain roe without being detected? Mr. Macgillivray has shown that the salivary crypts of the Swifts are largely developed, and he believes that the substance with which the materials of its nest are fitted together, is secreted by their glands. I cannot doubt that this is the origin of the similar and more copious substance in the nest of the North American Swift, and in those of the *Collocalia esculenta.* We can thus understand its vesicular and laminated structure, and the curious reticulated structure of the Philippian Island species. The only change required in the instinct of these several birds is that less and less foreign materials should be used. Hence I conclude that the Chinese make soup of dried saliva.*

In looking for a perfect series in the less common forms of birds' nests, we should never forget that all existing birds must be almost infinitely few compared with those which have existed since footprints were impressed on the beach of the New Red Sandstone formation of North America.

If it be admitted that the nest of each bird, wherever placed and however constructed, be good for that species

* [It is almost needless to observe that we must remember the date at which this was written; but it may be remarked that as early as 1817 it was pointed out by Home (*Phil. Trans.*, p. 332) that the proventriculus of Collocalia is a peculiar glandular structure probably suited to secrete the substance of which the nest consists.—G. J. R.]

under its own conditions of life; and if the nesting-instinct varies ever so little, when a bird is placed under new conditions, and the variations can be inherited, of which there can be little doubt—then natural selection in the course of ages might modify and perfect almost to any degree the nest of a bird in comparison with that of its progenitors in long past ages. Let me take one of the most extraordinary cases on record, and see how selection may possibly have acted; I refer to Mr. Gould's observation* on the Australian Megapodidæ. The *Talegalla lathami* scrapes together a great pyramid, from two to four cart-loads in amount, of decaying vegetable matter; and in the middle it deposit its eggs. The eggs are hatched by the fermenting mass, the heat of which was estimated at 90° F., and the young birds scratch their way out of the mound. The accumulation propensity is so strong that a single unmated cock confined in Sydney annually collected an immense mass of vegetable matter. The *Leipoa ocellata* makes a pile forty-five feet in circumference and four feet in height, of leaves thickly covered with sand, and in the same way leaves its eggs to be hatched by the heat of fermentation. The *Megapodius tumulus* in the northern parts of Australia makes even a much larger mound, but apparently including less vegetable matter; and other species in the Malayan Archipelago are said to place their eggs in holes in the ground, where they are hatched by the heat of the sun alone. It is not so surprising that these birds should have lost the instinct of incubation, when the proper temperature is supplied either from fermentation or the sun, as that they should have been led to pile up beforehand a great heap of vegetable matter in order that it might ferment; for, however the fact may be explained, it is known that other birds will leave their eggs when the heat is sufficient for incubation, as in the case of the Fly-catcher which built its nest in Mr. Knight's hot-house.† Even the snake takes advantage of a hot-bed in which to lay its eggs; and what concerns us more, is that a common hen, according to Professor Fischer, "made use of the artificial heat of a hot-bed to hatch her eggs."‡ Again Réaumur, as well as Bonnet,

* *Birds of Australia*, and *Introduction to the Birds of Australia*, 1848, p. 82.

† *Yarrel's British Birds*, vol. i, p. 166.

‡ Alison, article "Instinct" in *Todd's Cyclop. of Anat. and Physiol.*, p. 21.

observed* that ants ceased their laborious task of daily moving their eggs to and from the surface according to the heat of the sun, when they had built their nest between the two cases of a bee-hive, where a proper and equable temperature was provided.

Now let us suppose that the conditions of life favoured the extension of a bird of this Family, whose eggs were hatched by the solar rays alone, into a cooler, damper, and more wooded country: then those individuals which chanced to have the accumulative propensity so far modified as to prefer more leaves and less sand, would be favoured in their extension; for they would accumulate more vegetable matter, and its fermentation would compensate for the loss of solar heat, and thus more young birds would be hatched which might as readily inherit the peculiar accumulative propensity of their parents as our breeds of dogs inherit a tendency to retrieve, another to point, and another to dash round its prey. And this process of natural selection might be continued, till the eggs came to be hatched exclusively by the heat of fermentation; the bird, of course, being as ignorant of the cause of the heat as of that of its own body.

In the case of corporeal structures, when two closely allied species, one for instance semi-aquatic and the other terrestrial, are modified for their different manners of life, their main and general agreement of structure is due, according to our theory, to descent from common parents; and their slight differences to subsequent modification through natural selection. So when we hear that the thrush of South America (*T. Falklandicus*), like our European species, lines her nest in the same peculiar way with mud, though, from being surrounded by wholly different plants and animals, she must be placed under somewhat different conditions; or when we hear that in North America the males of the kitty wrens,† like the male of our species, have the strange and anomalous habit of making "cock-nests," not lined with feathers, in which to shelter themselves;—when we hear of such cases, and they are sufficiently numerous in all classes of animals, we must attribute the similarity of the instinct to inheritance from common progenitors, and the dissimilarity, either to

* Kirby and Spence, *Introd. to Entomol.*, vol. ii, p. 519.

† Peabody in *Boston Journ. Nat. Hist.*, vol. iii, p. 144. For our British species see Macgillivray, *Brit. Birds*, vol. iii, p. 23.

selected and profitable modification, or to acquired and inherited habit.. In the same manner, as the northern and southern thrushes have largely inherited their instinctive modification from a common parent, so no doubt the thrush and blackbird have likewise inherited much from their common progenitor, but with somewhat more considerable modifications of instinct in one or both species, from that of their ancient and unknown ancestor.

We will now consider the variability of the nesting-instinct. The cases, no doubt, would have been far more numerous, had the subject been attended to in other countries with the same care as in Great Britain and the United States. From the general uniformity of the nests of each species, we clearly see that even trifling details, such as the materials used and the situation chosen on a high or low branch, on a bank or on level ground, whether solitary or in communities, are not due to chance, or to intelligence, but to instinct. The *Sylvia sylvicola*, for instance, can be distinguished from two closely allied wrens more readily by its nest being lined with feathers than by almost any other character. ("Yarrell's British Birds.")

Necessity or compulsion often leads birds to change the situation of their nests: numerous instances could be given in various parts of the world of birds breeding in trees, but in treeless countries on the ground, or amongst rocks. Audubon (quoted in "Boston Journ. Nat. Hist.," vol. iv, p. 249) states that the Gulls on an islet off Labrador, "in consequence of the persecution which they have met with, now build in trees," instead of in the rocks. Mr. Couch ("Illustrations of Instinct," p. 218) states that three or four successive layings of the sparrow (*F. domesticus*) having been destroyed, "the whole colony, as if by mutual agreement, quitted the place and settled themselves amongst some trees at a distance—a situation which, though common in some districts, neither they nor their ancestors had ever before occupied here, where their nests became objects of curiosity." The sparrow builds in holes in walls, on high branches, in ivy, under rooks' nests, in the holes made by the sand-martins, and often seizes on the nest made by the house-martin: "the nest also varies greatly according to the place" (Montague, "Ornitho. Dict.," p. 482). The Heron (Macgillivray, "Brit. Birds," vol. iv, p. 446: W. Thompson, "Nat. Hist. Ireland," vol. ii, p. 146) builds in trees, on precipitous sea-cliffs, and amongst heath on the ground. In the United States the *Ardea herodias* (Peabody in "Boston Journal Nat. Hist.," vol. iii, p. 209) likewise builds in tall or low trees, or on the ground; and, which is more remarkable, sometimes in communities or heronries, and sometimes solitarily.

Convenience comes into play: we have seen that the Taylor-bird in India uses artificial thread instead of weaving it. A wild Gold-finch (Bolton's *Harmonia Ruralis*, vol. i, p. 492) first took wool, then cotton, and then down, which was placed near its nest. The common Robin will often build under sheds, four cases having been observed in one season at one place (W. Thompson, "Nat. Hist. Ireland," vol. i, p. 14). In Wales the Martin (*H. urbica*) builds against perpendicular cliffs, but all over the lowlands of England against houses; and this must have prodigiously increased its range and numbers. In Arctic America in 1825 *Hirundo lunifrons* (Richardson, "Fauna Boreali-Amercani," p. 331) for the first time built against houses; and the nests, instead of being clustered and each having a tubular entrance, were built under the eaves in a single line and without the tubular entrance, or with a mere ledge. The date of a similar change in the habits of *H. fulva* is also known.

In all changes, whether from persecution or convenience, intelligence must come into play in some degree. The Kitty-wren (*T. vulgaris*), which builds in various situations, usually makes its nest to match with surrounding objects (Macgillivray, vol. iii, p. 21); but this perhaps is instinct. Yet when we hear from White (Letter 14) that a Willow-wren (and I have known a similar case), having been disturbed by being watched, concealed the orifice of her nest, we might argue that the case was one of intelligence. Neither the Kitty-wren nor Water-ouzel (" Mag. of Zool.," vol. ii, 1838, p. 429, invariably build domes to their nests, when placed in sheltered situations Jesse describes a Jackdaw which built its nest on an inclined surface in a turret, and reared up a perpendicular stack of sticks ten feet in height—a labour of seventeen days: families of this bird, I may add (White's "Selborne," Letter 21), have been known regularly to build in rabbit-burrows. Numerous analogous facts could be given. The Water-hen (*G. chloropus*) is said occasionally to cover her eggs when she leaves her nest, but in one protected place W. Thompson ("Nat. Hist. Ireland," vol. ii, p. 328) says that this was never done. Water-hens and Swans, which build in or near the water, will instinctively raise their nest as soon as they perceive the water begin to rise (Couch "Illustrations of Instinct," p. 223–6). But the following seems a more curious case:—Mr. Yarrell showed me a sketch of the nest of a Black Australian Swan, which had been built directly under the drip of the eaves of a building; and, to avoid this, male and female conjointly added semicircular * to the nest, until it extended close to the wall, within the line of drip; and then they pushed the eggs into the newly added portion, so as to be quite dry. The Magpie (*Corvus pica*) under ordinary circumstances builds a remarkable, but very uniform nest; in Norway they build in churches, or spouts under the eaves of houses, as well as in trees. In a treeless part of Scotland, a pair built for several years in a gooseberry bush, which they barricaded all round in an extraordinary manner with briars and thorns, so that "it would have cost a fox some days' labour to have got in." On the other hand, in a part of Ireland, where a reward had been offered for each egg and the magpies had been much persecuted, a pair built at the bottom of a low thick hedge, 'without any large collection of materials likely to attract notice." In Cornwall, Mr. Couch says he has seen near each other, two nests, one in a hedge not a yard from the ground and "unusually fenced in with a thick structure of thorns;" the other "on the top of a very slender and solitary elm—the expectation clearly being that no creature would venture to climb so fragile a column." I have been struck by the slenderness of the trees sometimes chosen by the magpie; but, intelligent as this bird is, I cannot believe that it foresees that boys could not climb such trees, but rather that, having chosen such a tree, it has found from *experience* that it is a safe place.†

Although I do not doubt that intelligence and experience often come into play in the nidification of Birds, yet both often fail: a Jackdaw has been seen trying in vain to get a stick through a turret window, and had

* [A word is here accidentally omitted in the MS.—G. J. R.]

† For Norway, see in *Mag. of Zool. and Bot.*, 1838, vol. ii, p. 311. For Scotland, Rev. J. Hall, *Travels in Scotland*, see Art. "Instinct" in *Cyclop. of Anat. and Physiol.*, p. 22. For Ireland, W. Thompson, *Nat. Hist. of Ireland*, vol. ii, p. 329. For Cornwall, see Couch, *Illustrations of Instinct*, p. 213.

not sense to draw it in lengthways: White (Letter 6) describes some martins which year after year built their nests on an exposed wall, and year after year they were washed down. The *Furnarius cunicularius* in S. America makes a deep burrow in mud-banks for its nest; and I saw ("Journal of Researches," p. 216) these little birds vainly burrowing numerous holes through mud-walls, over which they were constantly flitting, without thus perceiving that the walls were not nearly thick enough for their nests.

Many variations cannot in any way be accounted for: the *Totanus macularius* (Peabody, "Boston Journ. Nat. Hist.," vol. iii, p. 219) lays her eggs sometimes on the bare ground, sometimes in nests slightly made of grass. Mr. Blackwall has recorded the curious case of a yellow Bunting (*Emberiza citrinella*) given in "Yarrell's British Birds," which laid its eggs and hatched them on the bare ground: this bird generally builds on or very close to the ground, but a case is recorded of its having built at a height of seven feet. A nest of a Chaffinch (*Fringilla cœlebs;* "Annals and Mag. of Nat. History," vol. viii, 1842, p. 281) has been described, which was bound by a piece of whipcord passing once round a branch of a pine tree, and then firmly interwoven with the materials of the nest: the nest of the chaffinch can almost be recognized by the elegant manner with which it is coated with lichen; but Mr. Hewitson ("British Oology," p. 7) has described one in which bits of paper were used for lichen. The Thrush (*T. musicus*) builds in bushes, but sometimes, when bushes abound, in holes of walls or under sheds; and two cases are known of its having built actually on the ground in long grass and under turnip-leaves (W. Thompson, "Nat. Hist. of Ireland," vol. i, p. 136: Couch, "Illustrations of Instinct," p. 219). The Rev. W. D. Fox informs me that an "eccentric pair of blackbirds" (*T. merula*) for three consecutive years built in ivy against a wall, and always lined their nest with black horse-hair, though there was nothing to tempt them to use this material: the eggs also were not spotted. The same excellent observer has described (in "Hewitson's British Oology") the nests of two Redstarts, of which one alone was lined with a profusion of white feathers. The Golden-crested Wren (Mr. Sheppard in "Linn. Trans.," vol. xv, p. 14) usually builds an open nest attached to the under side of a fir-branch, but sometimes on the branch, and Mr. Sheppard has seen one "pendulous with a hole on one side." Of the wonderful nest of the Indian Weaver-bird (*Ploceus Philippensis*, "Proc. Zool. Soc.," July 27, 1852), about one or two in every fifty have an upper chamber, in which the males nest, grooved by the widening of the stem of the nest with a pent-house added to it. I will conclude by adding two general remarks on this head by two good observers (Sheppard in "Linn. Trans.," vol. xv, p. 14, and Blackwall quoted by Yarrell, "British Birds," vol. i, p. 444). "There are few birds which do not occasionally vary from the general form in building their nests." "It is evident," says Mr. Blackwall, "that birds of the same species possess the constructive powers in very different degrees of perfection, for the nests of some individuals are finished in a manner greatly superior to those of others."

Some of the cases above given, such as the Totanus either making a nest or building on the bare ground, or that of the Water-ouzel making or not making a dome to its nest, ought, perhaps, to be called a double instinct rather than a variation. But the most curious case of a double instinct which I have met with, is that of the *Sylvia cisticola* given by Dr. P. Savi ("Anns. des Sc. Nat.," tome ii, p. 126). This bird in Pisa annually makes two nests; the autumnal nest is formed by leaves being sewn together with spiders' webs and the down of plants, and is placed in marshes; the vernal nest is placed in tufts of grass in corn-fields, and the leaves are not sewn together; but the sides are thicker and very different materials are used. In such cases, as was formerly remarked with respect to corporeal structures, a great and *apparently* abrupt change might be effected in the instinct of a bird by one form alone of the nest being retained.

In some cases, when the same species ranges into a different climate, the nest differs; the *Artamus sordidus* in Tasmania builds a larger, more compact, and neater nest, than in Australia (Gould's "Birds of Australia"). The *Sterna minuta*, according to Audubon ("Anns. of Nat. Hist.," vol. ii, 1839, p. 462), in the southern and middle U. States merely scoops a slight hollow in the sand; "but on the coast of Labrador it makes a very snug nest, formed of dry moss, well matted together and nearly as large as that of the *Turdus migratorius*." Those individuals of *Icterus Baltimore* (Peabody in "Boston Journ. of Nat. Hist.," vol. iii, p. 97) "which build in the south make their nests of light moss, which allows the air to pass through, and complete it without lining; while in the cool climate of New England they make their nests of soft substances closely woven with a warm lining."

Habitations of Mammals.—On this head I shall make but few remarks, having said so much on the nests of Birds. The buildings erected by the Beaver have long been celebrated; but we see one step by which its wonderful instincts might have been perfected, in the simpler house of an allied animal, the Musk Rat (*Fiber zibethicus*) which, however, Hearne* says is something like that of the Beaver. The solitary Beavers of Europe do not practise, or have lost the greater part of their constructive instincts. Certain species of Rats now uniformly inhabit the roofs of houses,† but other species keep to hollow trees—a change analogous to that in swallows. Dr. Andrew Smith informs me that in the uninhabited parts of S. Africa the hyænas do not live in burrows, whilst in the inhabited and disturbed parts they do.‡ Several mammals and birds usually inhabit burrows made by other species, but when such do not exist, they excavate their own habitations.§

In the genus Osmia, one of the Bee family, the several species not only offer the most remarkable differences, as described by Mr. F. Smith‖ in their instincts; but the individuals of the same species vary to an unusual degree in this respect; thus illustrating the rule, which certainly seems to

* *Hearne's Travels*, p. 380. Hearne has given the best description (pp. 227–236) ever published of the habits of the Beaver.

† Rev. L. Jenyns in *Linn. Trans.*, vol. xvi, p. 166.

‡ A case sometimes quoted of Hares having made burrows in an exposed situation (*Anns. of Nat. Hist.*, vol. v, p. 362), seems to me to require verification: were not the old rabbit-burrows used?

§ *Zoology of the Voyage of the* Beagle, "Mammalia," p. 90.

‖ *Catalogue of British Hymenoptera*, 1855, p. 158.

hold in corporeal structures, namely, that the parts which differ most in allied species, are apt also to vary most in the same species. Another Bee, the *Megachile maritima*, as I am informed by Mr. Smith, near the sea makes its burrows in the sand-banks, whilst in wooded districts it bores holes in posts.*

I have now discussed several of the most extraordinary classes of instincts; but I have still a few miscellaneous remarks which seem to me worth making. First for a few cases of variation which have struck me: a spider which had been crippled and could not spin its web, changed its habits from compulsion into hunting—which is the regular habit of one large group of spiders.† Some insects have two very different instincts under different circumstances, or at different times of life; and one of the two might through natural selection be retained, and so cause an apparently abrupt difference in instinct in relation to the insects' nearest allies: thus the larva of a beetle (the *Cionus scrophulariæ*), when bred on the scrophularia, exudes a viscid substance, which makes a transparent bladder, within which it undergoes its metamorphosis; but the larva when naturally bred, or transported by man, on to a verbascum, becomes a burrower, and undergoes its metamorphosis within a leaf.‡ In the caterpillars of certain moths there are two great classes, those which burrow in the parenchyma of leaves, and those which roll up leaves with consummate skill: some few caterpillars in their early age are burrowers, and then become leaf-rollers; and this change was justly considered so great, that it was only lately discovered that the caterpillars belonged to the same species.§ The Angoumois moth usually has two broods: the first are hatched in the spring from eggs laid in the autumn on grains of corn stored in granaries, and then immediately take flight to the fields and lay their eggs on the standing corn, instead of on the naked grains stored all round them: the moths of the second brood (produced from the eggs laid on the standing corn) are hatched in the granaries, and then do not leave the granaries, but deposit their eggs on the grains around them; and from these eggs proceed the vernal brood which have the

* [Here follows a section on the instincts of Parasitism, Slave-making, and Cell-making, which is published in the *Origin of Species*.—G. J. R.]

† Quoted on authority of Sir J. Banks in *Journal Linn. Soc.*

‡ P. Huber in *Mém. Soc. Phys. de Genève*, tome x, p. 33.

§ Westwood, in *Gardeners' Chronicle*, 1852, p. 261.

different instinct of laying on the standing corn.* Some hunting spiders, when they have eggs and young, give up hunting and spin a web wherewith to catch prey: this is the case with a Salticus, which lays its eggs within snail-shells, and at that time spins a large vertical web.† The pupæ of a species of Formica are *sometimes*‡ uncovered, or not enclosed within cocoons; this certainly is a highly remarkable variation; the same thing is said to occur with the common Pulex. Lord Brougham§ gives us a remarkable case of instinct, namely, the chicken within the shell pecking a hole and then "chipping with its bill-scale till it has cut off a segment from the shell. It always moves from right to left, and it always cuts off the segment from the big end." But the instinct is not quite so invariable, for I was assured at the Eccalobeion (May, 1840) that cases have occurred of chickens having commenced so close to the broad end, that they could not escape from the hole thus made, and had consequently to commence chipping again so as to remove another and larger rim of shell: moreover occasionally they have begun at the narrow end of the shell. The fact of the occasional regurgitation of its food by the Kangaroo‖ ought, perhaps, to be considered as due to an intermediate or variable modification of structure, rather than of instinct; but it is worth notice. It is notorious that the same species of Bird has slightly different vocal powers in different districts; and an excellent observer remarks that "an Irish covey of Partridges springs without uttering a call, whilst on the opposite coast the Scotch covey shrieks with all its might when sprung."¶ Bechstein says that from many years' experience he is certain that in the nightingale a tendency to sing in the middle of the night or in the day runs in families and is strictly inherited.** It is remarkable that many birds have the capacity of piping long and difficult tunes, and others, as the Magpie, of imitating

* Bonnet, quoted by Kirby and Spence, *Entomology*, vol. ii, p. 480.

† Dugès in *Anns. des Sci. Nat.*, 2nd series, tome vi, p. 196.

‡ F. Smith in *Trans. Ent. Soc.*, vol. iii, N.S., Pt. iii, p. 97; and De Geer, quoted by Kirby and Spence, *Entomology*, vol. iii, p. 227.

§ *Dissertation on Natural Theology*, vol. i, p. 117.

‖ W. C. Martin in *Mag. of Nat. Hist.*, N.S., vol. ii, p. 323.

¶ W. Thompson, in *Nat. Hist. Ireland*, vol. ii, p. 65, says that he has observed this, and that it is well known to sportsmen.

** *Stuben-vögel*, 1840, s. 323. See on different powers of singing in different places, s. 205 and 265.

all sorts of sounds, and yet that in a state of nature they never display these powers.*

As there is often much difficulty in imagining how an instinct could first have arisen, it may be worth while to give a few, out of many cases, of occasional and curious habits, which cannot be considered as regular instincts, but which might, according to our views, give rise to such. Thus, several cases are on record† of insects which naturally have very different habits having been hatched within the bodies of men—a most remarkable fact considering the temperature to which they have been exposed, and which may explain the origin of the instinct of the Gad-fly or Œstrus. We can see how the closest association might be developed in Swallows, for Lamarck‡ saw a dozen of these birds aiding a pair, whose nest had been taken, so effectually that it was completed on the second day; and from the facts given by Macgillivray§ it is impossible to doubt that the ancient accounts are true of the Martins sometimes associating and entombing alive sparrows which have taken possession of one of their nests. It is well known that the Hive-bees which have been neglected "get a habit of pillaging from their more industrious neighbours," and are then called corsairs; and Huber gives a far more remarkable case of some Hive-bees which took almost entire possession of the nest of a Humble-bee, and for three weeks the latter went on collecting honey and then regorged it at the solicitation, without any violence, of the Humble-bee.|| We are thus reminded of those Gulls (*Lestris*) which exclusively live by pursuing other gulls and compelling them to disgorge their food.¶

In the Hive-bee actions are occasionally performed which

* Blackwall's *Researches in Zoology*, 1834, p. 158. Cuvier long ago remarked that all the passeres have apparently a similar structure in their vocal organs; and yet only a few, and these the males, sing; showing that fitting structure does not always give rise to corresponding habits. [Concerning birds which imitate sounds when in captivity not doing so in a state of nature, see p. 222, where there is evidence of certain wild birds imitating the sounds of other species.—G. J. R.]

† Rev. L. Jenyns, *Observations in Nat. Hist.*, 1846, p. 280.

‡ Quoted by Geoffry St. Hilaire in *Anns. des Mus.*, tome ix, p. 471.

§ *British Birds*, vol. iii, p. 591.

|| Kirby and Spence, *Entomology*, vol. ii, p. 207. The case given by Huber is at p. 119.

¶ There is reason to suspect (Macgillivray, *British Birds*, vol. v, p. 500) that some of the species can only digest food which has been partially digested by other birds.

we must rank amongst the most wonderful of instincts; and yet these instincts must often have been dormant during many generations: I refer to the death of the queen, when several worker-larvæ are necessarily destroyed, and being placed in large cells and reared on royal food, are thus rendered fertile: so again when a hive has its queen, the males are all infallibly killed by the workers in autumn; but if the hive has no queen, not a single drone is ever destroyed.* Perhaps a ray of light is thrown by our theory on these mysterious but well ascertained facts, by considering that the analogy of other members of the Bee family would lead us to believe that the Hive-bee is descended from other Bees which regularly had many females inhabiting the same nest during the whole season, and which never destroyed their own males; so that not to destroy the males and to give the normal food to additional larvæ, perhaps is only a reversion to an ancestral instinct, and, as in the case of corporeal structures reverting, is apt to occur after many generations.†

I will now refer to a few cases of special difficulty on our theory—most of them parallel to those which I adduced when discussing in Chapter VIII corporeal structures. Thus we occasionally meet with the same peculiar instinct in animals widely remote in the scale of nature, and which consequently cannot have derived the peculiarity from community of descent. The Molothrus (a bird something like a starling) of N. and S. America has precisely the same habits with the Cuckoo; but parasitism is so common throughout nature that this coincidence is not very surprising. The parallelism in instinct between the White Ants, belonging to the Neuroptera, and ants belonging to the Hymenoptera, is a far more wonderful fact; but the parallelism seems to be very far from close. Perhaps as remarkable a case as any on record of the same instinct having been independently acquired in two animals very remote from each other in relationship, is that of a Neuropterous and a Dipterous larva digging a conical

* Kirby and Spence, *Entomology*, vol. ii, pp. 510–13.

† [Concerning the question why there are so many drones as to require killing, see *Animal Intelligence*, p. 166, where I suggest that among the ancestors of the Hive-bee the males may have been of use as workers. But possibly the drones may even now be of use as nurses to the larvæ, for I am told by an experienced bee-keeper that he believes this to be the case.—G. J. R.]

pit-fall in loose sand, lying motionless at the bottom, and if the prey is about to escape, casting jets of sand all round.*

It has been asserted that animals are endowed with instincts, not for their own individual good, or for that of their own social bodies, but for the good of other species, though leading to their own destruction: it has been said that fishes migrate that birds and other animals may prey on them: † this is impossible on our theory of natural selection of self-profitable modification of instinct. But I have met with no facts in support of this belief worthy of consideration. Mistakes of instinct, as we shall presently see, may in some cases do injury to a species and profit another; one species may be compelled, or even apparently induced by persuasion, to yield up its food or secretion to another species; but that any animal has been specially endowed with an instinct leading to its own destruction or harm, I cannot believe without better evidence than has hitherto been adduced.

An instinct performed only once during the life of an animal appears at first sight a great difficulty on our theory; but if indispensable to the animal's existence, there is no valid reason why it should not have been acquired through natural selection, like corporeal structures used only on one occasion, like the hard tip to the chicken's beak, or like the temporary jaws of the pupa of the Caddis-fly or Phryganea, which are exclusively used for cutting open the silken doors of its curious case, and which are then thrown off for ever.‡ Nevertheless it is impossible not to feel unbounded astonishment, when one reads of such cases as that of a caterpillar first suspending itself by its tail to a little hillock of silk attached to some object, and then undergoing its metamorphosis; then after a time splitting open one side and exposing the pupa, destitute of limbs or organs of sense and lying loose within the *lower* part of the old bag-like split skin of the caterpillar: this skin serves as a ladder which the pupa ascends by seizing on portions between the creases of its abdominal segments, and then searching with its tail, which is provided with little hooks, thus attaches itself, and

* Kirby and Spence, *Entomology*, vol. i, pp. 429–435.

† Linnæus in *Amœnitates Academicæ*, vol. ii; and Prof. Alison on "Instinct" in *Todd's Cycl. of Anat. and Physiol.*, p. 15.

‡ Kirby and Spence, *Entomology*, vol. iii, p. 287.

afterwards disengages and casts off the skin which had served it for a ladder.* I am tempted to give one other analogous case, that of the caterpillar of a Butterfly (*Thekla*), which feeds within the pomegranate, but when full fed gnaws its way out (thus making the exit of the butterfly possible before its wings are fully expanded), and then attaches with silk threads the point to the branch of the tree, that it may not fall before the metamorphosis is complete. Hence, as in so many other cases, the larva works on this occasion for the safety of the pupa and of the mature insect. Our astonishment at this manœuvre is lessened in a very slight degree when we hear that several caterpillars attach more or less perfectly with silken threads leaves to the stems for their own safety; and that another caterpillar, before changing into a pupa, bends the edges of a leaf together, coats one surface with a silk web, and attaches this web to the footstalk and branch of the tree; the leaf afterwards becomes brittle and separates, leaving the silken cocoon attached to the footstalk and branch; in this case the process differs but little from the ordinary formation of a cocoon and its attachment to any object.†

A really far greater difficulty is offered by those cases in which the instincts of a species differ greatly from those of its related forms. This is the case with the above mentioned Thekla of the pomegranate; and no doubt many instances could be collected. But we should never forget what a small proportion the living must bear to the extinct amongst insects, the several orders of which have so long existed on this earth. Moreover, just in the same way as with corporeal structures, I have been surprised how often when I thought I had got a case of a perfectly isolated instinct, I found on further enquiry at least some traces of a graduated series.

I have not rarely felt that small and trifling instincts were a greater difficulty on our theory than those which have so justly excited the wonder of mankind; for an instinct, if really of no considerable importance in the struggle for life, could not be modified or formed through natural selection. Perhaps as striking an instance as can be given is that of the worker of the Hive-bee arranged in files and ventilating, by a peculiar movement of their wings, the

* Kirby and Spence, *Entomology*, vol. iii, pp. 208–11.
† J. O. Westwood in *Trans. Entomol. Soc.*, vol. ii, p. 1.

well-closed hive: this ventilation has been artificially imitated,* and as it is carried on even during winter, there can be no doubt that it is to bring in free air and displace the carbonic acid gas: therefore it is in truth indispensable, and we may imagine the stages—a few bees first going to the orifice to fan themselves—by which the instinct might have been arrived at. We admire the instinctive caution of the hen-pheasant which leads her, as Waterton remarked, to fly from her nest and so leave no track to be scented out by beasts of prey; but this again may well be of high importance to the species. It is more surprising that instinct should lead small nesting birds to remove their broken eggs and the early mutings, whereas with partridges, the young of which immediately follow their parents, the broken eggs are left round the nest; but when we hear that the nests of those birds (Halcyonidæ) in which the mutings are not enclosed by a film, and so can hardly be removed by the parent, are thus "rendered very conspicuous;"† and when we remember how many nests are destroyed by cats, we cannot any longer consider them instincts of trifling importance. But some instincts one can hardly avoid looking at as mere tricks, or sometimes as play: an Abyssinian pigeon when fired at, plunges down so as to almost touch the sportsman, and then mounts to an immoderate height:‡ the Bizcacha (Lagostomus) almost invariably collects all sorts of rubbish, bones, stones, dry dung, &c., near its burrow: Guanacoes have the habit of returning (like Flies) to the same spot to drop their excrement, and I saw one heap eight feet in diameter; as this habit is common to all the species of the genus, it must be instinctive, but it is hard to believe that it can be of any use to the animal, though it is to the Peruvians, who use the dried dung for fuel.§ Many analogous facts could probably be collected.

Wonderful and admirable as most instincts are, yet they cannot be considered as absolutely perfect: there is a con-

* Kirby and Spence, *Entomology*, vol. ii, p. 193.

† Blyth in *Mag. of Nat. Hist.*, N.S., vol. ii.

‡ *Bruce's Travels*, vol. v, p. 187.

§ See my *Journal of Researches*, p. 167 for the Guanaco; for the Bizcacha, p. 145. Many odd instincts are connected with the excrement of animals, as with the wild Horse of S. America (see *Azara's Travels*, vol. i, p. 373), with the common House Fly and with Dogs; see on the urinary deposits of the Hyrax, Livingston's *Missionary Travels*, p. 22.

stant struggle going on throughout nature between the instinct of the one to escape its enemy and of the other to secure its prey. If the instinct of the Spider be admirable, that of the Fly which rushes into its toils is so far inferior. Rare and occasional sources of danger are not avoided: if death inevitably ensues, and creatures cannot have learnt by seeing others suffer, it seems that no guardian instinct is acquired: thus the ground within a solfortara in Java is strewed with the carcases of tigers, birds, and masses of insects killed by the noxious exhalations, with their flesh, hairs, and feathers preserved, but their bones entirely consumed.* Migratory instinct not rarely fails, and the animals, as we have seen, are lost. What ought we to think of the strong impulse which leads Lemmings, Squirrels, Ermines,† and many other animals which are not regularly migratory, occasionally to congregate and pursue a headlong course, across great rivers, lakes, and even into the sea, where vast numbers perish; and ultimately it would appear that all perish? The country being overstocked seems to cause the original impulse; but it is doubtful whether in all cases scarcity actually prevails. The whole case is quite inexplicable. Does the same feeling act on these animals which causes men to congregate under distress and fear; and are these occasional migrations, or rather emigrations, a forlorn hope to find a new and better land? The occasional emigrations of insects of many kinds associated together, which, as I have witnessed, must perish by countless myriads in the sea, are still more remarkable, as they belong to families none of which are naturally social or even migratory.‡

* Von Buch, *Descript. Phys. des Iles Canaries*, 1836, p. 423, on the excellent authority of M. Reinwardts.

† L. Lloyd, *Scandinavian Adventure*, 1854, vol. ii, p. 77, gives an excellent account of the migration of Lemmings: when swimming across a lake, if they meet a boat, they crawl up one side and down the opposite side. Great migrations took place in 1789, 1807, 1808, 1813, 1823. Ultimately all seem to perish. See Högström's account in *Swedish Acts*, vol. iv, 1763, of ermines migrating and entering the sea. See Bachman's account in *Mag. of Nat. Hist.*, N.S., vol. iii, 1839, p. 229, of the migration of squirrels; they are bad swimmers and get across great rivers.

‡ Mr. Spence in his Anniversary address to the Entomological Society, 1848, has some excellent remarks on the occasional migration of insects, and shows how inexplicable the case is. See also Kirby and Spence, *Entomology*, vol. ii, p. 12; and Weissenborn in *Mag. of Nat. Hist.*, N.S., 1834, vol. iii, p. 516, for interesting details on a great migration of Libellulæ, generally along the course of rivers.

The social instinct is indispensable to some animals, useful to still more for the ready notice of danger, and apparently only pleasant to some few animals. But one cannot avoid thinking that this instinct is carried in some cases to an injurious excess: the antelopes in S. Africa and the Passenger Pigeons in N. America are followed by hosts of carnivorous beasts and birds, which could hardly be supported in such numbers if their prey were scattered. The Bison of N. America migrates in such vast bodies, that when they come to narrow passes in the river-cliffs, the foremost, according to Lewis and Clarke(?),* are often pushed over the precipice and dashed to pieces. Can we believe when a wounded herbivorous animal returns to its own herd and is then attacked and gored, that this cruel and very common instinct is of any service to the species? It has been remarked† that with Deer, only those which have been much chased with dogs are led by a sense of self-preservation to expel their pursued or wounded companion, who will bring danger on the herd. But the fearless wild elephants will "ungenerously attack one which has escaped into the jungles with the bandages still upon its legs."‡ And I have seen domestic pigeons attack and badly wound sick or young and fallen birds.

The cock-pheasant crows loudly, as everyone may hear, when going to roost, and is thus betrayed to the poacher.§ The wild Hen of India, as I am informed by Mr. Blyth, chuckles like her domesticated offspring, when she has laid an egg;

* [The note of interrogation is in the MS.—G. J. R.]

† W. Scrope, *Art of Deer Stalking*, p. 23.

‡ Corse, in *Asiatic Researches*, vol. iii, p. 272. This fact is the more strange as an Elephant which had escaped from a pit was seen by many witnesses to stop and assist with his trunk his companion in getting out of the pit (*Athenæum*, 1840, p. 238). Capt. Sulivan, R.N., informs me that he watched for more than half an hour, at the Falkland Islands, a Logger-headed Duck defending a wounded Upland Goose from the repeated attacks of a Carrion Hawk. The upland goose first took to the water, and the duck swam close alongside her, always defending her with its strong beak; when the goose crawled ashore, the duck followed, going round and round her, and when the goose again took to the sea the duck was still vigorously defending her; yet at other times this duck *never* associates with this goose, for their food and place of habitation are utterly different. I very much fear, from what we see of little birds chasing hawks, that it would be more philosophical to attribute this conduct in the duck to hatred of the carrion hawk rather than to benevolence for the goose.

§ Rev. L. Jenyns, *Observations in Natural History*, 1846, p. 100.

and the natives thus discover her nest. In La Plata the Furnarius builds a large oven-like nest of mud in as conspicuous a place as possible, on a bare rock, on the top of a post, or cactus-stem; * and in a thickly peopled country, with mischievous boys, would soon be exterminated. The great Butcher-bird conceals its nest very badly, and the male during incubation, and the female after her eggs are hatched, betray the nest by their repeated harsh cries.† So again a kind of Shrew-mouse at the Mauritius continually betrays itself by screaming out as soon as approached. Nor ought we to say that these failures of instinct are unimportant, as principally concerning man alone; for, as we see instinctive wildness directed towards man, there seems no reason why other instincts should not be related to him.

The number of eggs of the American Ostrich scattered over the country, and so wasted, has already been noticed. The Cuckoo sometimes lays two eggs in the same nest, leading to the sure rejection of one of the two young birds. Flies, it has often been asserted, frequently make mistakes, and lay their eggs in substances not fitted for the nourishment of their larvæ. A Spider‡ will eagerly seize a little ball of cotton when deprived of her eggs, embedded as they are in a silken envelope; but if a choice be given her, she will prefer her own eggs, and will not always seize the ball of cotton a second time: so that we see sense or reason here correcting a first mistake. Little birds often gratify their hatred by pursuing a Hawk, and perhaps by so doing distract its attention; but they often mistake and persecute (as I have seen) any innocent and foreign species. Foxes and other carnivorous beasts often destroy far more prey than they can devour or carry away: the Bee Cuckoo kills a vast number more bees than she can eat, and "unwisely pursues without interruption this pastime all the day long."§ A queen Hive-bee confined by Huber, so that she could not lay her eggs in worker cells, would not deposit, but dropped them, upon which the workers devoured them. An unfertilized queen can lay only male eggs, but these she deposits in worker and royal cells—an aberration of instinct not surprising under the circum-

* *Journal of Researches*, p. 95.

† Knapp, *Journal of a Naturalist*, p. 188.

‡ These facts are given by Dugès in *Anns. des Sc. Nat.*, 2nd series tome vi, p. 196.

§ *Bruce's Travels in Abyssinia*, vol. v, p. 179.

stances; but "the workers themselves act as if they suffered in their instinct from the imperfect state of their queen, for they fed these male larvæ with royal jelly and treat them as they would a real queen."* But what is more surprising, the workers of Humble-bees habitually endeavour to seize and devour the eggs of their own queens; and the utmost activity of the mothers is "scarcely adequate to prevent this violence."† Can this strange instinctive habit be of any service to the Bee? Seeing the innumerable and admirable instincts all directed to rear and multiply young, can we believe, with Kirby and Spence, that this strange aberrant instinct is given them "to keep the population within due bounds?" Can the instinct which leads the female spider savagely to attack and devour the male after pairing with him‡ be of service to the species? The carcase of her husband no doubt nourishes her; and without some better explanation can be given, we are thus reduced to the grossest utilitarianism, compatible, it must be confessed, with the theory of natural selection. I fear that to the foregoing cases a long catalogue could be added.

Conclusion.—We have in this chapter chiefly considered the instincts of animals under the point of view whether it is possible that they could have been acquired through the means indicated on our theory, or whether, even if the simpler ones could have been thus acquired, others are so complex and wonderful that they must have been specially endowed, and thus overthrow the theory. Bearing in mind the facts given on the acquirement, through the selection of self-originating tricks or modification of instinct, or through training and habit, aided in some slight degree by imitation, of hereditary actions and dispositions in our domesticated animals; and their parallelism (subject to having less time) to the instincts of animals in a state of nature: bearing in mind that in a state of nature instincts do certainly vary in some slight degree: bearing in mind how very generally we find in allied but distinct animals a gradation in the more complex instincts, which show that it is at least possible that a complex instinct might have been acquired by successive steps; and

* Kirby and Spence, *Entomology*, vol. ii, p. 161 (3rd ed.).

† *Ibid.*, vol. i, p. 380.

‡ *Ibid.*, vol. i, p. 280. A long list of several insects which either in their larval or mature condition will devour each other is given.

which moreover generally indicate, according to our theory, the actual steps by which the instinct has been acquired, in as much as we suppose allied instincts to have branched off at different stages of descent from a common ancestor, and therefore to have retained, more or less unaltered, the instincts of the several lineal ancestral forms of any one species: bearing all this in mind, together with the certainty that instincts are as important to an animal as their generally correlated structures, and that in the struggle for life under changing conditions, slight modifications of instinct could hardly fail occasionally to be profitable to individuals, I can see no overwhelming difficulty on our theory. Even in the most marvellous instinct known, that of the cells of the Hive-bee, we have seen how a simple instinctive action may lead to results which fill the mind with astonishment.

Moreover it seems to me that the very general fact of the gradation of complexity of instincts within the limits of the same group of animals; and likewise the fact of two allied species, placed in two distant parts of the world and surrounded by wholly different conditions of life, still having very much in common in their instincts, supports our theory of descent; for they are explained by it: whereas if we look at each instinct as specially endowed, we can only say that it is so. The imperfections and mistakes of instinct on our theory cease to be surprising: indeed it would be wonderful that far more numerous and flagrant cases could not be detected, if it were not that a species which has failed to become modified and so far perfected in its instincts that it could continue struggling with the co-inhabitants of the same region, would simply add one more to the myriads which have become extinct.

It may not be logical, but to my imagination, it is far more satisfactory to look at the young cuckoo ejecting its foster-brothers, ants making slaves, the larvæ of the Ichneumidæ feeding within the live bodies of their prey, cats playing with mice, otters and cormorants with living fish, not as instincts specially given by the Creator, but as very small parts of one general law leading to the advancement of all organic bodies —Multiply, Vary, let the strongest Live and the weakest Die.

INDEX TO MENTAL EVOLUTION IN ANIMALS.

A.

B.

C.

D.

E.

F.

G.

H.

J.

K.

L.

M.

N.

O.

P.

R.

S.

T.

U.

V.

W.

Y.

Z.

INDEX TO MR. DARWIN'S POSTHUMOUS ESSAY.

A.

B.

C.

D.

E.

F.

G.

H.

I.

J.

K.

L.

M.

N.

O.

P.

Q.

R.

S.

T.

W.

Y.

HARRISON AND SONS, PRINTERS IN ORDINARY TO HER MAJESTY, ST. MARTIN'S LANE.

For EU product safety concerns, contact us at Calle de José Abascal, 56–1°,
28003 Madrid, Spain or eugpsr@cambridge.org.

www.ingramcontent.com/pod-product-compliance
Lightning Source LLC
LaVergne TN
LVHW090300210926
848480LV00001B/85

* 9 7 8 1 1 0 8 0 3 7 8 7 7 *